D0024788

DEMCO

WESTERN EUROPE:
geographical perspectives
THIRD EDITION

ICELAND
Reykjavik

SWEDEN

FINLAND

NORWAY

Helsinki

Oslo   Stockholm

ESTONIA

LATVIA

IRELAND

LITHUANIA

DENMARK

Copenhagen

Dublin

UNITED
KINGDOM

NETHERLANDS

POLAND

BELORUS

London

The Hague

Berlin

Brussels

BELGIUM

GERMANY

CZECH
REP.

UKRAINE

LUXEMBOURG

SLOVAK REP.

Paris

Vienna

HUNGARY

Berne

AUSTRIA

ROMANIA

SWITZERLAND

SLOVENIA

FRANCE

CROATIA

BOSNIA

SERBIA

BULGARIA

PORTUGAL

ITALY

MONTENEGRO

MACEDONIA

Madrid

Rome

ALBANIA

Lisbon

SPAIN

GREECE

Athens

0   500

km

WESTERN EUROPE:

geographical perspectives

THIRD EDITION

HUGH CLOUT
MARK BLACKSELL
RUSSELL KING
DAVID PINDER

Longman
Scientific &
Technical

Longman Scientific & Technical
Longman Group UK Limited
Longman House, Burnt Mill, Harlow
Essex, CM20 2JE, England
and Associated Companies throughout the world.

Copublished in the United States with
John Wiley & Sons, Inc, 605 Third Avenue, New York, NY
10158

First published 1985
Second edition 1989
Third edition   1994

British Library Cataloguing in Publication Data
A catalogue record for this book is available from the British
Library

ISBN 0–582–09283–3

Library of Congress Cataloging-in-Publication Data
A catalog record for this book is available from the Library
of Congress.

ISBN 0–470–22147–X (USA only)

Set 8 in Times 9/11 pt.
Printed in Malaysia

Contents

List of figures

List of tables

Preface

At the beginning of this century, Paul Vidal de la Blache (1903) observed that the land of France resembled a medal, struck in the likeness of its people. In so doing, he introduced a methodology for comprehending both the interrelationship of people and nature and the spatial differentiation of long-settled countries of Europe. His approach was to prove long lasting and influenced generations of geographers who produced highly detailed textbooks that explored the intricacies of place. Major works dealing with component parts of Europe were produced by British geographers (e.g. Houston 1964; Monkhouse 1959; Mutton 1961), while writers in North America tended to embrace a country-by-country approach to cover the continent of Europe as a whole (e.g. Gottmann 1950; Hoffmann 1953; Pounds 1953). These and other comparable texts went through several editions and fashioned the education of countless students.

Then came a phase in the late 1960s and 1970s when geographers tended to cast their thoughts about European area studies within a frame of applied (or applicable) geography and examined economic trends, social processes and management problems in the context of individual nations or official planning regions (e.g. House 1978; Thompson 1970). By emphasizing the importance of process, rather than accentuating the uniqueness of place, the legacy of the past or the endowments of nature, this approach was much more problem orientated than that which preceded it. Geographers stressed the role of planners and the activities of planning institutions and their writings reflected the belief that wise management could manipulate resources in the best interests of society at large. Such assumptions had, of course, to be rethought in the light of the depression that followed the oil crisis of 1973. It became clear that cheap energy, low unemployment, political compromise and racial harmony no longer formed the scenario against which the human geography of Western Europe was enacted. Authors subsequently recast their thoughts in a more pessimistic way and sometimes adopted a much more political approach to regional problems (e.g. Carney, Hudson and Lewis 1980).

The present book does not propose such an overtly political standpoint but has certainly been written with the current economic and social challenges of the 1980s in mind. It is presented in a thematic way and, rather than emphasizing the uniqueness of each of the eighteen states of Western Europe, attempts to draw out common characteristics and shared problems, while always recognizing spatial variations in their expression. Geography, whatever else it may be, remains the study of areal differentiation; and hence maps and case studies represent important features of this book. The essays within it have been written by four university geographers who have wide experience of teaching and researching European matters. Three obtained their doctorates for research in Continental Europe and each has his own distinctive areal, linguistic and thematic expertise. The collection is edited by Hugh Clout, who also wrote the introduction and dealt with population and social change, tertiary activities, urban conditions, regional development and Western Europe in the 'new Europe.' Mark Blacksell authored the sections on political evolution, recreation and conservation, and marine resources; Russell King wrote on migration and on agricultural matters; while David Pinder covered energy and manufacturing.

It has to be stressed that the book has been written in a selective, generalized way and at an introductory level, with a British undergraduate audience in mind. Only a minority of bibliographical items in languages other than English have been included. Each essay is concluded with some suggestions for further reading while the consolidated bibliography at the end of the book lists a much fuller range. Place-names are normally expressed in their conventional English form, except where they refer to planning regions (e.g.

Bretagne) rather than broader geographical areas (e.g. Brittany). The illustrations have been prepared by members of the Cartographic Units at University College, London and at the universities of Exeter and Plymouth, while manuscripts have been typed by many hands in different cities.

Acknowledgements

We are grateful to the following for permission to reproduce copyright material:

Almqvist and Wiksell International for Fig. 10.1 redrawn from Fig. 1 (Kariel 1982); *Annales de Géographie* and the author Prof. Alain Metton for Fig. 7.6 (redrawn from Metton 1982); the author, Mrs Marianne Carey for Fig. 9.1 (redrawn from van Valkenburg and Held 1952); the Délégation à l'Aménagement du Territoire et à l'Action Régionale, Paris for Fig. 7.7 (redrawn from DATAR originals); Mr Hervé Théry for Fig. 8.4 (redrawn from a map in Informations RECLUS, no. 19, 1990, Montpellier); Professor Georges Cazès for Fig. 7.4 (redrawn from a map in the *Travaux de l'Institut de Géographie de Reims*, 1988); The Economist for Figs 3.6 and 3.7 (redrawn from The Economist, 1.5. 82, 18.4.87); the Population Council for Fig. 4.5 (redrawn from a figure, p. 35 in Bourgeois-Pichat 1981); the Office of Population, Censuses and Surveys for Fig. 8.2 (redrawn from a Crown Copyright figure, p. 12 in *1991 Census Preliminary Report for England and Wales*, HMSO); the Organization for Economic Cooperation and Development, Paris for material in Tables 5.11, 5.12 and 5.13 (*OECD Energy Statistics* 1970–85) and 4.1, 5.9, 10.3, 10.4; Professor Richard Munton for material in Table 9.6.

1

Optimism and uncertainty

Fundamental features

More attention has been devoted by geographers to Europe than to any similarly sized portion of the earth's surface. Generations of scholars have described and analysed its diverse natural environment, the numerous human activities that it sustains and the complex imprints that they impart to its evolving cultural landscapes. Research monographs, textbooks and learned articles, replete with maps and statistical series, record these facts and interpretations in every language in use across the continent. The majority of these works are expressed at the scale of the state, the region or the precise locality; only rarely do authors look to wider horizons and try to capture the fundamental essence of Europe (Mellor and Smith 1979; Hoffman 1989).

Long and intimate knowledge of the continent and penetrating understanding of the minds of its geographers enabled W. R. Mead (1982) to produce such a distillation in a challenging and wide-ranging essay entitled 'The discovery of Europe'. It is his belief that the distinctiveness of Europe as a whole derives from an amalgam of physical geography, ethnography, technology, values and territorial organization. The physical resource base of this continent of peninsulas, isthmuses and islands, stretching from Mediterranean latitudes to beyond the Arctic Circle, is intricate and remarkably diverse, with localized resources often complementing those available in neighbouring areas. The assembled interdigitation of land and sea, successions of mountains, plains and valleys, arrays of soils, and gradations from maritime to continental climatic regimes are without parallel.

Nowhere else in the world is there comparable cultural complexity across such a relatively compact area, whether this is expressed with regard to tangible features, such as traditional house styles or field patterns, or in terms of language or religious tradition (Fig. 1.1).

In this respect, sharp lines on maps must be recognized as deceptive, since significant minority groups are to be found within apparently homogeneous cultural domains. Migration flows during many centuries have produced far more complicated meshes of language and religious tradition than Fig. 1.1 could possibly convey, with the recent arrival of migrants from North Africa, Asia and many other distant parts of the world providing striking examples.

For two centuries, Western Europe was the focus of technological innovation, cradling the Industrial Revolution, diffusing technical skills throughout the world, and experiencing the emergence of post-industrial civilization more intensely than elsewhere. West Europeans flourished by virtue of their technical acumen and commercial expertise, with several nations amassing great empires by land or sea. The fruits of technology, colonialism and trade added even more advantages to those enjoyed by the dwellers of this already well-endowed part of the globe. Imperialism and commerce ensured that methods and techniques developed in Western Europe were adopted widely throughout the world. The same processes contributed to the diffusion of shared European values, including inventiveness and democracy.

In addition, Western Europe bears the distinctiveness of an intense proliferation of states of varying ethno-political status, ranging from one-nation federations (such as Austria and Germany) and multi-ethnic units (such as Belgium and Switzerland) to one-nation states (such as France and Italy) (Fig. 1.1c). However, several one-nation states contain areas with autonomous status (e.g. Alto Adige, Sardinia, Sicily and the Val d'Aosta in Italy) or with special status with regard to bilingualism (e.g. Friesland in the Netherlands) or legal system (e.g. Scotland in the UK). Western Europe's states and their boundaries have undergone complex evolution through time, with the territorial imperative being at the heart of hundreds of

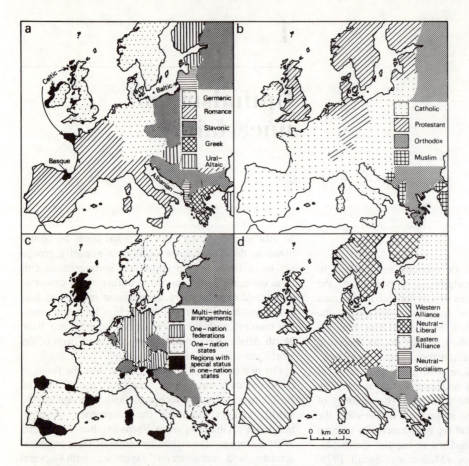

Fig. 1.1
Cultural indicators:
(a) language groups;
(b) religious tradition;
(c) ethno-political
situation; (d) political
blocs pre-1989.

the continent's wars, including the two world wars of our own century. Europe has a more elaborate patchwork of political boundaries and national 'systems' of administration and organization of ordinary life than any other continent (Jenkins 1986).

Nationality implies peripherality as well as centrality and Western Europe is endowed with areas of marginality along all of its political boundaries. For four and a half decades after the Second World War none of these was sharper or harsher than the line that partitioned states and component regions of the old continent, stretching from the Atlantic to the Urals, into two fundamental blocs: the Eastern and Western alliances, with relatively few countries maintaining neutral positions (Fig. 1.1d) (Wallace 1990). Of course, the East/West divide was not impervious; Western travellers and tourists visited Eastern Europe but significantly fewer East Europeans travelled westwards. Radio and television programmes transmitted in West Germany and Austria reached substantial parts of East Germany, Czechoslovakia, Hungary and Yugoslavia, while Western radio programmes were received much further eastward, and continue to be so.

None the less, in many respects, the 'real division of Europe' was between territories that looked ultimately to Moscow or to Washington DC and it was given most acute expression along the line of the 'Iron Curtain', regardless of whether that was composed of brick, concrete, barbed wire or minefield (Fontaine 1971; Ritter and Hajdu 1989). Twelve of the chapters of this book concern themselves with economic and social conditions in the states located to the west of that critical line and with the lives of the Europeans who inhabit them. The focus of the final chapter is shifted to the 'new Europe' that is fast emerging.

Seen from the global perspective, Western Europe is composed of a set of 'old', predominantly urbanized and industrialized nations which are served by dense networks of transport and communication (Jordan 1973). Indeed, the population of Western Europe is older than that of any other continent. In Germany there are already more than four pensioners for every ten workers; by the year 2030 their numbers may well be almost equal. West Europeans are rich, well fed (even over fed) and well educated. They live long, healthy and relatively comfortable lives, and display

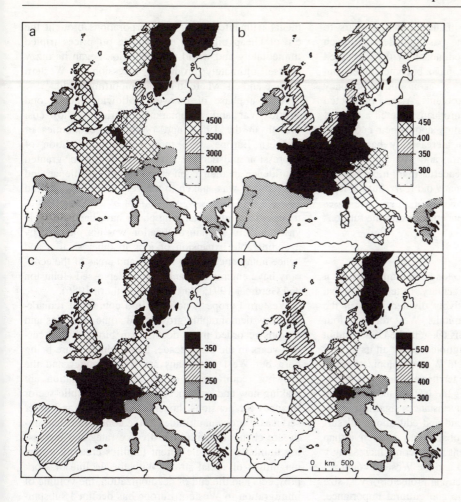

Fig. 1.2
Indicators of well-being:
(a) consumption of energy
(tonnes of oil equivalent per
capita), 1988; (b) motor cars
per thousand population,
1988; (c) television sets per
thousand population, 1988;
(d) telephones per thousand
population, 1988.

low levels of natural increase. Their agricultural systems are highly productive, regularly generating 'mountains' of surplus butter, grain and fruit and 'lakes' of surplus milk and wine which could relieve the suffering of the world's under-fed millions.

While acknowledging the general validity of this array of facts and perceptions, the mythical 'average' West European would insist that attention be drawn to at least three major qualifications. First, the countries to the west of the long-taken-for-granted East/West divide through Europe display important spatial differences with regard to the affluence and well-being of their inhabitants (Ilbery 1986). Such features emerge clearly from analyses of per capita income, energy consumption, housing quality, health care, car ownership and a host of other indicators (Fig. 1.2).

The broad dimensions of these differences are well known, with an affluent West European 'core' contrasting with a poorer Atlantic and Mediterranean 'periphery' (Ilbery 1984).

Second, these differences are simply spatially packaged expressions of social differences with regard to employment, class and access to financial resources, educational capital and political power. 'Regional problems' do not exist of themselves, in splendid spatial isolation; they are just geographical expressions of more deeply rooted social conditions which have grown in each specific political economy (Massey 1979). In Western Europe, they cannot be divorced from the working of capitalism in the late twentieth century, while in Eastern Europe they reflect the painful recent espousal of capitalist principles (Carney, Hudson and Lewis 1980; Peet 1980; Williams 1987).

Third, many of the indicators that defined the strength and general well-being of Western Europe in the eyes of the world underwent a drastic reversal in recent years. The 1950s and 1960s had been a period of unprecedented economic growth and political stability and West Europeans had been swept along in a tide of rising expectations, generally believing that they would enjoy even greater affluence in the years to come (Landes 1969). In the past two decades, such assumptions have become redundant. Widely accepted facts of economic strength have become fables and commonly held principles of economic supremacy have been converted into problems. After a protracted phase of optimism, West Europeans entered a time of disillusionment and uncertainty about their economic future. By the late 1980s increasingly harmonious relations between Washington and Moscow, coupled with reform and largely peaceful revolution beyond the 'Iron Curtain', lessened many old fears, but the fragmentation of the former USSR and of Yugoslavia raises new economic and political uncertainties for the future.

Changing conditions

In many ways the world is no longer as Eurocentric as it was earlier in the twentieth century. In terms of global strategy, Europe was effectively divided between the Communist East and the capitalist West for over four decades, with policy decisions taken in the USSR and in the USA being of ultimate significance for all its inhabitants. Since the Second World War, decolonization has brought independence to the territories that made up the empires of Belgium, France, the Netherlands, Portugal and the UK, although in many instances significant contacts with respect to trade, aid and education still survive. European culture is probably less distinctive than in the past since the continent has become increasingly open to the rest of the world and the Western countries are in the shadow of a partner and protector of vast size and unprecedented economic and cultural importance, namely the USA.

In the economic context, multinational corporations, often based in North America or the Far East, have come to exercise control of numerous branches of manufacturing and business life in Western Europe (Lee and Ogden 1976). In very many respects, the dominating influence in each West European country during many of the post-war years was that of the USA, rather than that of another European country or of Europe as a whole (Aron 1964). However, the completion of the Single European Market of the EC at the end of 1992 may impart more of a unified (West) European image to the wider world. Dramatic changes in the world energy market since the end of 1973 drove home even more forcibly the fact that West European nations were no longer the masters of their own destinies. As Sampson (1968) had stressed just a few years earlier, Europe was more than ever at the mercy of events, discoveries and decisions outside its frontiers. The decision by the Organization of Petroleum Exporting Countries (OPEC) members to abandon price-fixing arrange-

ments with international oil corporations brought the era of cheap oil to an end. Crude oil prices tripled immediately and world-wide inflation began its dizzy course, ultimately producing recession in Western Europe and elsewhere in the oil-importing world.

After a phase of recovery following the Second World War and a remarkable period of low-cost growth in the 1950s and 1960s, the economies of Western Europe were thrown into a condition of depression and decline after 1970. Taken-for-granted confidence in their economic strength, resilience and adaptability was shaken to the very roots (Dicken and Lloyd 1981). Firms went bankrupt by the hundreds of thousand and manufacturing plants have been closed in vast numbers. Deindustrialization was not matched by a comparable expansion of jobs in other sectors and hence many once-strong regions and areas of the economy have entered the ranks of the depressed (Johnston and Gardiner 1991; Martin and Rowthorn 1986).

Western Europe is now set in a context of remarkably low demographic growth and some countries and regions recorded more deaths than births during certain years in the past decade. Fewer children are being born but West Europeans are living longer and the growing requirements of the retired population are making new and quite necessary calls on public-sector finances for pensions, housing, health care and other services. European attitudes to the concept of the family are changing, with high divorce rates and increasing numbers of single-parent families creating a new magnitude of social strain. With the decline of work prospects and the end of decolonization, the volume of immigration to Western Europe has declined substantially by comparison with the peak years of the 1960s; indeed, attempts have been made by several reception countries to redirect labour migrants back home and the threat of repatriation gives rise to profound political concern. The presence of so many 'new Europeans' has introduced a striking measure of cultural diversity to many areas with, for example, Islam becoming an important new component in the religious complexion of Western Europe (Trébous 1980). To some extent the countries of the West have already become significantly multicultural, but this ongoing process continues to be surrounded by serious social tensions which have their most painful expressions in inner-city environments and post-war housing estates. The recent arrival of large numbers of asylum-seekers, refugees and 'ethnic Germans' has created new strains in many regions.

It may be argued that West Europeans have become more politically conscious than in earlier decades and certainly there is an increasing trend for all aspects of planning and public administration to be enveloped in political challenge and debate. A growing number of people are expressing their concern for social justice and are attempting to exercise their political muscle in order to change the distribution of scarce resources to the benefit of those in need. Concern for the rights (actual or claimed) of specific sectors of society and for regional minority groups is being expressed more forcefully and sometimes violently. The rise of 'green' political parties, whose aims are imbued with the principles of ecology and conservation, reflects a growing concern for the world at large; and the critical moral issues surrounding the use of nuclear energy and the possession of nuclear deterrents are receiving wider public debate than ever before. A worrying new feature has been the rise of far-right political groups in many countries, whose members preach the dangerous message of national and ethnic 'purity'.

In the 1990s, Western Europe finds itself in the apparently paradoxical situation of becoming both more united and more fragmented at the same time. The EC enlarged from six original signatories of the Treaty of Rome (1957) to twelve member states, with Turkey having applied and other Mediterranean, Scandinavian and Alpine states expressing intense interest. However, similarities of habit and behaviour are also accompanied by profound antagonisms, and the Community is far from unanimous on numerous financial and policy issues. Entry of the Iberian countries in 1986 unleashed a formidable range of new challenges for the Community; even more will surround further enlargement in the years ahead.

Within individual states many territorial minorities (the so-called 'old nations' of Europe) are expressing their desire for greater power for the management of their own affairs (Williams 1980). The very existence of such organizations manifests a basic lack of faith in the ability of national political systems to perceive and cope with economic, social and cultural questions in the outer regions during the painful transition to post-industrial conditions. Autonomist or nationalist sentiments have emerged from Basques, Bretons, Catalans, Corsicans, Flemings, Scots, Walloons, Welshmen and from members of various other groups. It may be argued that regionalization and devolution of power in Belgium, France, and Spain have reduced the grasp of some centralized national administrations; but demands for greater regional power continue to be pressed in many parts of Western Europe and the problems of Northern Ireland remain particularly desperate (Boal and Douglas 1982). Repeated outbreaks of violence in Corsica since the early 1980s reflect the divergence of views between relatively liberal members of the island's regional assembly and those held by political extremists belonging to the now outlawed libera-

tion front. In Spain, the violent activities of ETA, the armed wing of the Basque separatist movement, provoke widespread public outrage. French authorities have made serious efforts to dislodge suspected Basque guerrillas from their traditional sanctuaries in neighbouring parts of the French Basque country along the Pyrenean border (see Fig. 2.2)

In many ways, Western Europe is more open and accessible than in the past. West Europeans now travel to neighbouring countries more easily than before, with growing prosperity, improved holiday benefits for workers, cheap air travel, package holidays and rising rates of car ownership in the 1950s and 1960s having guided the trend and fuelled the demand for new vacation experiences (Williams and Shaw 1991; Urry 1990) (see Ch. 10). Despite recent economic and social ills, the number of cars in Western Europe continued to rise from 69 million in 1971 to 142 million in 1992. High rates of acquisition characterized the new consumer societies of Greece, Portugal and Spain, while much lower rates were found among the fairly 'saturated' car-owning societies of Denmark, Sweden and the UK.

In spite of unemployment and profound economic difficulties, the number of tourists visiting the countries of Western Europe continued to rise each year during the 1970s, and in some respects tourism was considered as one of the few growth activities at a time of recession. In 1985 56 per cent of the inhabitants of the Twelve went away on holiday, with proportions rising as high as two-thirds of Dutch and Danes but falling to two-fifths of Portuguese, Irish and Belgians. Two-thirds of West Europeans spend their holidays in their home country and this share is far exceeded in Mediterranean states (Greece, Spain, Portugal each 92 per cent, Italy 87 per cent). Perhaps surprisingly only 3 per cent of residents of the EC take their holidays outside Western Europe (Eurostat 1991a). As a result of these demands the environments of many European localities continue to undergo dramatic remodelling to cater for the needs of tourists and many regional economies are highly dependent on their presence and purchasing power (Naylon 1967; Morris 1985; Morris and Dickinson 1987). Over 3 per cent of the workforce of the EC depends entirely on tourism for its livelihood and a further 7 per cent derives seasonal or part-time employment from it. However, it must be admitted that in the last few years a duality of holiday taking has emerged. Many West Europeans economize on their holiday budgets and satisfy themselves with shorter vacations or destinations close to home. By contrast, those in secure, well-paid employment take several holidays or visit increasingly distant destinations just as an extrapolation of the outward-thrusting trends in the 1960s would lead one to expect

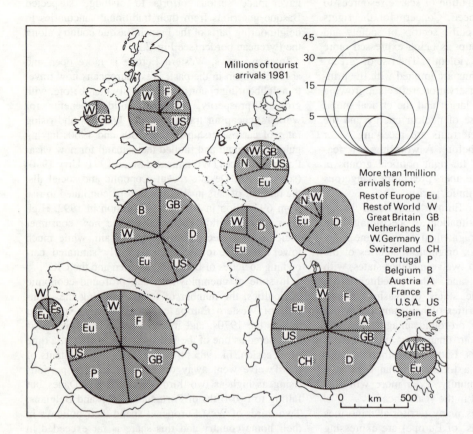

Millions of tourist arrivals 1981

More than 1 million arrivals from:

Rest of Europe	Eu
Rest of World	W
Great Britain	GB
Netherlands	N
W.Germany	D
Switzerland	CH
Portugal	P
Belgium	B
Austria	A
France	F
U.S.A.	US
Spain	Es

Fig. 1.3
Top ten West European
nations for the arrival of
foreign tourists, 1981.

(Williams and Zelinsky 1970). Over half of the citizens of the EC go away on holiday at least once each year and tourism represents just over 5 per cent of the GDP of the countries of the Twelve.

In the early 1980s, Italy, Spain and France occupied the leading places among the top ten West European states for the arrival of foreign tourists and together accounted for two-thirds of foreign inflows (Fig. 1.3). A quarter of tourists travelling to a destination in the top ten countries came from West Germany, with a further 21 per cent originating in France. Many tourist trips tend to be dominated by a dual quest for sunshine and 'the familiar' (especially in matters of food) but none the less, each generation of West Europeans is undoubtedly visiting (if not always really experiencing) other parts of its native continent more frequently than members of every generation that preceded it. The European Commission is keen to assist further development of tourism in the Twelve by easing border checks, improving transport facilities, extending traditional holiday 'seasons', and encouraging rural, cultural and other less familiar types of tourism, thereby boosting this form of activity in localities where economic initiatives are greatly needed.

The harsher than expected economic, social and political realities that many West Europeans are undergoing at this time of recession are literally brought into our gaze and into our living rooms by news and current affairs reporting on television and in the columns of the more serious newspapers. The geographical background to these issues forms the substance of this book which ranges across eighteen countries and does not attempt to focus exclusively on the EC. The following chapters do not claim to be all-embracing thematically, nor can they dwell on the specificities of place or of organizational system in the countries of the West. Instead, they are written at a higher level of generalization, in full recognition of both the advantages and the limitations of such an approach.

Further reading

Economist 1992 *Atlas of the New Europe*. The Economist, London

Hudson R, Rhind D, Mounsey H 1984 *An Atlas of EEC Affairs*. Methuen, London

Lee R, Ogden P E (eds) 1976 *Economy and Society in the EEC*. Saxon House, Farnborough

Mead W R 1982 The discovery of Europe. *Geography* **67**: 193-202

Williams A 1987 *The Western European Economy*. Hutchinson, London

2

Political evolution

Security, democracy and trade

The political geography of the twentieth century in Western Europe has been characterized by a series of dramatic upheavals, triggered by the gradual disintegration of the empires that were so dominant over the previous 200 years in the face of assertive nationalism. The fragmentation of the Austro-Hungarian, the German and the Ottoman empires into a multitude of independent nation-states, not to mention the subsequent loss of world-wide colonial territories by Belgium, the UK, France, Germany, Italy, the Netherlands, Spain and Portugal, made a fundamental realignment and reassessment of Europe's global role in the latter half of the twentieth century inevitable (Harris 1977; Heater 1992).

After 1945, the most obvious manifestation of the changed circumstances was the ideological schism that resulted in the continent being physically divided into two by the Iron Curtain for more than forty years. Western Europe became closely allied to the political and economic ideals championed by the USA, while Eastern Europe languished under the Communist ideology of the USSR. For more than a generation this division stood clear and seemingly immutable, symbolizing an uneasy political balance, not only in Europe but generally across the globe.

During this period Western Europe was guided politically by three overwhelming concerns. First, in the years immediately after the end of the Second World War, when the cold war was at its height, the need for collective security against both a resurgent Germany and the rapacious expansionism of the USSR was the prime objective. Although there was some subsequent relaxation in the tensions caused by the threat of imminent attack as the years went by, the perceived need for collective security remained a powerful cohesive force throughout Europe, in the East as well as in the West (Gorbachev 1987).

The second concern was the widespread desire to see democratic governments in all the individual states in Western Europe (Everling 1980). Great care was taken by the Western Allies to ensure that democratic institutions replaced the paraphernalia of the Fascist dictatorships in both Germany and Italy after 1945, and the Western European nations as a group exerted pressure on any of their number with a totalitarian regime to ensure its demise. In Portugal, the successive dictatorships of Salazar and Caetano were overthrown by a military *coup d'état* in 1974, to be followed rapidly by free elections, and in Spain the death of Franco in 1975 was the signal for an almost immediate return to democracy. In both cases, a desire to end the isolation imposed by their West European neighbours was important in contributing to the change. Equally in Greece where, after a *coup d'état* in 1967, a military junta ruled until 1974, international disapproval and actual exclusion from most areas of European decision-making in the European Community, the Council of Europe and the North Atlantic Treaty Organization were critical factors in ensuring that the suspension of democratic rule was only an interlude (Georgakopoulos 1980). Subsequently, in Turkey where in 1980 there was also a successful *coup d'état*, international pressure eventually persuaded the government to hold democratic elections.

The third unifying concern was a general determination to reduce and, if possible, abolish the multiplicity of trade quotas and tariffs that existed between European states in the years immediately after the end of the Second World War. Throughout the 1950s and 1960s, negotiations were entered into at a number of different levels within the framework of the General Agreement on Tariffs and Trade, the Organization for Economic Cooperation and Development (and its predecessor the Organization for European Economic Cooperation), the European Economic Community and the European Free Trade Association (EFTA) with the result that by the 1970s there was almost completely free movement of industrial goods within Western Europe and, indeed, among all the major Western trading nations (Hechter and Brustein 1980). While the individual economies of all the West European states were expanding in the affluent years of the 1960s, successful extensions to the principles of free trade were relatively easy to achieve. But it is indicative of the strength of the commitment that even in recession it has never seriously wavered. The trebling in the world market price of crude oil in the years after 1973 put great pressure on West European economies and ever since their fortunes have fluctuated, with severe downturns in both the early 1980s and the early 1990s; but, despite these vicissitudes, the fundamental structure of free trade within Western Europe has remained virtually unscathed. Indeed, it has been progressively strengthened. In 1991 the seven members of EFTA joined with the twelve members of the EC to form the European Economic Area, almost certainly the first step towards a single community (and market) of 375 million people. Meanwhile, within the EC the whole concept of free trade has been further developed with the creation of the single internal market from the beginning of 1993.

Since 1989, the tense certainties of the cold war era have been shattered by the collapse of the Soviet Empire and the resulting political power vacuum in Eastern Europe. As far as Western Europe is concerned the new situation offers great opportunities, but also many dangers. Without exception, the former client states in the East, as well as the former Soviet republics, have rejected Communism and in a variety of different ways looked to the West to provide a role model and economic support for their future development. For Western Europe the open invitation to extend its economic and political philosophy into the area that had previously been most vehemently opposed to it has been both flattering and heady, but it has also raised serious questions about its capacity to deliver. The major problem is that the political map in much of Eastern Europe has disintegrated along with the USSR. Unification in 1990 saw the two Germanies merged into a single Federal Republic; Yugoslavia has collapsed into civil war and is gradually being reconstituted as a mosaic of independent republics; and Czechoslovakia divided in January 1993 into separate Czech and Slovak states. As far as Western Europe is concerned all these internal changes make it very difficult to develop any coherent strategy to address the massive social and economic problems facing its eastern neighbours.

To describe the desire in Western Europe for collective security, democracy and free trade as three separate forces is to some extent misleading; essentially they are different manifestations of a remarkable sense of purpose that has broadly united the region for nearly half a century. The key to this unity has been the overwhelming perceived threat from the USSR in the East, which traditionally has enabled economic, and other, differences in Western Europe to be subsumed in the interests of regional security. Now that this threat has been removed, it may well prove more difficult to sustain the general commitment to unity. There have always been substantial economic discrepancies between countries and regions in Western Europe and they have tended to become more marked and institutionalized as the region as a whole has become more integrated (Wallace 1990). When the EC comprised just six mainland states, it was entirely reasonable to think in terms of a single economic space. Now it embraces twelve countries, some of which like Ireland, Portugal, Spain and Greece are on the Atlantic and Mediterranean peripheries and have economies that are substantially weaker than those of the industrialized core, and it may well prove more difficult to sustain political unity in the face of such economic discrepancies. The opening up of Eastern Europe has made the task even harder. This new eastern periphery is even poorer and the sheer scale of its economic problems means that the West European core will be unable to generate sufficient wealth to satisfy all the demands put upon it. Seeds of dis-content between rich and poor could encourage political rivalry and dissension, threatening the unity of purpose that has so impressively characterized Western Europe since 1945.

The North Atlantic Treaty Organization and collective security

The desire throughout Western Europe for a common defence front against any threat, either external or internal, to the democratically elected governments in the region is embodied in NATO. For the greater part of its existence since 1949 the overwhelming danger was seen as the USSR, but since 1989 with the demise and eventual collapse of the Soviet Empire the alliance has been forced to reassess its orientation and purpose.

Somewhat surprisingly, given the ferocity and destruction of the Second World War, the need for a formal expression of solidarity in the form of a common defence treaty was not immediately apparent to the victorious Western Allies in 1945. In the euphoria that prevailed at the end of the six years of conflict it was rather naïvely assumed that the United Nations

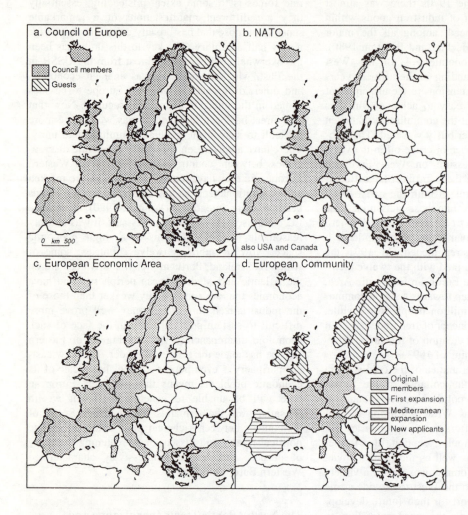

Fig. 2.1
European institutions

and its charter, to which all the European states were signatories, provided sufficient guarantees of protection. Such hopes were of course short-lived; the repeated attempts at destabilization by the Communist Party, not only in Eastern Europe, but also in Greece and Turkey and even Belgium and France, caused great alarm both in Western Europe and in the USA and evoked a decisive strategic response. The United States Congress accepted the Truman Doctrine in 1947, thus pledging American support for free peoples anywhere who were resisting subjugation by armed minorities or outside pressure. The first beneficiaries of the new commitment were Greece and Turkey, where Communist *putsches* were effectively extinguished thanks to military and economic aid from the USA. In north-west Europe, Belgium, France, Luxembourg, the Netherlands and the UK signed the Brussels Treaty in 1948, guaranteeing the support of

all should any one of their number be attacked.

Wherever the Communist Party tried to secure support in free elections, as in Berlin in 1946, it fared badly and this fact, together with the growing military cohesion between West European states, drove the USSR into taking pre-emptive action. On 24 June 1948, it imposed the Berlin Blockade and threatened to precipitate just the kind of confrontation that the Brussels Treaty was designed to combat. For their part the Americans sought to widen the concept of collective security beyond Western Europe to include Canada and the USA as well, thus encompassing the whole of the North Atlantic area. The idea was widely welcomed and NATO was born in the spring of 1949. The original members were the five signatories of the Brussels Treaty, together with the USA and Canada and five further European nations: Denmark, Iceland, Italy, Norway and Portugal. In 1952, Greece and

Turkey were admitted, along with West Germany in 1955. Since then, formal membership has changed little. France withdrew from the military committee, though not the alliance itself, in 1966 and Spain became a member in 1981 (Fig. 2.1).

First and foremost NATO has always been a military alliance, built around the concept of collective security. An armed attack against any one of the signatories in either Europe or North America would be considered as an attack against them all. The geographical scope of the treaty is defined very precisely as the territory of any one of the member nations and islands administered by them north of the Tropic of Cancer. It also covers attacks on ships and aircraft in the North Atlantic and Mediterranean areas.

The organization is, however, more than just a military pact: it provides for cooperation in a much wider field and has always been seen as a means of defending a way of life, not only by military might, but also by joint action in political, social and economic matters. One of the most important instances of such cooperation was the attitude of the organization to the membership of West Germany. When agreement was reached in 1954 that the new state should be allowed to join as a full member, the other signatories to the treaty implicitly accepted West Germany's own, then controversial, view of its status as the only legitimate German state with the right of representation for the whole of the German nation, east as well as west. It meant that East Germany was not recognized in any shape or form by NATO and its members, a crucial factor in formalizing the political confrontation in Europe and the divisive influence of the Iron Curtain (Blacksell 1982). Eventually, the revolution in 1990 which led to the unification of the two Germanies in an enlarged Federal Republic went a long way towards vindicating West Germany's traditional stance (Harris 1991).

Until the middle of the 1960s, NATO was overwhelmingly concerned with containing the political and military threat posed by the USSR. There was little open dissension among the member states and private misgivings were put aside in the face of brutal acts of repression, such as the invasion of Hungary by Soviet troops in 1956. This unity was extremely important, because it supported the prevailing orthodoxy that world politics in the middle of the twentieth century was a matter for negotiation between the two superpowers – the USA and the USSR.

Nevertheless, there was always a degree of unease in Western Europe about the wisdom of identifying so closely with the USA, and gradually a view began to gain ground that NATO was a positive impediment to the relaxation of international tension. Pre-eminent among the doubters was France under the forceful leadership of President de Gaulle. He had an alternative vision of the distribution of political power, preferring a European axis linking Paris to Moscow, rather than what he saw as the Anglo-Saxon domination of NATO. In March 1966, when all French forces were withdrawn from the NATO command, even though France formally remained a member of the alliance, the administrative facilities in France had to be moved, the military headquarters going to Brussels, the Defence College to Rome and the Central European Command to Brassum in the Netherlands. Although French disenchantment had been apparent for some years, the actual withdrawal came as a shock to the other members, undermining the fundamental unity of purpose and weakening the self-confidence of the whole alliance.

Further evidence of reassessment elsewhere in Western Europe was provided by the *Ostpolitik* of the SPD/FDP coalition headed by Chancellor Brandt, which came to power in 1969 in West Germany (Sallnow and John 1982; Blacksell and Brown 1983). It was the first time the socialists had ever been the party of government and they were eager to demonstrate both their independence and their capacity for new initiatives. A whole series of treaties and non-aggression pacts was negotiated with the countries of Eastern Europe, including one with the USSR in 1972. A dialogue was also started with East Germany which led to the signing of a treaty, the first step in normalizing relations between the two states. While it was always strenuously denied that these developments were intended in any way to undermine the status of NATO, inevitably they represented a softening of the antagonistic stance that had prevailed hitherto towards the regimes in Eastern Europe.

It is probable that NATO's cohesion would have been more seriously undermined had it not been for the invasion of Czechoslovakia in 1968 by armies from the Warsaw Treaty Organization. The ruthless suppression of the tentative attempts by the Czechoslovaks to broaden their contacts with their Western European neighbours was a sharp reminder to the NATO allies that talk at that stage of political and military realignment in Europe as a whole was still very premature.

Even so, the 1970s was a period when NATO's influence was relatively weak. The military junta that took power by force in Greece in 1967 severely dented the general belief in the ultimate, inevitable triumph of democracy. It also meant that Greece was no longer able to play a full part in NATO's defensive commitments. In 1976, a rumbling territorial dispute between Greece and Turkey over Cyprus and various other islands in the Aegean Sea was only narrowly prevented from erupting into war (Rustow and Penrose 1981). It was another crack in the façade of unity, which had been such a feature of the early years of the alliance. Further, more minor evidence of internal conflict was the so-called Cod War of 1972–73 over fishing rights in territorial waters between Iceland, the UK and to a less degree West Germany (Mitchell 1976). Perhaps the most serious blow to NATO's prestige, however, was the increasingly frequent bilateral negotiations between the USSR and the USA over reductions in nuclear armaments, which took place outside the framework of the alliance. At the time it was organized in 1975, the European Security Conference in Helsinki was heralded as the greatest single step towards military *détente* in Europe during this century.

After 1980, NATO somewhat unexpectedly again assumed a more central role in the development of Western Europe. The invasion by the USSR of Afghanistan in 1980 and its shadowy and sinister role in the imposition of martial law in Poland in 1981 succeeded in reawakening many of the fears of the cold war in relations between Western and Eastern Europe and certainly rekindled interest in NATO among its member countries. A measure of the revival was the fact that Spain, after much agonizing, became the thirteenth member in May 1981 (Minet, Siotic and Tsakaloyannis 1981). Such a move had been debated ever since the end of the Franco dictatorship in 1975, but fears of being too closely identified with the foreign policy of the USA had always caused the Spanish government to draw back from actually signing the treaty. The growing tension in the relations between East and West decided them in favour of joining and although subsequently there was a referendum on the issue, Spain remains a full member of the alliance. The succession of bilateral agreements between the USA and the USSR in the mid-1980s limiting the numbers of strategic nuclear weapons also served to strengthen, rather than weaken NATO. They shifted the emphasis in defence to modern conventional weapons as the guarantee of long-term security in Europe.

The disintegration and eventual demise of the USSR in 1991 have inevitably called into question the whole future of the NATO alliance. The Warsaw Treaty Organization has been disbanded and, rather than facing a monolithic Communist empire across the Iron Curtain to the east, NATO now has the task of forging links with multiplicity of newly independent, nationalist states both in Eastern Europe and what used to be the USSR. The old division remains in that the sphere of influence of NATO is still restricted to Western Europe and its forces may not even be stationed in what was East Germany. However, Russia

and a number of East European countries are now represented at some NATO meetings and the whole structure of the alliance is slowly being rethought.

The difficulties attending the redefinition are starkly illustrated by events in what was Yugoslavia. The former federation began to fragment into separate states early in 1991, the process of separation becoming increasingly bloody as it proceeded. Full-blown civil war is now raging in Bosnia-Herzegovina yet NATO is powerless to intervene, as are most other pan-European bodies, such as the EC and the Western European Union (WEU). It is now clear that NATO, as originally conceived, was organized primarily to counter Communist aggression from the East and that it is largely impotent when it comes to internal conflict, such as is occurring along its eastern border, even though there is a very real possibility that the fighting may spill over into some of its member countries, notably Greece and Turkey.

A common culture and the search for a European identity

The Council of Europe

Fear of Soviet aggression was not the only pressure drawing the countries of Western Europe closer together; in the wake of the Second World War there was a strong, and enduring, sense of the need to develop a European identity and stress the positive features of a shared culture. Over the years the movement has developed a number of institutional forms with varying degrees of success. Most have been primarily concerned with economic cooperation, but others have taken wider terms of reference. The most ambitious conceptually is the Council of Europe, set up in 1949 with a view to achieving unity between its members so as to safeguard their common heritage and facilitate social and economic progress. The original members were Belgium, Denmark, France, Italy, Luxembourg, the Netherlands, Norway, Ireland, Sweden and the UK, quickly followed by Greece, Iceland, Turkey, the Saar and West Germany (1950) and Austria. Cyprus, Switzerland and Malta all joined during the 1960s; Portugal, Spain and Liechtenstein during the 1970s; and Finland and San Marino during the 1980s. By that time the Council encompassed the whole of Western Europe, except for the micro states of Andorra, Monaco and the Vatican (Catudal 1975) (Fig. 2.1).

The broad scope of the Council was one of its strengths, for it could truly claim to speak for Western Europe as a whole, but it also proved one of its severest limitations. The areas of agreement and cooperation among the twenty-three members tended, almost inevitably, to be general and unspecific, making difficult the formulation of detailed policies. There was, however, universal acceptance of the need for democracy and the protection of individual human rights, and states proved sensitive to censure by the Council in these matters. Greece's membership was suspended between 1969 and 1974 while the country was ruled by the military junta and neither Portugal nor Spain was able to join while governed by a dictatorship. Equally violations of human rights were regularly investigated and, if necessary, criticized by the European Court of Human Rights established in 1958. For the most part, member countries were quick to remedy any shortcomings identified.

The progress made by the Council towards institutional initiatives leading to formal integration was disappointing and gradually caused the organization to be edged out of the mainstream of political reorganization in Western Europe. The Council never acted as a forum, let alone a focus, for the debate about economic cooperation, and the sensitivity of some members to any reference to national defence resulted in the subject being the only one specifically excluded from consideration by the founding statute. This was in deference to the same neutral nations – Austria, Ireland, Sweden and Switzerland – which found it impossible to join NATO and which would have refused to have anything to do with the Council of Europe had there been the slightest suspicion that it might encompass military matters. The lack of agreement about a substantive agenda for discussion and the tacit acceptance that controversial matters should not be debated if there were a chance that they might cause members to leave were the main limitations on the Council's effectiveness. On the other hand, it meant that there was always a forum in Western Europe where any state could air its concerns at government level, a virtue well illustrated by the Council's unwavering support for democracy in Greece.

From its inception the Council always deplored the cold war division of Europe, but was universally rejected by the Communist regimes in Eastern Europe. Events since 1989, however, have eventually provided the opportunity for it to extend its horizons eastwards. Czechoslovakia, Hungary and Poland are now all full members and 'special guest status' in the Parliamentary Assembly has been conferred on Albania, Bulgaria, Estonia, Latvia, Lithuania, Romania, Russia and Slovenia. It also seems certain that it is only a matter of time before most of the other new states that have emerged, or are emerging, are similarly included. The Council is an important civilizing influence and the countries of Western Europe, and now Eastern Europe and beyond, have looked to it to legitimize their standards of economic, social and political behaviour for nearly half a century. Nevertheless, it remains marginalized; in the ferment of change now occurring it is still basically a forum and not the real focus for the new European order.

Nordic initiatives

For the four Scandinavian states – Denmark, Finland, Norway and Sweden – together with Iceland, the Council of Europe has never offered a wholly satisfactory, or sufficient, answer to their desire for closer political cooperation (Smidelius and Wiklund 1979). Nordic unity has a sporadic history reaching back to the late fourteenth century, but it really acquired popular appeal, if not tangible success, in the nineteenth century. The first really significant recent development was the setting up of the Nordic Council in 1952. It was a loose-built, ministerial forum that was convened about once a year and most of its achievements have been the result of the endeavours of permanent committees of civil servants working under the Council's general guidance. Not surprisingly, the most progress has been made on detailed issues such as customs control, railway freight tariffs, vehicle insurance and the free movement of labour; larger and politically more sensitive issues have proved much more intractable. Nevertheless, citizens of all five member countries can move freely throughout any part of Scandinavia and have the right to seek work wherever they wish. For unskilled and semi-skilled jobs, this principle was quickly established but, in common with other parts of Europe, it proved much more difficult to gain agreement about the interchangeability of professional qualifications. It was not until the late 1960s that all restrictions were finally removed (Turner and Nordquist 1982).

Even within this relatively homogeneous group of countries with a long history of cooperation, progress towards integration has not always been smooth (Smidelius and Wiklund 1979). The political constraints placed upon Finland as part of the price for the ending of Soviet wartime occupation prevented it from joining the Nordic Council until 1955. Strains have also been caused by the differing approaches adopted by the Scandinavian countries towards closer economic cooperation with the rest of Europe and the EC.

In the 1960s, when all but Iceland belonged to EFTA but none to the EC, a separate customs union in Scandinavia seemed a desirable and achievable objective. Nordek was agreed in principle in 1969, but as the details began to be debated the whole idea lost favour as the disadvantages of economic union to individual countries began to be apparent. Eventually, Finland withdrew in 1970 and the concept foundered when Denmark became a member of, and all the other Scandinavian countries concluded separate trade agreements with, the EC in 1973. After protracted negotiations, Nordek was rendered even more irrelevant in 1991 when the members of EFTA and the EC combined to form a single free trade zone, the European Economic Area (Fig. 2.1.)

Scandinavia as it has existed for the past half-century is in the process of being altered fundamentally by the collapse of the USSR. Historically, there have always been close cultural and economic links between the Baltic states, but Latvia, Lithuania and Estonia and, to a lesser extent, Finland were all forcibly integrated into the Soviet bloc throughout the cold war. Now that they are once again independent states all are seeking to re-establish their traditional contacts in the Baltic region. As yet, these feelers have produced only tentative results, but it seems certain that the boundaries of Europe will shift decisively eastwards here to create an enlarged Scandinavian or, more accurately, Baltic region. Whether it remains as closely orientated towards Western Europe as Scandinavia has been in the post-Second World War era is a matter of speculation, but no region is better equipped to exploit the opportunities in the emerging market economies of the republics of the former USSR.

Boundary disputes and regional autonomy

The predominant political trend in Western Europe in the latter part of the twentieth century has been closer cooperation, but this has been by no means exclusively the case. Given the large, and steadily growing, number of small states and the turbulent history of the continent in the past two centuries, it would have been surprising had the political boundaries and resulting allegiances been universally accepted without dissent (Williams 1980). Minority languages such as Basque, Catalan, Flemish, Gaelic and Welsh have all been the catalyst for struggles to prevent cultural identities being swamped by the seductive attractions of large-scale political groupings (Stephens 1978; Teumis 1978). It is all too easy for national governments to dismiss these movements, but it should be remembered that many, such as for example the Scots in Scotland, represent populations and areas larger than some of the independent states in Western Europe. Even so, the only recent example in Western Europe of a completely new independent state emerging from a regional struggle is Ireland in 1922, but Iceland did not become fully independent from Denmark until 1944. In contrast, the break-up of the USSR and the demise of the Communist governments in Eastern Europe has resulted in a widespread upsurge in regional nationalism, creating a plethora of new states, with the probability that more will follow. Many, such as Slovenia and Croatia abut, or are close to, the eastern boundary of Western Europe and it is almost inconceivable that they will not soon be viewed as an integral part of the region (Fig. 2.2).

Within Western Europe the Second World War left a number of unresolved border issues in its wake, though most have subsequently been settled. It was uncertain whether the Saar should be incorporated into France or West Germany; whether the South Tyrol should be part of Austria or Italy; and where the boundary between Italy and what was Yugoslavia (now Slovenia) should run. Eventually, after a plebiscite, the population of the Saar elected to become part of West Germany; after a number of years of intermittent guerrilla activity, the existing boundary between Austria and Italy in the Tyrol was accepted; and in 1979 a final line was agreed for the boundary between Italy and Yugoslavia ensuring that the much disputed city of Trieste was incorporated into Italy. Most intractable of all was the future of Germany, split into two states, with the status of its former capital in limbo administered jointly by the American, the British, the French and the Soviet governments. In the space of a year between the autumn of 1989 and the autumn of 1990, as part of what now can be seen as the collapse of the USSR and its influence in Europe, the Berlin Wall was demolished and Germany was unified. The state of what had been West Germany expanded to encompass both East Germany and Berlin, in the process consigning those two symbols of the cold war in Europe, East Germany and West Germany, to the dustbin of history (Wallace and Spät 1990).

There are a number of other disputes where a lasting solution has proved more elusive. Most involve a small minority with a distinctive culture and language in danger of being swamped by the national majority. In the northern part of Belgium the population is predominantly Flemish speaking, whereas in the south French is the norm (Ashworth 1980). Although the state is officially bilingual, the conflict between the two factions has been extremely acrimonious, threatening the stability of the national government (Stephenson 1972; Hill 1974).

The United Kingdom, as its name implies, is an amalgam of various national groups, and the Scots, the Welsh and the Irish have each felt at some time or other that their culture and people have been poorly served by the predominantly English government in

Fig. 2.2
Selected regional minorities
in Western Europe

London. In Northern Ireland an overlay of bitter religious conflict between the Roman Catholic minority and the Protestant majority makes the divisions particularly deep and acrimonious, illustrating the inadequacies of partition as a political solution to such disputes. The establishment of the Irish Free State in 1922 was designed to provide a state where aims of Irish nationalism could be realized, but however carefully the boundaries were drawn minorities on both sides were always inevitable. The result has been continuing conflict and, since 1969, there has been an intermittent state of undeclared war in Northern Ireland, the object of which has been to transfer the province from the UK to Ireland. Ironically, the short-term result has been the exact reverse. For fifty years from 1922 Northern Ireland exercised autonomy over most of its internal affairs, but the growing civil unrest caused nearly all these powers to be suspended in 1972 in favour of direct rule from London. Ever since, repeated attempts to restore a devolved government have

been frustrated by the continuing conflict (Boal and Douglas 1982).

Spain is another country where the regional divisions are deep and seem unlikely to be solved either peacefully or quickly. The return of democratic government in 1975 after more than forty years of Franco's dictatorship revived the political aspirations of several of the country's regions (Ling 1973). In northern Spain the people of the Basque country have long hankered after independence and have pursued a persistent and often bloody campaign to gain acceptance for their cause. Similar, though less violent movements have also begun to flourish in Catalonia in the north-east and in Andalusia in the south.

Elsewhere, the calls for autonomy are more spasmodic. In the UK, Scotland and Wales have flirted intermittently with independence, but in both countries when matters came to a head in 1978 and 1992 the people rejected any weakening of the union, the first time in a referendum, the second in a general election. In France, Breton nationalism briefly threatened to become a political force to be reckoned with in the 1970s, but the only enduring regional discontent is on the island of Corsica (Kofman 1982).

It is important not to over exaggerate the wider significance of these regional power struggles. If their cases are to receive any general attention, then minorities have to exploit every possible means of attracting publicity for their causes. However, in Western Europe as a whole over the past five decades, governments have been at pains to pursue European unity, believing that the things binding the various countries in the region together are more important than those that divided them.

Free trade and the road to economic integration

If cultural unity in the sense that Jean Monnet used the term when he was trying to gain support for his concept of a United States of Europe in the 1940s has remained elusive, real progress has been at the rather more prosaic level of economic integration. In the late 1940s there was widespread frustration and dissatisfaction with the complex network of tariffs and quotas which seriously inhibited trade. A general consensus emerged that if they could be dismantled and replaced by a system of free trade, then this would provide the stimulus for a much-needed economic revival.

In fact, the dissatisfaction and general recognition of the need for change can be traced back to well before the outbreak of the Second World War, but it was only in its aftermath that significant progress began to be made. Throughout the years of conflict the USA had poured aid across the Atlantic with a view to freeing Europe as a whole rather than any individual county. Despite some initial reluctance and hesitation, the American government also agreed with a generosity, unparalleled before or since, to secure the peace as it had financed the war by continuing the flow of economic aid in the form of both money and raw materials. The main instrument was the European Recovery Programme, widely referred to as the Marshall Plan, mooted in 1947 and put into operation the following year. Between 1948 and 1952, the equivalent of $US13 500 million was distributed under the aegis of the programme and eventually these resources did provide the intended stimulus necessary for the revival of the shattered industrial economies of Western Europe (Blacksell 1981).

Initially, however, considerable problems were raised by the American gesture, because no Europe-wide institution existed for managing the distribution of monies among the individual states. The impasse was finally broken by the creation of the Organization for European Economic Cooperation (OEEC), the main task of which was to devise a formula for sharing out Marshall Aid. In common with virtually all the other umbrella organizations founded at this time, the OEEC also had a broader and more grandiose objective of re-creating a sound and viable European economy. It sought to achieve this by facilitating free trade and, as far as possible, removing all barriers to the movement of goods among its member countries. The strategy had two distinct, though closely linked, aspects: the removal of tariffs and quotas on trade and the maintenance of stable currency exchange rates.

Progress on both was spectacular and by the late 1950s there was almost total free trade in industrial goods and the currencies of all the member states were freely convertible. The latter in particular was a momentous achievement, due almost entirely to the success of a subsidiary of the OEEC, the European Payments Union (EPU), which provided for the multilateral settlement of trading accounts and also provided guarantees to ensure the stability of the weaker currencies. In practice, the EPU meant that rather than having to settle every trade transaction with any other in an acceptable (hard) currency, it could calculate the sum of all its transactions with the whole group and settle the account on a monthly basis. The problems caused by fluctuations in the value of European currencies were overcome by the EPU guarantees. The system provided precisely the kind of stability and reduction in bureaucracy which enabled international trade in an expanding economy to flourish. The average increase in the volume of trade in real terms throughout the OEEC countries in the decade 1950–60 was 47 per cent, and all the members, except for Iceland and Turkey, shared in the improvement. There was, however, considerable variation, with West Germany (182 per cent) and Austria (104 per cent) recording increases well above the norm. There is little doubt that the success of the EPU was one of the outstanding achievements in the whole process of political and economic integration, and the speed and lack of acrimony with which it was created contrast starkly with the wrangling and long-drawn-out argument that preceded many later developments.

By the end of the 1950s, the OEEC had accomplished all its major objectives and its exclusively West European orientation was coming to be something of an impediment. The member countries wanted to see the principles of free trade extended world-wide and to be in a position to deal with the other major Western trading nations on a more equal footing. In

1958 the EPU was superseded by the European Monetary Agreement which extended the scope of the EPU arrangement beyond Europe to the whole of the pound sterling and French franc areas, while, at the same time, most West European currencies became freely convertible with the US dollar. The changes meant that barriers to international trade between all the main industrial nations of the non-Communist world were reduced to a level that had never previously been possible. In 1961, the OEEC itself was wound up and replaced by the Organization for Economic Cooperation and Development (OECD) which eventually numbered the USA, Finland, Canada, Japan and Australia among its members. This new organization provided the framework for sustaining economic growth through trade and sought to spread the benefits of industrialization to the developing world by drawing the economically less developed countries more closely into the system of international trade. Throughout the 1960s the OECD was generally judged to be highly successful, but more recently it has proved only moderately effective in protecting developing countries from the effects of economic depression in the industrial nations and at ensuring a fairer distribution of wealth between rich and poor countries. The OECD is also not an exclusively European organization; the focus of its work is global, rather than concentrating on promoting the economic health of one particular area.

It would be misleading to give the impression that the enthusiasm for European integration came mainly from the USA, because it wanted to see economic recovery without the political complications of having to deal with a multiplicity of different governments. The attractions of closer cooperation have been obvious to most shades of European opinion for a long time, and the Western European governments themselves have made by far the most significant long-term contribution to integration in the form of the EC.

The seeds of this initiative go back as far as 1921 when the Belgian and Luxembourg governments agreed to remove all restrictions on trade and commerce between their two countries. In 1944, the Dutch government in exile in London decided to join the arrangement, although the decision did not take effect until 1948, well after the end of the Second World War. The Benelux agreement, as it was called, was only a small beginning among three of the smallest states in Western Europe, but it was nevertheless of considerable importance. The emphasis on free trade, primarily in industrial goods, did much to set the tone of the subsequent more comprehensive initiatives which flourished in the general atmosphere of economic

buoyancy and expansion that pervaded the 1950s and 1960s.

The European Community

Early moves

The origins of the EC as such have their roots in the chaotic political and economic aftermath of the Second World War. The industrial nations of north-west Europe had been severely disrupted by the conflict and one of the essential prerequisites for revival was the creation of a more integrated and mutually interdependent industrial structure. The critical stumbling block was finding a way of readmitting Germany to the industrial system without allowing it to develop the political means once more to threaten the peace. The urgency of finding a solution was increased, because by 1950 the USA was heavily involved in another war under the auspices of the United Nations in Korea, and there was a world-wide shortage of steel for manufacturing military supplies. In this situation it was somewhat perverse not to be exploiting German and other European capacity to the full for rather doctrinaire political reasons.

The deadlock was broken by the French, a number of whose leading politicians had been advocating greater political and economic integration in Europe for a number of years. Their Foreign Minister, Robert Schuman, proposed the setting up of a European Coal and Steel Community (ECSC) in 1950, primarily to manage jointly the coal, iron and steel production of France and Germany, but open to any of the industrial countries in north-west Europe. Enthusiasm for the plan varied. The newly formed federal republic in West Germany was very enthusiastic, as it rightly considered that membership of such an organization would confer a measure of much-needed political respectability. The Benelux countries all welcomed the idea as an extension of their own supranational initiatives. Italy was also more than ready to take part because of the implied international approval that membership would bring in its wake. On the other hand, the UK, then the largest industrial country in Europe, was distinctly lukewarm, fearing, in particular, that its recently nationalized coal and steel industries might find difficulty in adjusting to such an ambitious multilateral venture.

Despite the reservations of the UK, the ECSC was set up by the Treaty of Paris, with Belgium, France, West Germany, Italy, Luxembourg and the Netherlands as members, coming into operation on 1 August 1952. It rapidly set about removing restrictions on production and trade in coal and iron and steel products and watched over a rapid expansion in the volume of trade in all the industries involved. There is considerable doubt abut the extent to which the ECSC actually managed the revival of heavy industry or was simply a mechanism for formally removing barriers to trade which would have fallen in any case. Certainly the organization found it difficult to resist protective measures used by member governments who felt the 'free trade' in their own products would benefit from judicious subsidies. Freight rates on the mainly nationalized European railway systems were, and still are regularly manipulated to confer advantages on domestic industries. Equally, price-fixing among the major producers was also difficult both to detect and control. In the latter part of the 1950s, the apparent inability of the ECSC to manage the declining markets for coal, iron and steel in the member countries and its failure to take remedial measures of sufficient scale to avoid the consequences of depression in the traditional heavy industrial regions, such as the Borinage in Belgium and the Nord–Pas de Calais in France, raised serious doubts about its true effectiveness.

In the early 1950s, however, the initial expansion of both production and trade led to something approaching euphoria among the member states and a desire to see further steps towards European integration. Their attention turned to defence and the possibility of establishing a European Defence Community but, after a first flush of enthusiasm for the idea in 1952, political objections to such a pooling of strategic resources began to multiply. Eventually, both Italy and France, the latter having been the prime mover in the original proposal, failed to ratify the draft treaty and the whole project was abandoned in 1954.

The Treaties of Rome

The collapse of the European Defence Community was a serious blow and at the time it appeared that the prospects for progress towards unity were dim, yet within two years a much more ambitious set of proposals had been agreed. In 1957, the six members of the ECSC signed two Treaties of Rome to take effect from 1 January 1958, one creating the European Economic Community (EEC), the other the European Atomic Energy Community (Euratom). Once again overtures to the UK to join had proved unsuccessful and the Communities therefore retained their rather narrow geographical focus.

Of the two new organizations, one, the EEC, was to provide the springboard for a process of integration, both economic and political, that is still in full spate (Williams 1991). The other, Euratom, was to prove

something of an irrelevance, as the early promise of nuclear energy was beset with technical problems and burgeoning competition from oil.

As its name suggests, the EEC was concerned primarily with reducing still further the barriers to trade within Western Europe and, like the ECSC, was born as much out of the immediate economic requirements of France and West Germany as out of any lofty commitment to the ideal of integration. France was eager to have freer access for its large farming industry to the markets in the industrial cities of West Germany, while the latter wanted a larger share in the French consumer market for its manufactured goods. The Treaty of Rome reflects these interests, but contains a very wide range of provisions for creating a broadly based, integrated economy in the Community, covering such matters as a customs union, common policies for agriculture and transport, joint initiative for social security and development aid, and the harmonization of the different legal systems. In the event, however, it was in the promotion of free trade and the formulation of a common agricultural policy that the EEC was most active in its early years.

Trade

In striving for the abolition of tariffs and quotas within the EEC, the members were reflecting the more general moves to free trade being pursued in Western Europe as a whole by the OEEC and on an even wider basis by the General Agreement on Tariffs and Trade. The EEC policy was, however, much more comprehensive in that its objective was a customs union, which meant that not only would restrictions on trade within the Community be abolished, but also that there would be a common tariff wall to the world outside. As far as industrial goods were concerned, the common tariff posed little problem because of the more general moves to reduce restrictions, but the situation was very different for agricultural products. The majority of West European countries had traditionally enacted strong measures to protect their domestic farming economies and anything approaching free trade would have wrought havoc with this carefully manipulated section of the economy. For this reason the creation of a customs union had to go forward hand in hand with the formulation and implementation of a common agricultural policy.

The initial timetable agreed by the EEC suggested that the customs union should be introduced in a series of stages and be fully operational by the beginning of 1970. The problem was that after a good start, the parallel agreement on a common agricultural policy proved much more difficult to achieve. Rather than simply removing a set of restrictions, in the case of agriculture it was necessary to agree a framework of subsidy and support for farmers in six different countries and to devise a policy for imports and exports for the whole Community. For the French, who laid the greatest store by a common agricultural policy, the delays caused first frustration and then fury and, in 1965, they threatened to withdraw from the EEC completely if agreement was not reached. The threat was obviously effective, because by the middle of 1967 common guaranteed price levels were in operation for most of the major agricultural products. The impasse was solved by the relatively simple stratagem of agreeing guaranteed minimum price levels for all products no less than the highest minimum price guarantees previously operating in any individual country. It was to prove an effective but extremely expensive solution which, although it ensured the continued survival of the EEC, meant that the bulk of the budget has always been committed to agricultural support. Inevitably, this in turn has meant that other positive policies have had much less chance of becoming successfully established, a fact that in the long term has seriously blunted the effectiveness of the EC as a whole (Inglehart 1971).

Once agreement on agriculture had been reached, the implementation of the customs union posed few problems and by 1 July 1968 all tariffs within the Community had been completely abolished, eighteen months ahead of the original timetable. The EEC was by this juncture far more than just a statement of intent by six countries since they had transferred major aspects of their economic policy-making from a national to a Community level. It was not surprising that there should have been a concurrent desire to strengthen the structure and, on 1 July 1967, the ruling authorities of all three communities (the ECSC, the EEC and Euratom) were merged into a single European Community (the EC) thus tightening yet further the bonds between member countries.

Beyond the Community

Elsewhere in Western Europe, the emerging cohesion of the EC was watched with a mixture of concern and envy. While other countries had chosen not to join rather than being directly excluded, the success of the new organization clearly altered their relationship both with the Community's members and with each other (Puchala 1971). Largely to counter the diversion of trade that seemed likely to result from the establishment of the customs union, a second group of countries constituted themselves as the European Free Trade Association (EFTA) in 1960. The driving force behind the initiative was the UK which, though it

declined to join the EC for a combination of political and economic reasons, was none the less alarmed at the prospective consequences of being excluded from European markets. The single overriding objective of EFTA was to stimulate free trade between its members. It was not interested in creating a customs union or interfering in any other way with the relationships of member countries towards other states and it claimed to have no supranational ambitions. This enabled it to embrace a wide spectrum of opinion, from countries like Sweden, Switzerland and Austria which were politically neutral, to the UK which had long-standing preferential trade agreements for agricultural products with non-European Commonwealth countries. In addition, EFTA was able to accept easily small countries such as Norway and Denmark, as well as a country with a poorly developed economy like Portugal, because the association was only concerned with free trade in industrial goods among the members, who were therefore free to protect their economies from other sources of competition. Through the ploy of devising associate member status, it was even possible to include Finland in the agreement, despite the fact that in the 1960s it was extremely constrained when it came to joining any international agreement because of its treaty obligations with the USSR.

Given the scale of the realignments implied by the new groupings and the somewhat arbitrary political and economic logic behind the actual membership of the two organizations, it is hardly surprising that there have subsequently been significant changes. For the most part, these have benefited the EC at the expense of EFTA; Denmark and the UK transferred their allegiance in 1973, Portugal in 1986. The EC has also attracted more new members from elsewhere in Europe, with Ireland joining in 1973, Greece in 1981 and Spain in 1986. For its part, the story of EFTA has been one of gradual decline; the only new members it has managed to attract are Iceland and Liechtenstein. In order to survive at all the association has had to draw closer to the EC. In 1973, in the wake of its first haemorrhage of members, the remainder all signed individual trade agreements with the EC, bringing the two bodies much closer together. In 1991, after a protracted series of negotiations going back over a decade, the process took a decisive step further forward with the announcement of the European Economic Area, formally linking the two bodies into a single free trade grouping.

New members of the European Community

In its initial form, the EC was a relatively cohesive and close-knit group of north-west European states, sharing common land borders and with overwhelmingly industrial economies. Undoubtedly, the experience of increasing cooperation in the years after 1952 drew the six original members steadily closer together, but in the 1960s there was a reluctance, particularly on the part of France, to disrupt the emergent political structure of the Community by admitting additional members. Only once the mould had been set did it become easier to countenance an expansion of membership, for it was by then clear that any newcomers would have to accept the Community on its own terms. In 1971 the UK, together with Denmark, Ireland and Norway, applied to join and all were successful, although Norway subsequently withdrew its application. From 1 January 1973, therefore, the Community had nine members and had to begin to adjust to a new set of political and economic pressures. The concept of a single free trade area with minimal government intervention to allow for regional economic discrepancies was just about tenable with six original members, a close-knit coterminous grouping on the mainland of Europe, but the three new members were on the fringe, with only Denmark sharing a land border with any of the founder members. There was considerable apprehensiveness in some areas of the applicant countries, especially in the declining heavy industrial regions in the north of England, Scotland and Wales, about the dangers of being marginalized economically. The UK wanted to see much more scope for positive economic intervention in the form of a comprehensive regional policy, rather than having to rely on the EC's existing narrow and overwhelming concentration on agriculture. It is a measure of how firmly the EC had become set in its ways during the 1960s that, while lip service was paid to the need for a regional policy in the form of the European Regional Development Fund (ERDF) created in 1974, there was little fundamental alteration in the general pattern of expenditure (Croxford, Wise and Chalkley 1987).

Until the end of the 1970s, the whole tenor of the EC was that of an organization geared primarily to serving the development needs of the industrial nations of Western Europe and it was frequently upbraided for not showing sufficient concern for the problems facing both the less developed parts of Europe and the developing world. Although treaties of association were concluded with Greece, Turkey, Malta and Cyprus in the early 1960s, they fell far short of even a form of associate membership of the Community. None of these countries enjoyed much protection from the barriers to free trade with their industrial neighbours created by the customs union, nor did they receive significant amounts of aid to help them modernize their economies so that they would be able to compete

on a more equal footing (Commission of the European Communities 1982a).

Much the same was also true of the developing countries outside Europe. A series of conventions was signed, mainly with former colonial territories, but the effect has been to protect traditional markets for these countries' raw material exports in Europe, rather than providing a vehicle for redistributing wealth between the developing and industrial worlds. In fairness, it should be said that all these agreements were eagerly sought after by the developing countries concerned and they have always been keen to renegotiate when any particular programme came to an end. After 1975, there was also a marked, if not fundamental, change of policy on the part of the Community itself. It heralded the first of a continuing series of Lomé conventions, the fourth of which came into operation in 1991, which were very much more concerned with providing resources to improve the infrastructure of developing nations, though the scale of the aid is small in relation to the enormity of the task in most of the recipient countries.

Since the mid-1970s, political considerations have forced the EC to adopt a wider view when considering new applications for membership. Under the terms of the Treaty of Rome, any European country has the right to apply and in the past two decades Greece (1981), Portugal and Spain (1986) have all joined, thus shifting the economic focus decisively southwards towards the Mediterranean. All three would have applied earlier, but they were left in no doubt that on account of the structural weakness of their economies alone they would not have been accepted (Deubner 1980). Gradually, however, other considerations began to press their claims; it became clear that one of the best ways of preserving the fragile new democratic governments in Greece, Portugal and Spain was by drawing them more closely into the institutional framework of Western Europe (Pepelasis et al 1980). The EC was faced with the very real dilemma that to reject their overtures would simply herald the return of some form of totalitarian regime (Leigh and van Praag 1978).

Welcome as the Mediterranean expansion was in political terms, it posed very real difficulties for the Community economically. Incomes generally were some 40 per cent below the Community average so that the accession of the new members, in particular Spain which has a population of 36 million, nearly one in ten of the EC total, generated insistent demands for a substantial transfer of resources from the north to the south. Agriculture posed a very intractable problem, because the industry was dominant in all three countries, but different in kind from that practised in much of northern Europe. It meant that not only would the burden on the Common Agricultural Policy be further increased, but that the new crops typical of the Mediterranean would have to be included alongside those grown further north (Williams 1984)

The effect on EFTA of the growing membership of the EC has been traumatic. When Denmark and the UK joined, well over half the customers covered by its free trade agreement were suddenly removed and the loss was hardly compensated for by the acquisition of two new members, Iceland in 1970 and Liechtenstein in 1978. As a matter or urgency, the EFTA members had to negotiate a formal agreement with the EC and because of the association's lack of any institutional structure, each had to do so separately. To some extent this undermined the authority of EFTA, for every agreement was slightly different and underlined the impossibility of it ever evolving into a more cohesive organization. Nevertheless, EFTA's work was far from finished; it continued to lobby on behalf of its members in their dealings with the EC, in most cases preparing the way for eventual membership. So far, the most important step in this process has been the creation of the single European Economic Area, discussed above, but Portugal swapped camps and joined the EC in 1986. Austria, Sweden and Switzerland lodged formal applications to join the EC but in a referendum late in 1992 the Swiss voted against entry to the European Economic Area which would have been the first stage.

Renegotiating the Treaty of Rome

Politically the EC has found it difficult to live up to its liberal economic success and has struggled to evolve an acceptable governmental structure. One of the major difficulties has been the way in which it is hemmed in by its founding treaties, which determine the areas of national life over which the Community has jurisdiction and the way in which it may exercise its power. In the 1960s, despite painfully slow progress in many areas, development proceeded on the coat-tails of the global movement towards free trade, but the changed world economic climate since the latter part of that decade, together with the growing size and complexity of the Community itself, have necessitated fundamental changes in both policy and institutional structure. The instability caused by the collapse of the post-Second World War financial system caused recurrent crises in the EC throughout the 1970s and blighted most of what it wanted to do until the impasse was broken by the creation of the European Monetary System (EMS) in 1978. This system of agreed exchange rates has gradually gained acceptance among the members of the Community, the gist of the system being that individual national currencies are only allowed to fluctuate in value against each other within predetermined limits, unless there is a formal agreement to adjust their relative values. The EMS proved rather less successful than had been hoped, with the UK withdrawing in 1992. Now it is less sure that the EMS may be used as the launching pad for full monetary integration, and probably a single EC currency, the ECU, by the end of the century. If this happens, it will represent the most important development since the Community's foundation, for it will formally have transferred the bulk of economic decision-making away from individual member states to the level of the EC, thus incomparably strengthening its economic and political role.

In policy terms, the three founding treaties were very restrictive; they allowed cooperation and integration in certain areas, notably trade, but effectively precluded it in others. Unless there is a treaty commitment, no area of government activity falls within the jurisdiction of the EC. As time has elapsed and the Community itself has grown, it has been necessary to modify the nature and extent of the treaty obligations. The mechanism employed to do this has been to modify the most comprehensive of the founding treaties, the Treaty of Rome establishing the EEC. So far there has only been one modification, the Single European Act 1986, which has been agreed and ratified by the member states. The most important provision of the Act is the creation of a single internal market throughout the Community, free of all restrictions on the movement of goods, persons and services by the beginning of 1993, a goal that has been realized more or less on time. The Act also extends the scope of EC policy-making to new areas, notably the environment and research and development (R & D), as well as introducing the concept of qualified majority voting. Previously all decisions had to be taken unanimously, always a recipe for inaction and one which threatened complete paralysis as the Community grew in size. Qualified majority voting distributes votes among the member states in accordance with their size and then allows certain matters to be decided by a majority of votes cast, thus removing the power of veto which had so severely limited development in the past.

In 1992 a further modification to the Treaty of Rome, the Maastricht Treaty, was agreed, though as yet it has still to be ratified by the parliaments of all the member states. The most important provisions are for making possible the monetary integration and the single currency described above, as well as for the creation of a central reserve bank, but it also encompasses many other issues, such as social policy and the environment. Another significant aspect of the treaty is the redefinition of political power within the EC. On the one hand, it extends the scope of qualified majority voting, but it also seeks to limit power at the centre by introducing the concept of subsidiarity. This is a requirement that all decisions should be taken at the lowest appropriate level of government, so that they are as close as possible to the people affected. In practice it is a means of stemming the centralist tide in decision-making away from EC institutions and towards national and regional governments. It is important to note that the precise meaning of the commitment varies between the different member states. In a largely unitary state like the UK the devolution is to the national government; in a federal state like Germany it is the provinces, just as much as the national government that benefit.

From the outset the political structure of the EC lacked democratic credibility, relying until 1979 on an appointed advisory assembly to ensure that the voice of the people was heard directly, rather than through their member governments. To improve the level of public accountability a directly elected European Parliament was chosen for the first time in 1979 and its 518 Euro-MPs are now an integral part of the political scene. Unfortunately, the Parliament has had little real power delegated to it and remains primarily a debating forum, despite regular promises that it will assume a key role in Community decision-making (Brewin 1987). The problem is that the governments of member states have ultimately proved reluctant to derogate powers away from their own national parliaments to the EC; indeed, they have moved decisively in the past decade to strengthen their position through increased emphasis on the Council of Ministers. In theory this has always existed, but its role has now become very much more formalized. Member states in turn assume the presidency of the Council for six months at a time, during which period they chair ministerial meetings from the level of head of state downward and it is these committees that exert the real political control over the civil servants in the European Commission. The European Parliament remains effectively sidelined.

The European Court is, however, one Community institution that has succeeded in carving out for itself a significant role at the expense of individual member states. It is independent and has the role of interpreting and ensuring compliance with the founding treaties, its decisions taking precedence over those of national courts. Although its jurisdiction is limited to the treaties, as the scope of these and the extent of EC activities have grown, so has the Court's influence and importance. It now finds itself in an almost unchallengeable position, although it is still dependent on the governments of the member states to enforce its decisions.

The shape of things to come

The future of the EC, and Western Europe as a whole, is very much in the balance. The Iron Curtain, the traditional frontier in the east, has been swept away since 1989 to be replaced by a multitude of states and an even greater multitude of people all wanting to join the economic, political and social system that has evolved since the end of the Second World War. The process of expansion is already well under way. What was East Germany has been absorbed into a unified and enlarged Germany and the EC has concluded association agreements with Czechoslovakia, Hungary and Poland. Unfortunately, the political structures in virtually all the newly independent countries in the east are extremely fragile and lack public acceptance. Civil conflict, notably in what was Yugoslavia and some of the former republics of the USSR, has so far been fierce and bloody, but so far sporadic. The worry is that where the EC has tried to intervene to preserve peace it has so far proved ineffective and there is little hope that any of the other pan-West European organizations, such as NATO or the WEU, will be any more successful. A further problem is the parlous state of the economies in all these countries and the difficulties they are experiencing in adjusting to the market economy, which is the linchpin of the economic system in Western Europe (Weidenfeld and Janning 1991).

Here lies the dilemma. Western Europe having worked unremittingly for the downfall of the Communist hegemony in the east cannot simply abandon those countries now that this has been achieved. Even if it wanted to do so, it would be unthinkable to allow internecine warfare to rage on its borders without even attempting to bring it under control. However, intervening economically to integrate these countries into its economic fold will place enormous strains on its own economic system and could jeopardize much of the progress, economic, political and social, of the past fifty years. It is a challenge that will require a degree of European statesmanship far in excess of anything required to create NATO, the EC, or any of the other pan-European institutions which are the foundations of present-day Western Europe.

Further reading

Blacksell M 1981 *Post-war Europe: a political geography* 2nd edn. Hutchinson, London

Heater D 1992 *The Idea of European Unity*. Leicester University Press, Leicester

Urwin D 1989 *Western Europe since 1945, a Political History* 4th edn. Longman, Harlow

Urwin D W, Paterson W E (eds) 1990 *Politics in Western Europe Today. Perspectives, policies and problems.* Longman, London

Williams A M 1991 *The European Community*. Blackwell, Oxford

3

Demographic and social change

Population

From growth to stagnation

Since the Second World War, the population of Western Europe has undergone a complicated transformation with regard to numbers, distribution and composition. The many marriages contracted soon after hostilities ended in 1945 produced a veritable 'baby boom' that was strikingly different from the demographic depression of the 1930s and the massive loss of life that occurred between 1939 and 1945 as a direct result of fighting or because of general disruption and poor conditions among the civilian population. Parts of the political map of Europe were redrawn once war had ended and consequently the second half of the 1940s witnessed an important redistribution of people. For example, in 1950, refugees from the Sudetenland, former eastern provinces of Germany, and from minority areas in Eastern Europe made up 16 per cent of the population of what was to become West Germany (George 1972). A comparable process was to take place during the 1960s as roughly 2 million people moved from East Germany to the West. Even in the early 1990s, the arrival of refugees, from East European and non-European countries, posed severe problems for some West European governments.

By mid-century, the eighteen countries of Western Europe contained 284.7 million people, some 11.3 per cent of the world total. Their demographic growth continued during the 1950s and 1960s, albeit with decreasing vigour. By contrast, the labour-hungry economies of Western Europe were growing fast during those decades and large numbers of labour migrants had to be enlisted from elsewhere since the products of the late 1940s baby boom were not yet old enough to enter the labour market. In addition, the processes of decolonization and repatriation, together with major changes in world affairs, dispatched yet more migrants to swell Western Europe's total. For example, decolonization

and especially the granting of independence to Algeria in 1962 resulted in the repatriation of 1,160,000 French nationals between 1954 and 1963. A miniature baby boom did occur in some countries of Western Europe in the mid-1960s but the general tendency during that decade was for natural increase to decline and continuing population growth to become increasingly reliant on immigration.

In the 1970s, the relationship between demographic factors and migration was modified again. Economic recession, with consequent job loss on a massive scale, and the presence of the baby-boom generation on the labour market meant that new supplies of foreign labour were not needed. Many 'guestworkers' were sent home as soon as their short-term work permits expired. At the same time, rates of natural increase continued to fall in most of Western Europe, even reaching the point of more deaths than births being recorded in some countries and especially so during the mid-1970s. Fertility picked up fractionally in a few areas in the late 1970s and early 1980s, but there is no doubt that Western Europe has now entered (or to be more accurate, re-entered) a population cycle that displays only modest growth and will soon be characterized by stability or even decline. The present chapter will focus on demographic trends and attendant modifications in employment and society; migratory processes will be examined in Chapter 4 and discussion of the urbanization of Western Europe since the Second World War will be reserved until Chapter 8.

Facts and figures

Outworking of the broad trends described above meant that by 1966 the population of Western Europe had risen by 14.2 per cent from its 1950 figure, to reach a total of 325.2 million. In the following twenty years the rate of growth decelerated very sharply to produce a total of 361.2 million in 1991 (i.e. an increase of 26

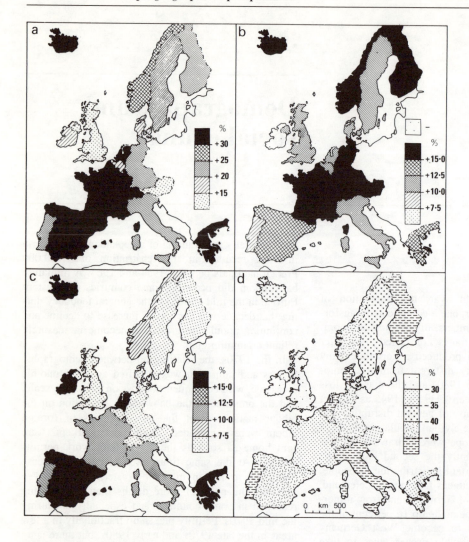

Fig. 3.1
Aspects of population change: (a) total growth, 1950–91; (b) change, 1950–66; (c) change (%), 1966–86; (d) decline in crude birth rates

per cent since 1950) but accounted for only 6.7 per cent of the world's greatly enlarged population. The interplay of differing rates of natural change and net migration generated enormous variations in this trend of growth both between countries and within them. During the four decades, net increases of more than 33 per cent were recorded in the Netherlands (+46.5 per cent), Switzerland (+40.4 per cent), Spain (+39.0 per cent), France (+34.8 per cent) and Greece (+33.3 per cent), but the relationship of demographic and migratory processes in those five countries was far from uniform (Fig. 3.1a)

Some of these nations displayed impressive rates of natural growth, but while the Netherlands received substantial numbers of repatriates and guestworkers, Spain dispatched large numbers of migrants to labour-hungry France and other industrial countries. Switzerland was the destination for large contingents

of migrants, coming especially from Italy. Spain's relative growth in population was certainly very high, but its absolute increase of 10.9 million in forty years was equalled by Italy and overtaken by West Germany (+12.1 million) and France (+14.4 million). Absolute increases of this kind of magnitude between 1950 and 1991 surpassed the individual totals of no less than ten West European countries in 1991. At the other extreme, modest rates of growth characterized Austria (+10.0 per cent), Belgium (+15.0 per cent) and the UK (+13.0 per cent), although in the latter case an absolute increase of no less than 6.9 million was involved.

Not surprisingly, national rates of growth were stronger between 1950 and 1966 than in the subsequent fifteen years in all West European countries save the Netherlands, Spain and the Republic of Ireland, with the latter country, long sapped by emigration and having official prohibition of birth control, switching

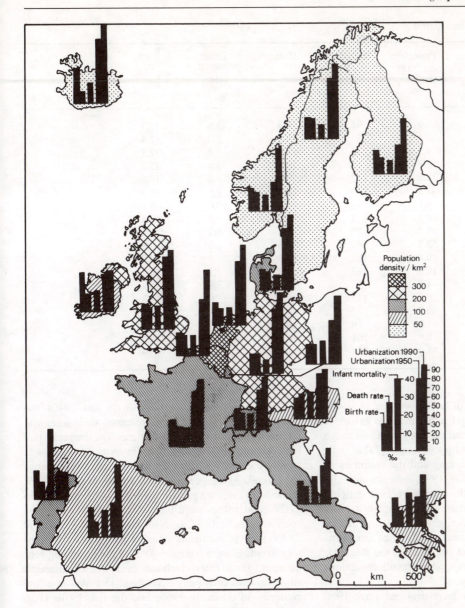

Population
density / km²

300
200
100
50

Urbanization 1990
Urbanization 1950
Infant mortality
Death rate
Birth rate

90
80
70
60
50
40
30
20
10

40

30

20

10

‰   %

Fig. 3.2
Demographic indicators,
1990 and measures of
urbanization.

0   km   500

from an initial loss of population to the fastest rate of
increase encountered anywhere in Western Europe.
Between 1950 and 1966, natural increase, immigration
and policies of labour recruitment combined to give
Switzerland (+29.8 per cent), West Germany (+19.6
per cent) and France (+18.5 per cent) the fastest rates
of growth and they were followed closely by some of
the Scandinavian states and the labour-exporting countries of Spain and Greece (Fig. 3.1b).

After the mid-1960s the pattern changed dramatically, with rates of growth between 1966 and 1986
falling to half the much lower West European average
(+8.8 per cent), the UK (+3.4 per cent) and West

Germany (+2.2 per cent), and a further four countries
also displaying below average growth (Fig. 3.1c). In
five countries the number of deaths exceeded births for
some period during the 1970s, with this trend being
most pronounced in West Germany where natural
decrease commenced in 1972.

Deaths and births

Death rates in Western Europe have declined since the
beginning of this century in response to continued
medical progress and are now remarkably similar in
the eighteen countries (Fig. 3.2). Higher rates now

Table 3.1 Demographic charactistics, 1992

	Total population (mill.)	Projection for 2000 (mill.)	Density per km²	Death rate ‰	Infant mortality rate ‰	Birth rate ‰
Austria	7.791	7.9	91	10.7	7.8	11.7
Belgium	9.978	9.3	325	10.5	8.0	12.4
Denmark	5.147	4.8	119	11.9	8.4	12.4
Finland	4.892	4.8	15	10.1	5.8	13.2
France	56.536	58.6	102	9.4	7.2	13.5
Germany	79.880	81.5	220	11.3	7.5	11.3
Greece	10.200	9.7	76	9.2	10.0	9.9
Iceland	0.253	0.3	2	6.7	5.3	18.8
Ireland (Rep.)	3.512	3.3	50	9.1	8.2	15.1
Italy	57.739	52.3	191	9.3	8.6	9.8
Luxembourg	0.381	0.4	142	10.0	9.9	12.9
Netherlands	15.010	16.3	363	8.6	7.1	13.2
Norway	4.250	4.3	13	10.7	7.8	14.4
Portugal	10.393	10.6	113	9.9	11.0	11.2
Spain	38.994	40.5	77	8.6	7.6	10.2
Sweden	8.591	9.0	19	11.1	5.6	14.5
Switzerland	6.751	6.9	161	9.5	7.3	12.5
UK	57.411	61.0	234	11.2	7.9	13.9

tend to reflect the success of health services in allowing populations to live longer rather than the existence of seriously inadequate medical provision, although there is always scope for improvement (Table 3.1). Thus in the UK, West Germany and the countries of Scandinavia, death rates for 1991 were fractionally higher than in the late 1940s simply because national populations had become more elderly. In most other parts of Western Europe, a slight decline was in evidence over the past four decades, with much more pronounced reductions of about one-third in the Republic of Ireland, Portugal and Spain, where health conditions have improved substantially.

Rates of infant mortality have declined since 1950 but considerable variations remain, most notably between the very low rates of Scandinavia and the Netherlands and the higher values recorded in some Mediterranean countries (Fig. 3.2). Infant mortality may be regarded as one of the most sensitive indicators of socio-economic conditions, since it reflects a whole host of factors, including quality of medical care, public health services, housing and income (Hall 1976). It will be difficult to reduce infant mortality beyond the current conditions in northern Europe since death within four weeks of birth (neonatal mortality) is strongly linked to pre-natal conditions and congenital weaknesses as well as antenatal and obstetric care.

After a steep rise in fertility in the late 1940s, the countries of Western Europe resumed their inter-war tendency for birth rates to decline. The increase that had occurred in the two decades prior to 1965 was merely a cyclical shift rather than an established long-term trend. At the peak of the baby boom of 1945–49, crude birth rates ranged from 14.2‰ (Luxembourg) to 27.2‰, but values were below 22.5‰ in all but three countries (Iceland 27.2‰, Netherlands 25.9‰, Portugal 25.6‰) to yield a mean index of 20.8 when the eighteen averages were summed. During the following four decades, birth rates declined everywhere, although slight upward fluctuations had been recorded in a few countries in the early 1960s and the late 1970s (Fig. 3.1d). By 1991 the range extended from 9.8‰ Italy to 18.8‰, but rates were below 14.0‰ everywhere except Norway (14.4), Sweden (14.5), the Republic of Ireland (15.1) and Iceland (18.8). The mean index was 12.6‰, one-third lower than in the late 1940s. In sharp contrast with past experience, southern Europe now records very low fertility rates: 1.5 children per mother in Greece and Portugal, and only 1.3 for Spain and Italy. This new taste for small families relates to uncertain economic prospects, lack of suitable housing, growth of job opportunities for women, and a shortage of good child care (Prioux 1989). Conversely, fertility in Sweden and some other northern countries started to rise after the mid-1980s.

Smaller families

Despite recent hints of recovery in some countries, the human situation summarized by these bald statistics is one of contracting family size across most of Western Europe. The proportion of couples with one child each or without children is increasing, while the share of families with several children is on the decline, and is particularly evident in the case of couples with four or more children. Calculations derived from patterns of family formation in West Germany estimated that a quarter of marriages in that country in the late 1970s would produce no children, a quarter would produce one child, 30 per cent two children and the remaining 20 per cent three or more children. Associated with this trend, the proportion of births to women aged 35 and over is falling, but contrasting behaviour is in evidence for birth patterns to younger women. Some complete family production by their mid-twenties and then return to employment; others first pursue their careers and postpone starting a family until they are well into their twenties.

No other demographic development in Western Europe has been so uniform in such a short period as this decline in fertility since the mid-1960s. During this time a range of contraceptive techniques has been widely diffused in most countries to allow the number and spacing of births to be planned more effectively. The economic, social and psychological processes at work to enable family planning to be widely accepted are complicated in the extreme but must include the following considerations. Traditional rural societies, in which large families were raised to supply farm labour and ensure security for parents in their old age, have largely disappeared as the agricultural workforce has declined throughout Western Europe. None the less, some regions that still have substantial agricultural employment still tend to display relatively high birth rates. Rural–urban migration has continued apace in such regions, as has the reach of mass communications, and it is clear that the role of the family is viewed differently by couples in urbanized, welfare-state societies than in more traditional farming economies. Old-established imperatives no longer hold sway and having children is increasingly seen as a way of 'completing the family', ensuring biological survival and finding new companionship. Many urban couples feel that cramped apartments or houses are quite unsuitable for raising more than a small family (of, say, one or two children); and, in any case, having more children involves added expenses which family allowances do little to compensate.

Even more important than these environmental and economic considerations is the evolving status of women in European society. The education and professional training of girls have improved substantially since the Second World War and their aspiration for an identity outside the home has been reflected by increased employment of women, with or without children, in industry and more especially in the service sector of the economy. Motherhood has become just one of several feminine roles and for many women pursuit of a career has taken preference to child rearing. Strict adherence to religious traditions which proscribe family limitation encourages larger families although many 'religious' West Europeans adopt liberal attitudes to the organization of their family life. Social and material objectives for personal fulfilment and well-being are now likely to be dominant in the minds of most couples.

The operation of factors such as these produced interesting fluctuations in national birth rates during the mid and late 1960s, by which time the baby boom of 1945–50 had reached reproductive age. In some countries (e.g. West Germany, Italy, Sweden, the UK) a second baby boom duly occurred but did not happen in others. The reasons for this difference were related in part to shortage of accommodation and other environmental factors, but the fundamental explanation was probably to do with young women having taken new opportunities to pursue careers, postponing both marriage and child bearing in consequence. Various social factors were also advanced for the slight upturn in the birth rate in the UK that commenced in the late 1970s. Women born shortly after the Second World War and still in their early thirties were catching up just as younger women born in the late 1950s were having their children. Many older mothers had postponed child bearing until their careers were well under way; others had been divorced and were starting new families with new partners. Other hypotheses included the fact that fewer British women were taking the contraceptive pill, preferring other less reliable methods after a medical report had linked deaths from thrombosis with the pill; and that many couples waited until they were better off and settled in their own home before embarking on reproduction.

Population policies

National policies regarding population vary both in broad terms and in highly significant details among the countries of Western Europe. At one extreme stands the Republic of Ireland where regulations indicate a pro-natalist attitude, while, at the other extreme, Germany and the Netherlands have adopted policies aimed at stabilizing population numbers. The other countries occupy various positions in the middle ground, either having no particular point of view or else remaining neutral and leaving such matters up to individuals. Official attitudes have become increasingly liberal in many of these middle-ground countries, as in France where legislation banning contraception was repealed in 1967 and abortion reform came in 1975, a year ahead of West Germany but nine years after Denmark and England and Wales.

The French anti-contraception law of 1920 had been devised for demographic purposes: to help compensate the deaths of 1914–18. It prohibited publicity for birth control and advisory clinics, and forbad the sale of contraceptives, except for a number of defined medical reasons. None the less, many French couples exercised their will-power and resorted to time-honoured means of birth control or turned to clandestine abortion.

It is probably impossible to establish a clear relationship between fertility trends and the myriad laws regarding contraception, abortion, sterilization, family benefits and medical care in the countries of Western Europe, since the marital behaviour of individuals depends on a complex web of social, cultural and economic variables rather than the dictates of law. However, that is not to deny that relative permissiveness or harshness of laws on family matters does condition general availability of information on family planning, provision of necessary materials for birth control, or procurement of safe abortions. For example, the situation in the Republic of Ireland, where artificial contraception is forbidden by both state and church and there is a strict interdiction on abortion, contrasts with that in Denmark, Germany, Sweden or the UK where the means of birth control are readily available and abortion, within certain periods of pregnancy, is legal. The remaining countries display various but generally increasing degrees of liberality on such matters (Blayo 1989).

None the less, national differences in legislation have been sufficiently great to generate 'abortion tourism' and smuggling of quantities of contraceptives into less liberal countries. Indeed, the latter practice continues into the Republic of Ireland to this day. During the early 1970s, about a quarter of abortions performed in England and Wales involved non-residents who came mainly from France and West Germany, prior to changes in their legislation in 1975 and 1976 respectively. Clinics in Denmark and the Netherlands were visited by large numbers of German and Belgian women, while London's clinics once formed one of the 'tourist attractions' for some Spanish women spending a short time in the UK. In recent years some 30 000 women went abroad each year for terminations; however, in 1983 the Spanish parliament approved a fiercely debated bill to permit abortion in cases where the mother's health is endangered or the foetus deformed. Estimates on the incidence of illicit terminations in the 1960s and early 1970s varied enormously. The average for Western Europe as a whole was thought to equal between 25 and 50 per cent of the number of live births, with annual estimates being in the order of 30 000–200 000 for Belgium, over 400 000 for France and over 650 000 for Italy. Estimates for West Germany suggested one abortion for every live birth, while in the Netherlands about 15 per cent of pregnancies were terminated illegally (Hall 1976).

Despite important national differences in legislation and significant variations in marital behaviour between socio-economic groups, a number of common trends have emerged in Western Europe and are likely to continue to operate in the future. Greater social and spatial mobility, together with the impact of mass communications, has brought about greater individualization of behaviour patterns and this trait will probably continue. More and more couples will plan the number and spacing of their offspring and it has been suggested that increased recourse to family planning may reduce the incidence of abortion. Wider education will encourage more women to seek employment outside the home and to try to continue economic activity after the birth of children. Conjugal fertility is likely to level out even further in all regions and through all social strata, so that the small family with one or two children will predominate throughout Western Europe. Life expectancies will probably not increase much more. The net result will be low rates of population growth for the remainder of the century, with recent projections raising the West European total by a mere 1.0 per cent to 381 million in 2025. The outcome of recent demographic trends will mean that the number of people of working age will decline in the 1990s and into the next century. Despite technological innovations, which perform work formerly dependent on human hands and brains, there will be many examples of mismatch between the supply of labour and the demand for it in the years ahead. Raising productivity, attracting more women into the workforce, raising the age of retirement, or (most controversial of all) allowing more immigration are possible means of adjusting to demographic decline.

Social characteristics

An ageing population

Widespread advances in living standards, combined with improved housing provision and medical and

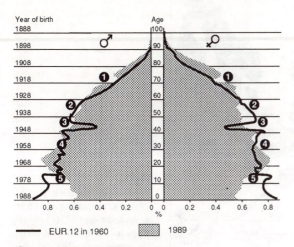

- ❶ Birth deficit due to the First World War
- ❷ Low birth-rate generations reach the age of fertility
- ❸ Birth deficit due to the Second World War
- ❹ Baby boom
- ❺ Non-replacement of generations

Fig. 3.3
Age pyramid for EC Twelve.

social services, are enabling more West Europeans to live longer while, at the same time, birth rates are much lower than in the 1940s and 1950s. The total population has become not only more numerous but also substantially older than at mid-century (Fig. 3.3). At mid-century, only 9 per cent of the population was aged 65 or over, but by 1991 the share had risen to 14.8 per cent, with the proportion of under 15-year-olds declining from 27.5 to 20.0 per cent over the same span of forty years. Sweden now has 18 per cent of its population aged 65 or over, with Denmark, Norway and the UK (each 16 per cent) close behind (Fig. 3.4a). By contrast, less than 12 per cent of the population of Iceland and the Republic of Ireland is aged 65 or more.

Life expectancy at birth (i.e. including infant mortality) now ranges from 74 years in Belgium and Portugal to 78 in Iceland, Sweden and Switzerland. These figures represent considerable advances since 1950. The average difference in male (72 years) and female (78 years) expectancy among West Europeans has widened by 3–4 years over the past four decades, with the lower increase for men being partly due to an upsurge of mortality rates in certain age groups associated with pathogenic behaviour and increased risk of accident. A clear tendency is for more people to live

into their eighties and, by virtue of the differences in life expectancy and the higher impact of wartime mortality on the male population, this means the survival of a growing number of elderly women. The decline of the traditional multi-generation household in urbanized Western Europe means that most elderly people live by themselves and may endure considerable social isolation. High divorce rates will accentuate this problem in the future. Many are in failing physical and mental health and depend on pensions and other forms of transfer payment for their survival. National pension schemes vary enormously in entitlement and amount but, despite the efforts made by some governments to provide all elderly people with an adequate minimum income, old folk who have no personal resources to supplement these minimum amounts must be recognized as being among the most deprived people in Western Europe. The worst off include the very old, who have contributed to pensions for only part of their working lives; widows who have never worked; and members of occupational categories that had no pension schemes until recently.

The need to provide and manage a range of types of accommodation for the elderly, according to their varying health requirements, is already one of the most urgent problems in Western Europe and will certainly grow in magnitude in the immediate future. The dimensions of the problem associated with an ageing society have only just started to be comprehended and it is to our collective shame that West European families and notions in general have shown themselves badly prepared to cope with a larger share of elderly people. This fundamental challenge remains very largely unmet (Warnes 1982; Noin and Warnes 1987). Clearly the consequences of an ageing population are very great. Business activities will have to adapt to meeting the requirements of the elderly; the health service and the caring professions will need to make profound adjustments; some forms of leisure or free-time activity may expand substantially; and builders and housing departments will have to construct more small houses and apartments for elderly households, as well as providing 'sheltered housing' (with staff always on the premises) and special residential homes. By 2020 a quarter of the population will be over 60 in most West European countries. Lobbies for larger pensions, better health services and more appropriate public transport will become more vociferous.

Changing household structure

The presence of greater numbers of old folk, the fact that different adult generations tend no longer to live

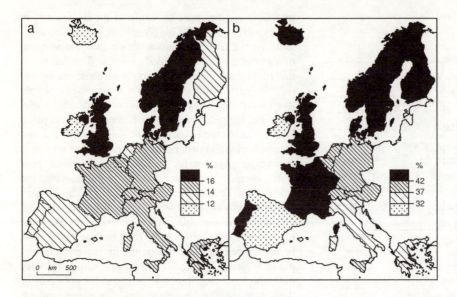

Fig. 3.4
Population characteristics,
1990: (a) proportion aged 65
years and over (%); (b)
women as proportion of civilian employment (%).

together, and the decline in the average number of children per family represent important explanatory factors behind a general reduction in household size (Hall 1988). Other causes include an increasing tendency for young people to move away from the parental home, as well as the dramatic increases in the number of divorces and one-parent families in recent years. In 1960, the average West European household contained 3.3 people, but by 1970 the figure had declined to 3.0 and now stands at 2.8. In 1990 two-person households represented the most common size category in most countries except for two groups. In the Republic of Ireland and Spain, larger households were most numerous (comprising 5 and 4 persons respectively), while one-person households formed the largest size category in Denmark and Norway (where they made up 40 per cent of the total), Sweden (33 per cent), Finland (28 per cent), West Germany (34 per cent) and Austria (32 per cent). Taking Western Europe as a whole, 26 per cent of households in 1990 were made up of lone persons compared with 18.2 per cent two decades earlier.

Information on divorce rates is far from complete, but there is clear evidence that numbers increased rapidly after 1960 in response to changes in national legislation. Little is known about broken marriages in strongly Roman Catholic nations, but for a dozen countries of Western Europe data are available which relate the annual number of divorces to each hundred marriages conducted during the year. We should observe at this point that, as the population has aged and attitudes to formal marriage have changed, the trend has been for the number of marriages to decline

with time. In 1965, the range was from 5.9 divorces per hundred marriages in Scotland to 18.2 in Denmark, with Sweden (17.8) and Austria (14.5) also displaying high ratios. In 1988 the values ranged from 6.5 divorces per hundred marriages in Italy (where divorce was legalized in 1974) to 45.2 in Denmark, 43.1 in Sweden and 41.5 in England and Wales.

In contrast with these figures, there are also signs of continuing attachment to 'family life', although such attachment is being expressed in other than conventional ways, for example remarriage after divorce of one or both partners and the fact that growing numbers of couples of the opposite sex and indeed the same sex are choosing to 'live together'. These trends may well represent a change in the concept and form of 'family life' rather than a rejection of the family as such (Coleman and Salt 1992). In any case, sizeable numbers of West Europeans, adults and children alike, are experiencing other kinds of tie. In some countries, alternatives to formal marriage have become widespread and this is especially so in northern Europe where, for example, two births in every five in Sweden were produced outside formal marriage in the late 1980s and long-established unmarried couples are by no means exceptional (Roussel 1987). In West Germany, roughly one birth in ten was outside formal marriage and in France and the UK one in every five (Decroly 1992) (Fig. 3.5). Demographers suggest that there seems to be a growing reluctance to marry, even when children are involved; age at marriage is rising and, failing substantial numbers of deferred marriages taking place, there has been a notable increase in one-person households in many parts of Western Europe.

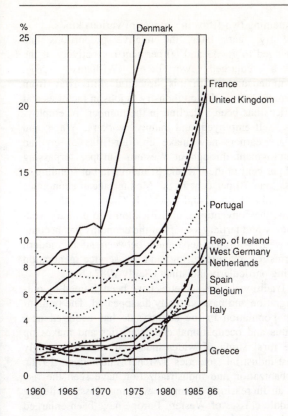

Fig. 3.5
Proportion of births outside marriage, by country 1960–86.

Life styles within families have changed considerably in response to the growing trend for women as well as their partners to take jobs outside the home. Since grandparents rarely live close at hand, this trend can cause problems in caring for young children and coping with domestic duties, but it has also had the very positive effect of playing down the rigid sex-stereotyped roles that past generations of husbands and wives were expected to perform. To some extent, male and female roles are converging; many women are spending less time than before on bringing up children and many men are willing to undertake domestic chores (Schmid 1982).

By virtue of the increasing occurrence of divorce and a host of other factors, the number of single-parent families has risen greatly so that in Belgium, Denmark, West Germany, the UK and the Scandinavian countries they now account for more than 15 per cent of all families with dependent children. In four cases out of five the single parent is a woman. In several countries the number of single-parent families is growing more rapidly than the number of 'complete' families. Unmarried mothers, divorced or separated women and

widows tend to work more than married mothers, but their jobs are often insecure and badly paid. They depend heavily on child allowances and supplements of various kinds to meet family needs and it is no surprise that single-parent families make up a disproportionate share of all low-income households. They deserve specialist assistance of various kinds, including nursery schools and a range of community facilities for older children.

Health and disease

West Europeans are certainly living longer and, in general terms, are healthier than ever before. Deaths from infectious diseases have declined dramatically since mid-century so that heart disease and cancer now form the leading causes. In the terminology of Eurostat (1991a) diseases of the circulatory system account for 45 per cent of deaths in the Twelve and malignant tumours for a further 24 per cent. The rising impact of socially induced disease gives rise to mounting concern; for example, chronic alcoholism occupies third position in the list of causes of serious illness. Death from cirrhosis of the liver is the most usual cause and its incidence has risen sharply among men aged 35–54 years. It must also be recognized that 'problem drinking' is also affecting growing numbers of women and young people; indeed, in several countries more than 5 per cent of the population is considered to be alcoholic. In addition, alcohol consumption represents an important risk factor in all kinds of accident, especially those on the roads. On an encouraging note, the total number of road accidents involving death or personal injury declined slightly in recent years even though the number of cars on the roads of Western Europe rose from 69 million in 1972 to 142 million in 1992. Speed restrictions and the wearing of safety belts may have helped this general reduction. Unfortunately, the number of young people, especially young men, involved in traffic accidents rose sharply and death on the road is now the main cause of premature death for the 15–24 age group. The steep rise in the number of motor cycles is part of this tragic story. Looking ahead, the ageing of population cohorts that have been accustomed to driving throughout adult life, combined with growing numbers of elderly pedestrians, may well lead to a disturbing rise in road accidents in the near future (Rallu 1990; Maffenini and Rallu 1991).

A higher proportion of West Europeans now smoke than they did in the 1950s with one man in every two and one women in every three being a regular smoker, despite the fact that smoking is a significant risk factor in the development of certain cancers and cardiovascular diseases. A heavy smoker may be considered to have 8–10 years trimmed off his or her life expectancy, with lung cancer being a growing cause of death among women as well as men in the 45–54 age group. Drug misuse represents a third form of socially induced disease and this has increased disturbingly in Western Europe since the 1960s, spreading from one area and social group to another and extending from young adults to juvenile age groups. Drug control poses major problems by virtue of the wide range of commodities involved and the fact that many medicines – which may be used as drugs – can be purchased across a pharmacist's counter. The link between certain forms of drug abuse and the incidence of AIDS is well known. Highest AIDS-related death rates have been recorded in large urban centres where liberal life styles find their niches and medical and care facilities are concentrated (Kostrubiec 1988; Jougla 1990). As of January 1990 drug addicts appeared to account for two-thirds of all AIDS cases in Italy and Spain, while in Denmark, the Netherlands and the UK four-fifths were homosexuals (Eurostat 1991a). At that time 90 per cent of the AIDS cases recorded in the EC involved men. France had almost one-third (8880) of all reported cases in the Twelve (followed by Italy, 5300 and Spain, 4630) and also the highest rate of AIDS victims per million inhabitants (158), followed by Spain (118) and Denmark (102). Far from being a 'gay plague', AIDS can affect straight and gay, married and single, rich and poor. It is no respecter of age, class, gender or sexuality, and as such is surely the greatest challenge that medical science, welfare services and West European society at large will have to face in the final years of the twentieth century.

Employment

Sectors and classes

Long-established forms of employment, such as agriculture, mining and traditional branches of manufacturing, have contracted massively during the past four decades as the balance of Western Europe's economy has been transformed by the emergence of modern industries, office work and a wide range of service activities. At mid-century there were highly important spatial variations in employment structure, with the agricultural economies of southern Europe contrasting with the greater development of manufacturing in the north-west, and urban economies everywhere being very different from those in the countryside. One-quarter of the total workforce was in farming and forestry, roughly one-third in manufacturing and the

remaining two-fifths in services of various kinds.

Forty years later, the relative proportions had changed to 7, 33 and 60 per cent respectively of the much enlarged workforce (Williams 1987). Commensurate with this structural shift away from farming and workshop industry to jobs in factories and offices has been a decline in the number of employers, self-employed and family workers. Wage and salary earners now make up four-fifths of civilian employment throughout Western Europe, surpassing 90 per cent in the UK and Scandinavia but standing at less than 70 per cent in the Mediterranean economies of Spain, Portugal and especially Greece. At the same time, the conventional classification into primary, secondary and tertiary sectors of the economy has become blurred, with the separation between some modern forms of manufacturing and of service generation being particularly indistinct (Bailly 1987a,b).

Traditional rural societies have been drained by out-migration and have virtually disappeared. The remaining population in the countryside has acquired new status and occupational characteristics and makes up the rural strata of wider industrial and increasingly post-industrial societies (Archer and Giner 1971). Urbanization and industrialization have seen the rise of an increasingly complex working class, while the middle classes of Western Europe have been enlarged and modified by the growing presence of managerial staff, bureaucrats, higher-grade clerical workers and teachers, all of whom represent 'hinge groups' in the processes of upward social mobility and typify the transition from industrial to post-industrial society (Mallet 1975). In the 1950s, many social scientists believed that West European countries were moving towards a condition in which class differences would be much less marked, barriers between classes less formidable, opposition between them less acute, and the consciousness of class membership itself less intense. Some even maintained that classes were virtually disappearing in response to rising living standards, changes in technology, wider spread of education and the impact of the welfare state, which served to redistribute a proportion of wealth through the taxation system.

Wisdom of hindsight suggests that these views were probably misplaced; social conflict has surely intensified since the late 1960s rather than declined. In most countries of Western Europe, access to really meaningful financial, cultural and educational capital has remained essentially in the hands of the 'establishment' (Marceau 1977). The socio-cultural influence of upper classes and a range of elite groups has endured during the post-war years and represents one of the most impermeable barriers to meaningful social mobility in Western Europe. The real and often limited significance of indices of social change needs to be appreciated; 'continuity' might be a more appropriate word to describe social relations in Western Europe over the past four decades rather than 'change'.

The changing workforce

It is arguable that the distribution of power in society may not have shifted substantially; however, the structure of the workforce has most certainly been modified by the recruitment of large numbers of guestworkers to perform less appealing jobs in every branch of work. For example, in West Germany, foreign workers comprise one-quarter of the underground workforce in the coal industry, one-half of the shiftworkers in foundries, one-quarter of the workforce in hotels and catering, and one-tenth of nurses and hospital staff. A controversial television programme attempted to determine what would happen to factories and public services in Stuttgart without guestworkers. It displayed pictures of streets piled high with refuse, factories and transport systems at a standstill, restaurants without service, and hospitals unable to function. Exactly the same kind of result would be produced in hundreds of cities throughout Western Europe. In addition, the presence of guestworkers and their families has introduced striking aspects of cultural and religious diversity in each of the reception countries, but such mass immigration has not occurred without tension arising.

Secondly, there has been a sharp rise in the number of women taking on paid employment, especially in modern branches of industry and in service activities (Commission of the European Communities 1991a). Women made up 39 per cent of the total workforce of Western Europe in 1989 but national proportions varied considerably from 32–35 per cent in rather more traditional societies such as Greece, Spain and the Republic of Ireland, where older generations believed a woman's place to be in the home, to 48 per cent in the more liberal societies of Sweden and Finland (Fig. 3.4b). In Scandinavia, two-thirds of all married women have paid employment and in several countries the proportion is almost one-half, but in the more traditional countries mentioned above and in the Netherlands far fewer women continue in paid employment after marriage. Expressed another way, some four-fifths of all women aged 25–54 years hold jobs in Sweden and Finland, while the proportion is closer to one-third in Italy, Greece, Spain, the Netherlands and the Republic of Ireland. However, it should not be forgotten that a notable proportion of female workers are engaged in part-time jobs that are neither secure nor well paid.

Other changes in the West European labour force involve a marked decline in participation by young adults, since many more pursue higher education than in earlier decades; and the widespread introduction of old-age pensions has brought in the concept of retirement and removed older workers from the labour force. Efforts have been made in Scandinavia and to a less extent elsewhere to replace the brutal break from work to retirement by a more gradual transition during which elderly staff work fewer hours and have the opportunity to prepare for their years after the world of work. Unfortunately, the mounting pace of job losses in recent years has overtaken many of these laudable initiatives.

The world of work

During the 1970s and 1980s, the employed population of Western Europe became more emphatically 'post-industrial', with 60.3 per cent of the workforce in service jobs in 1990 by comparison with 43.5 per cent twenty years previously. In absolute terms, this meant an increase in the service-sector workforce of 20 million people, roughly equal to the combined population of Greece and Portugal! Italy employed 4.5 million more people in the service sector in 1990 than in 1970, with increases in excess of 3 million being recorded in France and the UK. By 1990, the Benelux countries, Scandinavia and the UK were the most strongly service orientated, with over 65 per cent of their workforce in that sector, being closely followed by France (63.5 per cent). Service employment lagged behind the West European average especially in Portugal (45.7 per cent) and Greece (46.2 per cent) which still engaged important proportions of their workforces in agriculture, and in Austria (54.5 per cent) and West Germany (56.5 per cent), both of which had above average proportions of their employees in manufacturing (Table 3.2).

The industrial share of Western Europe's workforce declined by 9.5 million between 1970 and 1990, falling from 42.8 to 32.9 per cent of the total, not only because of the dynamism of the service sector but also in response to the severe economic difficulties experienced in so many branches of manufacturing. Contrary to the West European trend, both Greece (+21 per cent) and Portugal (+43 per cent) increased their manufacturing workforce between 1970 and 1990.

Employment in agriculture and forestry retreated by over 8.5 million between 1970 and 1990 and the primary sector of the workforce declined from 13.7 to 6.8 per cent. In Spain and Austria the agricultural workforce fell by one-half, while in Finland, France, Greece and the Republic of Ireland the loss was in

Table 3.2   Civilian employment by main sectors of economic activity, 1990 (%)

	Primary	Secondary	Tertiary
Austria	8.1	37.4	54.4
Belgium	2.8	28.9	68.3
Denmark	6.0	26.8	67.3
Finland	9.8	30.6	59.6
France	6.4	30.1	63.5
Germany (West)	3.7	39.8	56.5
Greece	26.6	27.2	46.2
Iceland	15.0	32.1	52.9
Ireland (Rep.)	15.1	28.4	56.5
Italy	9.3	32.4	58.2
Luxembourg	3.4	31.2	65.4
Netherlands	4.7	26.5	68.8
Norway	6.4	26.4	67.2
Portugal	18.9	35.3	45.7
Spain	13.0	32.9	54.0
Sweden	3.8	29.5	66.6
Switzerland	5.7	35.1	59.2
UK	2.2	29.5	68.4
EC 12	7.0	32.5	60.6
Western Europe	6.8	32.9	60.3

excess of 40 per cent. In absolute terms, the really big losses in farmworkers were in Spain (2.0 million), Italy (1.7 million) and France (1.4 million).

In short, by the early 1990s Western Europe was no longer composed of predominantly industrial neighbours; while, more frighteningly, a much larger number of workers were on the unemployment registers than held jobs in farming. In addition, it had become very clear that manufacturing some kinds of goods and generating some kinds of services represented a convergence of productive processes rather than a clear line of separation between distinctive sectors in the economy.

The quality of life

Just as the nature of work has changed, so have its pattern and duration. Overtime practices still vary from country to country, but the general trend has been for a reduction in the number of hours worked by each employee. After a period of very little change in the 1950s, increases in the duration of paid holidays, creation of part-time jobs and introduction of shorter working weeks all contributed to the same general trend. Employees now work a 38 to 40-hour, 5-day week in most countries, although 35 hours is usual in Scandinavia. In Belgium, Denmark, Italy and Norway

the number of hours worked by a jobholder in 1990 represented a one-third reduction on comparable figures for 1950. Flexitime, shortened lunch hours, three or four 'long' working days per week, and a number of short holidays during the year rather than one long break all represent substantial modifications to work patterns and enable workers to participate more fully in family and community life. The presence of so many married women on the labour market and the impact of unemployment have contributed to a rapid rise in part-time jobs. Four-fifths of such jobs are held by women and three-quarters of the total are in the service sector. In addition the incidence of shiftwork has doubled since the 1950s and has spread from manual work to computing, welfare services and many other sectors which employ large numbers of women. The growth of a 'black' or a 'shadow' economy, that goes unrecorded in official employment and tax statistics, reflects both this fundamental sectoral shift and the rise of unemployment. Economists have investigated multiple jobholding and what in the Third World would be called the 'informal sector' which carries no tax or social welfare obligations. They conclude that the black economy has grown in all West European countries since the mid-1960s and usually accounts for 7–10 per cent of the GDP of individual nations but stands at more than double that share in Italy and Spain.

Despite the undeniable hardship endured by some, the mythical 'average West European' now enjoys a much higher living standard than he or she did at mid-century simply in response to the fact that the average real GDP more than tripled during the four intervening decades. Family incomes have risen substantially and have given rise to a consumer society in which a declining proportion of income is spent on food and clothing, and a rising share on luxury goods, commercial services and holidays. Most houses are much better equipped than ever before, with nine out of ten households having a refrigerator, over half possessing a television set and over two-thirds having a car. The messages of the mass media and the greatly enhanced personal mobility afforded by car ownership offer the possibility of opening our minds to quantities of information and opinion and enriching our lives by new experiences in ways that earlier generations would have held inconceivable. Unfortunately, other features of the quality of life are less positive: violent crime is on the increase, especially offences against private property, as is juvenile delinquency. Equally disturbing is the fact that an ever-increasing share of West European households depends on social security transfers in the form of old-age pensions, unemployment monies and supplementary payments of various kinds to single-parent families and to the poor, who are most certainly still with us.

Unemployment

The numbers

At the start of 1990 some 14.2 million people were registered as unemployed in the EC – a figure very much greater than the 9.0 million in agricultural work at the time. For more than two decades after the Second World War, it seemed that mass unemployment had become a thing of the past over the greater part of Western Europe. Buoyant economic growth gave rise to millions of new jobs on building sites, in factories and in offices to employ an adult population that was expanding as the baby-boom generation came on to the labour market and guestworkers were recruited in vast numbers. Demographic and migratory processes interacted to raise the 15–64 years age group to some 200 million in 1980. In addition, many more women took paid employment and increased the potential labour force substantially. In the countries of the Nine, only about 2.2 per cent of the rapidly growing workforce was unemployed during the 1950s and 1960s, with the annual average falling to 1.9 per cent for 1964–66. In the eighteen countries of Western Europe, the proportion was a mere 1.4 per cent and a notable element in that figure was due to frictional unemployment in the short interval between jobs.

There were, of course, substantial national and regional deviations from these figures for the mid-1960s, with unemployment rates in the Republic of Ireland (4.5 per cent) and Italy (5.7 per cent) standing at more than double the average for the Nine. By contrast, unemployment involved less than 0.8 per cent of the workforce in Luxembourg, West Germany, the Netherlands, Denmark and France. The *Report on the Regional Problems of the Enlarged Community* (Commission of the European Communities 1973) showed that in the mid-1960s unemployment rates exceeded 6 per cent in a range of old industrial areas and more remote rural localities; namely, the southern half of Italy and the islands, the rural west of Ireland, Ulster, North Wales, the Highlands and Islands of Scotland, northern Jutland and parts of southern Belgium. Underemployment in traditional agriculture dampened down official rates in several of these regions and the same was undoubtedly true for Iberia and Greece. All 'old' industrial areas in Belgium and the UK, together with the remainder of the Irish Republic and Jutland, had between 4 and 6 per cent of their workforces unemployed but, by contrast, the general job picture in France, West Germany, the Netherlands and south-east England was unquestionably healthy.

By 1970, the west European total had risen to 2.5 million (1.8 per cent) and conditions deteriorated drastically after 1973–74, with the average unemployment rate for the Community in 1975 (4.3 per cent) being double that for the mid-1960s and the average more than doubling again to reach 12.0 per cent at the start of 1986, when 15.8 million were registered as being out of work in the Community – a figure somewhat greater than the total population of the Netherlands!

Definitions of unemployment and methods of collecting figures vary considerably between the countries of Western Europe but it is widely accepted that official figures underestimate the real situation. Household surveys and censuses reveal another large group of people – especially married women – who regard themselves as unemployed and would like to find work but are not on the formal unemployment register.

Unlike the pattern of the mid-1960s, Spain (with 16.1 per cent officially unemployed) and the Republic of Ireland (16.4 per cent) stood at the top of the list in 1990, with rates roughly double the EC average (8.3 per cent). Unemployment in the 1980s and 1990s has become a very different and much more diffuse phenomenon than that experienced in earlier years. It is no longer confined to the rural 'fringes' and the 'old' industrial areas of Europe. Nor is it experienced mainly by men, by the unqualified and those used to manual jobs. It is no longer a short-term problem but has become a generalized, long-term feature affecting many sectors of the labour force and all parts of Western Europe, albeit with differing degrees of intensity.

It is also a changing phenomenon, since job losses in any one sector of the economy do not necessarily imply higher unemployment on the whole. Even when GDP is declining, some new jobs are being created and at least a proportion of workers are changing jobs within and between sectors all the time. For example, jobs in service activities in the UK rose from 12.5 million in 1971 to 18.0 million in 1990 while employment in manufacturing contracted from 11.1 million to 7.8 million and in primary activities from 665 000 to 575 000. Because of differing dynamism in the varying sectors of the economy, the unemployed are not always made up of the same types of worker. The harsh truth of that statement is becoming increasingly apparent as redundancies are striking white-collar and professional workers who had come to regard their jobs as 'safe'.

The reasons

The conditions that have given rise to this crisis are both numerous and complex. At the heart of the problem stands the world recession which required greater economic efficiency from high-cost producers such as the countries of Western Europe. In addition, West European industry is experiencing fundamental structural changes, which are resulting from the transfer to low-cost developing countries of competitive advantages in production of low-technology and labour-intensive goods. Industrial firms in Western Europe have responded in various ways in their attempts to render themselves more competitive: by closing down capacity (newly recognized as being surplus to requirements), trimming costly labour inputs while reorientating production, and adopting new forms of industrial technology and marketing (Massey and Meegan 1982). In other instances, firms have ceased production completely. The net result has been the same: fewer jobs in manufacturing and in ancillary services as well. Routine paperwork is also retreating as a source of employment as new computerized techniques of information processing are widely applied to office jobs. Microelectronics has brought about a veritable revolution in this sector of the economy which was the backbone of many formerly healthy urban regions, where large numbers of white-collar workers, especially women, were taken on in the 1960s and 1970s.

Another major cause of employment problems is the fact that the workforce was swollen by the arrival on to the labour force of the largest-ever generation of school-leavers (Fig. 3.6). In every major West European country the number of children born rose each year from 1955 to 1964. From 1964 onwards the number declined, except in France and the Republic of Ireland. By the early 1980s the number of young people in their late teens and early twenties looking for their first jobs had reached an all-time peak, but this fell slowly as the decade progressed. But large numbers of young people continued to flood on to the labour market in the Republic of Ireland and they now face formidable problems as they look for work. For many of them, the choice is stark: emigrate or be unemployed. Thus in the early 1980s net migration losses involved about 1000 each year; in the late 1980s the annual figure was more like 30 000. The Irish birth rate remains the highest in Western Europe and, at 2.1 births per woman, the fertility rate has declined but is still very substantial.

A further contribution to widespread unemployment is the fact that over the past quarter century growing numbers of married women throughout Western Europe have sought paid work outside the home, often in the form of part-time employment which fits in with home-based activities. As we have seen, there are important variations in female participation rates between the more liberal societies of northern Europe and their more traditional Mediterranean counterparts. For example, only 30 per cent of Spanish women of working age had a job or wanted one (half the proportion in the UK). But that situation is changing fast as better-educated young women with fewer children are swelling the labour force and will surely inflate Spain's unemployment rate which is already in excess of 16 per cent. These two inputs have created severe strains in a situation where many employment opportunities are shrinking and are certainly failing to keep up with growing demand.

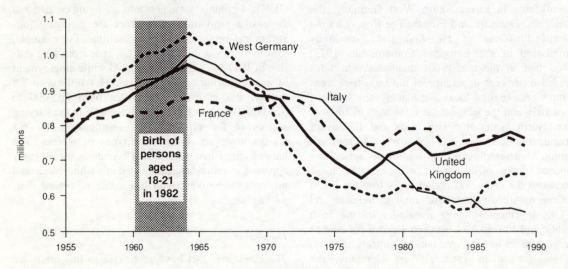

Fig. 3.6
Trend in births in four West European countries, 1955–90 (after *The Economist*).

Even more controversial is the question of whether generous social security payments discourage people from seeking work. It certainly has to be admitted that some nations are more benevolent than others in this respect. For example, the Dutch have developed a range of imaginative, caring but undoubtedly costly extensions of welfare payments. Social welfare and economic policy are bound together inextricably, but each nation in Western Europe ties the knots in a different way and with varying degrees of tension.

Those hardest hit

Four groups have traditionally been worst affected by unemployment: those living in depressed regions, the long-term unemployed, the unskilled and the young. They will continue to represent the core of the problem in the future, but it is clear that unemployment is spreading to all regions and to many sectors of the economy that have been recognized as buoyant until recently. In addition, women are bearing an increasing share of the unemployment burden and in most countries suffer a higher rate of unemployment than men. For a while, female workers were protected from the first effects of recession since they were heavily concentrated in less affected sectors of the economy such as service activities, but that period of grace did not last.

Old industrial regions with a legacy of declining 'sunset' industries typically display high unemployment rates. Textiles in north-west England and central Scotland and coal extraction in many European mining districts, all fit into that mould. In addition, relatively new industries in other regions have suffered the same fate in recent years, as the decline of the West Midlands' motor and engineering industries since 1970 bears clear witness. At a finer scale of analysis, 'inner cities' (as opposed to central business districts) have already been added to the list of areas with high unemployment, and suburbia and small-town Europe is being hit by redundancies as well.

Numbers of long-term unemployed have increased substantially in recent years and the longer a worker remains out of a job the more difficult it becomes to obtain one. Surveys in the late 1970s offered a sombre picture of workers who had been out of work for 12 months or more. About three-quarters were semi-skilled or unskilled manual workers; few had any formal qualifications; most came from declining industries; and many were in the older age range of the workforce and felt that their age told against them when they went in search of jobs. There is every reason to believe that the situation has deteriorated since then by the addition of younger, more qualified work-

ers to the list of the long-term jobless. Thus at the end of 1986, 52 per cent of the unemployed in the EC had been jobless for more than 12 months, with the proportion ranging from one-third in Denmark to two-thirds in Belgium, Italy and the Republic of Ireland. The hard truth is that a high long-term unemployment rate is dispiriting socially and useless economically; the longer someone is unemployed, the less able he or she is to compete for work.

Finally, and perhaps most tragically, there is the problem of youth unemployment which comprises school-leavers, those who pursued higher education and other young people who, for one reason or another, failed to obtain permanent employment. Not only are the cohorts of young people who flood on to the labour market remaining substantial each summer, but it is taking their members longer to find jobs. Individual countries have devised strikingly different formulae for organizing the lives of their young people. In each state of Western Europe, save the UK, young men undergo 'national' service for varying lengths of time. This normally involves military service but, under some circumstances, social or educational service may be undertaken and there are always grounds for exemption. Systems of apprenticeship and further and higher education also vary greatly from country to country. For example, young workers aged 16–18 in West Germany, Switzerland and Austria undertake apprenticeships on low pay for three years, while they continue education on a compulsory day-release basis in order to work towards a formal certificate of competence. Thus 70 per cent of all school-leavers in West Germany are absorbed by apprenticeship schemes, and two out of every five apprentices are girls. Teachers, manpower authorities and employers have clearly defined responsibilities in this process. Such an approach does not apply in the UK, where some teenagers on the open market receive higher rates of pay than their German counterparts. This economic fact works against the interests of British school-leavers since many employers are reluctant to take on 'expensive' and inexperienced young workers at this time of financial stringency. Youth training schemes do provide some opportunity for government-subsidized work experience for up to two years in the UK; however, many offer no formal qualification or promise of future employment.

Throughout Western Europe, older workers tend to be protected by job security laws, collective bargaining agreements and personnel practices which function to the relative disadvantage of the young. Unemployment rates among young people are normally in inverse relationship to their level of education or training. In broad terms that point still remains true but for the first time in several decades well-educated, middle-class young people are experiencing spells of unemployment and the sense of powerlessness and lack of control over their lives that has long been suffered by less educated members of society. Indeed, for almost everyone, prolonged unemployment is a seriously destructive experience that undermines personality and weakens the will and capacity to work.

As well as the four groups identified above, relatively recent immigrants and their children often experience high rates of joblessness. That problem, together with institutionalized discrimination, is shared by Surinamese and Antillans in the Netherlands, Turks in Germany, North Africans in France, and 'Black Britons' in the UK. But the legal status of these people is highly varied: some are technically foreigners, others have acquired the citizenship of a West European country, while others have held such status since birth. Unfortunately, all too many Europeans only see differences in skin colour and fail to appreciate complexities of legal status and all that this implies for right of abode (Condon and Ogden 1991).

Palliatives

Numerous attempts are being made to minimize the problem, including national measures to facilitate migration of workers to areas where appropriate jobs may still be available, artificial job-creation schemes, and a host of programmes for postponing the arrival of young people on the job market and perhaps even reducing the total supply of labour. Conscription and mass higher education are traditional and widely accepted mechanisms for keeping young people off the unemployment register and they are being supplemented by a host of special training schemes throughout Western Europe.

Repatriation of foreign workers when their work contracts come to an end is an effective but ruthless way of trimming a workforce that has no permanent right of abode in a particular country. Thus in West Germany, recruitment of new foreign workers was suspended in November 1973 and renewal of work permits was made increasingly difficult in cases where it was felt that guestworkers were competing directly with German nationals for jobs. Laws were enacted which imposed strict penalties on illegal immigration and a ruling of April 1975 entitled West German cities to turn away additional guestworkers (excluding EC nationals) once they made up 12 per cent of the local labour force. About 650 000 guestworkers duly left the country between 1973 and 1977 when their contracts expired, but an equal number of family members were allowed into the country during the following four

years to join guestworkers already living there. By the early 1980s, rates of unemployment among foreign workers were 50 per cent higher than among native West Germans.

Late in 1981, with social-welfare funds severely strained and unemployment still rising, the federal government approved a new batch of measures to restrict immigration. It recommended the *Länder* to keep out foreign workers' children aged 16 or more, as well as all children who had a parent living in their country of origin and all dependants of foreigners who were under contract to spend only a limited period in West Germany. Wives would be permitted to join their husbands only if their spouses had worked at least eight years in West Germany and if they had been married for at least one year. These measures reflected the general trend of public opinion regarding guestworkers. In 1978, public opinion polls had shown that one-third of West Germans were in favour of sending foreign workers home; by the summer of 1982 the proportion had risen to two-thirds. In 1981, two-thirds of those questioned said they would support a ban on the entry of guestworkers' dependants; in mid-1982, 90 per cent wanted such a ban. Early in that year, financial incentives were proposed to encourage guestworkers and their families to return to their home countries. West Germany then contained 4.6 million foreigners (comprising some 2 million workers and a rather larger total of dependants), or 7.7 per cent of the total population, while in West Berlin, Frankfurt, Munich and Stuttgart the proportion exceeded 20 per cent (O'Loughlin 1980).

In France, a provisional halt to labour migration was called in 1974 and was extended in 1975. Efforts to stop illegal immigration were stepped up and in 1977 the government introduced the offer of a special financial bonus plus a one-way air ticket for each guestworker who agreed to return home. By the summer of 1980 only 41 000 of the country's 1.8 million foreign workers had risen to the bait, with Portuguese making up more than half of that small total. One-third of the jobs thus vacated were taken by French workers, another third by previously unemployed foreign workers, while the remaining jobs were not filled. Belgium also banned further immigration of non-EC nationals (with the exception of certain skills), while Dutch authorities were forced to interpret work permit regulations very strictly and discourage the relatives of Mediterranean guestworkers from settling in the Netherlands. Dutch problems had been exacerbated by the arrival in 1974–75 of 120 000 migrants from Surinam who have fared less well on the labour market than the 300 000 who preceded them in the 1950s. The scope for continuing the politically sensitive process of repatriation is limited and could not be envisaged for workers and family members who possess residence permits. By contrast, West European governments have found scope for further curtailment of non-European immigration and in recent years have tightened their regulations on both entry and residence, thereby fuelling the notion of 'fortress Europe'.

Early retirement schemes offer the possibility of opening jobs for younger people, but pensions need to be financially attractive if such an approach is to work on a voluntary basis. The appearance of 'grey power' in the USA is a salutary manifestation of people's desire to continue working beyond retirement age. Limiting or prohibiting overtime, and working shorter days, weeks or even years offer ways of sharing out available work more widely, but schemes of this kind require adequate wages or social security payments for living standards to be maintained. Part-time work is already widespread in Scandinavia and the UK and the idea of job sharing has received considerable discussion but companies that have experimented with it have encountered problems. Daf, the Dutch vehicle firm, found that women staff in secretarial, medical and other service posts adapted well to work sharing and found that they could suit both their own and their employer's interests. Middle-aged, male factory workers in the southern Netherlands reacted less positively and resented being deprived of their full wage-earner status and having to bring home a slimmed-down wage packet, especially if their wives were fully employed. By contrast, the prejudice associated with this aspect of sexual stereotyping is breaking down among young workers of both sexes in the Netherlands and especially the open-minded citizens of Amsterdam. The hard truth is that fear of unemployment is a powerful stimulus for changing one's work routine. Doubtless this helps explain a surge of interest in flexible working arrangements by employees, companies and governments in recent years.

Whatever policies may be adopted to alleviate the hardships of unemployment and improve the chances of the long-term unemployed, the fact remains that there are insufficient jobs to absorb the total working population, while the supply of labour will continue to increase more rapidly than demand for workers into the 1990s. The impact of 'new technology' will continue to exacerbate the problem when demographic pressures start to slacken during this decade. In short, the age of full employment has passed into social history. Unemployment is not the preserve of scroungers; it extends throughout society, enmeshing executives, long-serving craftsmen, computer programmers and all other walks of life so that no job is secure. It will have to be recognized as an element of everyday life

for a large and growing section of the adult population in Western Europe during the remainder of the twentieth century. That condition will require a social revolution in attitudes, behaviour and self-esteem of almost unfathomable magnitude.

Looking further ahead the decline in the birth rate over recent years may mean that some parts of Western Europe may be labour-hungry again in the early years of the twenty-first century. For example, Germany may be fretting over labour shortages and may have to look to the nations of North Africa or Asia, despite the incorporation of 16 million new residents and the arrival of 'ethnic Germans' and others from Eastern Europe. However, that kind of demographic projection is of no consolation to those West Europeans who cannot find work now.

Inequalities

Poverty

The 'poor' are defined by the EC's Council of Ministers as people whose material, cultural and social resources are so limited as to exclude them from the 'minimum acceptable way of life' in the member state in which they live. The grinding, starving poverty of the inter-war years now barely exists in Western Europe but since 1975 the number of people with low income in the EC rose from 44 million to 50 million. These figures derive from family budget surveys where the 'poverty threshold' is recognized as half the average disposable income per capita in the country concerned. Table 3.3 compares the proportion of 'poor' residents in each member state of the EC with that country's share of the total population (Eurostat 1991a). Poverty in the Mediterranean south and along the Atlantic fringes (including the UK) is clearly in evidence. Rates of unemployment are still rising in some areas and among some groups, with social service provision in every country proving inadequate to cope with needs. The European Commission's report on the *First Programme to Combat Poverty* (1981) stressed that certain groups or categories of people – notably the elderly, immigrants, households headed by women, one-parent families, large households and the handicapped – were most vulnerable to relative poverty regardless of where they lived.

The evidence showed that economic growth did not of itself eliminate poverty, not even when accompanied by increased spending on welfare services. Thirty-five years ago, spending on these services in countries that were to become the Community of the Nine ranged from 8 to 15 per cent of the GNP; by 1980 the corresponding proportions were 20–30 per

Table 3.3   Poverty and population by member state of the EC, 1985. (%)

	Poverty	Population		Poverty	Population
Belgium	1.2	3.1	Italy	17.9	17.8
Denmark	0.8	1.6	Netherlands	3.3	4.5
France	17.5	17.1	Portugal	6.7	3.2
Germany (West)	12.2	19.0	Spain	14.6	11.9
Ireland (Rep.)	1.4	1.1			

cent. Spending on education rose from less than 4 per cent to 5–7 per cent of GNP and many EC programmes were developed to improve housing and to subsidize housing for the unemployed in Belfast and Mainz and resettlement help for return migrants to Greece. Despite schemes launched under three programmes, millions of people still live in substandard housing; in some countries, infant mortality rates in families of unskilled manual workers are still twice as high as those of professional families; health services are inadequate; and between an eighth and a quarter of young people in member countries leave school with few or no qualifications or vocational training. Trying to cope with problems associated with poverty has only just begun and the presence of large numbers of jobless adults, more elderly people and more one-parent families renders the task even more difficult.

Xenophobia

Rising unemployment rates served to stir up xenophobic tendencies among many West Europeans who feared that jobs which they believe should have been theirs have been purloined by guestworkers. For example, in 1964 the millionth guestworker to arrive in West Germany received a celebrity's welcome at Cologne station; now the talk is of *Überfremdung* – too many foreigners. Hundreds of thousands of second- and third-generation 'foreigners' have been born in West Germany and the birth rate among immigrant families is well above that among native Germans. The death rate is very much lower since at present the 'immigrant' population has a younger age profile than does the indigenous German population. In summer 1982, West Germany contained approximately equal numbers of Yugoslavs (637 000) and Italians (624 000), but the 1.5 million Turks comprised by far the largest single group of foreigners and have a strikingly different culture, religion and way of life from their hosts; only a fraction are interested in acquiring German nationality. They have experienced violent xenophobic attacks in some cities and graffiti such as 'Germany for the Germans', 'Foreigners out' and

'Turkish-occupied zone' have become increasingly commonplace.

At the start of 1986, France contained about 4 million foreigners, with that figure having contracted from 4.4 million in 1983 because of return migration and naturalization. Portuguese (765 000) formed the most numerous group, but it was North Africans (Moroccans 494 200, Tunisians 200,780, and especially Algerians 535 000) who were the most frequent targets of racial attacks. As in other West European countries, many of the 'foreigners' are in France to stay. The earlier phases of labour migration by lone males and later arrivals of wives and children have long passed (Castles 1984). Now the key issues are integration and assimilation of whole families, including many young people actually born in France and perhaps holding French nationality.

The popularity of Jean-Marie Le Pen and the *Front National* is a worrying demonstration of racism not only among the working class but also among some middle-class elements who respond to a philosophy of the purity of the French nation and of keeping France for the French. Le Pen insists that illegal immigrants should be sent home and even those in jobs should be encouraged to go, thereby helping to reduce unemployment among the 'French-French'. The *Front National* receives sizeable votes in de-industrializing working-class districts (with high unemployment and large immigrant communities), in southern towns (where many Europeans settled after leaving newly independent Algeria in 1962), and even in bourgeois neighbourhoods in Paris and other large cities. After a honeymoon of goodwill towards foreigners in the early 1980s, major political parties of both right and left adopted a tougher line in policies regarding immigrants from non-EC countries.

Guestworkers have different geographical origins in Belgium, the Netherlands, Switzerland and the UK. Many 'immigrants' share Islamic cultural traditions which are unfamiliar to most West Europeans (Gerholm and Lithman 1988). A walk through the back streets of any major city in Western Europe and a casual reading of graffiti reveals a widespread demonstration of racist attitudes. The slogans are expressed in various languages and daubed in differing colours, but the sentiments and the fundamental problems are very much the same. Even Brussels, arguably the 'capital' of Western Europe, is not immune. In three inner-city districts, poorer foreigners from the Mediterranean states, Asia and Africa make up between one-third and one-half of the total population and there the slogans are as embittered as anything to be seen in working-class Paris, London's East End or the poorer districts of Berlin. The immigration issue dominated campaigning during the 1985 election and some local authorities in the Belgian capital refuse to grant residence permits to new arrivals from Turkey or North Africa.

The latest part in this complicated story involves the arrival of refugees who are seeking asylum in Western Europe for fear of persecution in their home country on grounds of race, religion, social group or political opinion. In the early 1970s asylum-seekers numbered only a few thousand each year. Then numbers began to increase, with over 750 000 arriving from non-European sources between 1977 and 1986, including 200 000 in 1986 alone, and all at a time of mounting unemployment and xenophobia (Fig. 3.7). During 1984, 1985 and 1986, Sweden accepted about 5000 refugees for every million inhabitants, Denmark and Switzerland took in over 4000 per million and West Germany 3350, but the UK only 240 per million. West European governments have failed to agree a policy on refugees, partly because of the difficulty of judging the genuineness of asylum-seekers, including 50,000 Tamils arriving from Sri Lanka (1984–86), 40,000 Iranians (mostly young men avoiding military service) and thousands more from Ghana, Ethiopia, Pakistan, Zaïre and the Lebanon.

Increasingly, Austria, Greece and Italy have functioned as places of transit for those seeking asylum in the supposedly more comfortable and generous welfare states of north-west Europe. Each destination country operates in its own way with regard to refugees but all now attempt to discourage further arrivals. For example, once-generous Germany prevents non-European refugees from taking work for five years and requires them to live in camps. Agreement with East Germany in the mid-1980s halted the flow of Asians and Africans who had been arriving by plane to East Berlin and then been allowed to pass through the Wall. Belgium, France, the UK and formerly open-hearted Switzerland, Sweden and Denmark have tightened their border controls to ensure that asylum seekers who lack genuine claims are dispatched elsewhere or sent home.

During the summer of 1990 3000 Albanians arrived in Italy and were accepted as refugees fleeing from a particularly repressive regime. In the following year some 40 000 commandeered ships to cross the Adriatic, only to be turned back as unwelcome economic migrants. The Italian authorities justified their harsh response by arguing that many more young Albanians were desperate to follow. In addition, Italy already contained a large number of illegal, unskilled migrants from North Africa and Asia. Further north, West Germany accepted well over a million 'ethnic Germans' from Poland, Romania and the USSR, with numbers rising from 78 000 in 1987 to 400 000 in 1990. In accord with the country's Basic Law, citizenship is defined in ethnic terms and is automatic for those with evidence of German ancestry (Treasure 1991). Provision of housing, language training, social security payments and help to find employment has

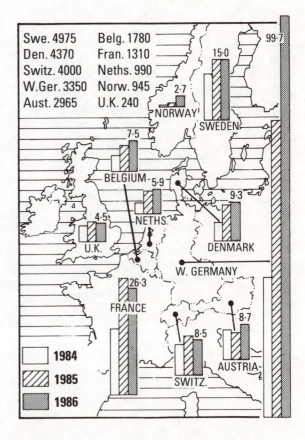

Fig. 3.7
Asylum-seekers in main West European countries—columns indicate annual arrivals in thousands (after *The Economist*).

proved expensive and has given rise to much social tension, for example in the cities of North Rhine-Westphalia where the greatest numbers of newcomers moved (Jones 1990). Reception conditions have become all the more strained because of the movement of large numbers of East Germans to the western *Länder*: over 340 000 arrived in 1989 alone. Freedom of movement for EC citizens within its borders is a fundamental element in the 1992 'package', but deciding who will be allowed to enter the Community, under what conditions, and how they will be treated once they have arrived will continue to give rise to vigorous debate.

Further reading

Archer M S, Giner S (eds) 1971 *Contemporary Europe: class, status and power*. Routledge and Kegan Paul, London

Bailey J (ed.) 1992 *Social Europe*. Longman, London

Castles S 1984 *Here for Good: Western Europe's new ethnic minorities*. Pluto, London

Kosiński L 1970 *The Population of Europe*. Longman, London

Warnes A M (ed.) 1982 *Geographical Perspectives on the Elderly*. Wiley, London

Wise M, Gibb R 1993 *Single Market to Social Europe: the European Community in the 1990s*. Longman, London.

4

Migration

Migration in context

Migration has been the main factor responsible for changing the population map of post-war Europe. Not only that, migration also lies at the root of, or is heavily involved in, several major trends shaping the contemporary human geography of Western Europe: processes like urbanization and counter-urbanization, regional economic growth and decline, rural depopulation and repopulation, inner-city decline and gentrification – to name but a few.

Human mobility can be classified in many ways. Generally, there is a broad correlation between distance moved and frequency of the movement. A continuum exists, ranging from short-distance daily commuting and weekly flows to second homes, through annual holiday trips and seasonal farm labour moves, to more permanent residential shifts which may be local, inter-regional or international in character. Before 1914, there were enormous outpourings of European population to the New World, chiefly the Americas. Since 1945 there have been large-scale transfers of migrants into the industrial countries of north-west Europe from the continent's Mediterranean fringe and from former 'colonial' territories like Algeria, the West Indies, India and Pakistan. In all, a bewildering variety of types and scales of movements!

This chapter is mainly concerned with migration *within* and *into* Western Europe. Historical migrations of Europeans overseas are mentioned only in passing. We shall also not dwell on local-scale intra-urban residential movement. The two main types of migration to be dealt with are international migration and inter-regional migration within individual countries.

As with many other socio-economic phenomena in Western Europe, the story of migration reached a kind of threshold in late 1973 with the onset of the oil crisis. The discussion is therefore structured around this 'tipping-point', although it will be shown that econ-

omic deterioration was by no means the only factor accounting for the sharp change in migratory trends in the early 1970s. The chapter contains four main parts. First, the evolution of international migration is traced up to 1973. Second, we examine the effects of the oil crisis on migratory patterns, with the freezing of new movements, the switch to the consolidation of foreign migrant stocks and the trend to return migration. The third section looks at new trends in international migration in Europe which have emerged since the mid-1980s and especially since the dismantling of the Iron Curtain in 1989. The final section explores trends in internal migration and makes a special study of France, which is the most instructive country to consider because of the variety of its internal migration processes. It is, however, worth noting that, with twelve of the eighteen major countries of Western Europe participating in the EC, the distinction between internal and international migration may become increasingly irrelevant.

Immigration has never been far from the European popular consciousness or political agenda. By 1990 there were an estimated 20 million foreigners living in Western Europe. Some have been settled for decades: these are mostly intra-European migrants such as the Irish in Britain, Portuguese in France and Greeks in Germany. Increasingly, however, Western Europe's migrant workers have been drawn from further afield: from Morocco and Turkey and from a variety of Asian and African countries. As elections in Belgium in November 1991 and in France and Germany in early 1992 have shown, extreme-right parties such as Jean-Marie Le Pen's *Front National* have gained political capital out of manipulating the immigration issue.

International migration: the picture to 1973

Two major influences lie behind international migration: political and economic. Much of Western

Europe's twentieth-century cross-national movement has been politically inspired, associated with the two world wars. Since 1945, economic reasons have taken over. The search for better jobs and higher incomes has guided millions of southern Europeans, Irish and Finns, mostly from rural backgrounds, to the great industrial conurbations and capital cities of north-west Europe. Most recently of all, in 'post-industrial' Europe, complex flows of high-level manpower have arisen between highly urbanized countries.

Kosiński (1970) has traced the earlier waves of European mobility and has shown how emigration has been a persistent theme in Europe's demography. Between 1820 and 1940 an estimated 55–60 million people left Europe for overseas, the annual rate peaking at over 1 million in the years immediately preceding the First World War. Most migrants – 38 million during 1820–1940 – were destined for the USA. Initially, they came from 'old' emigration countries like Great Britain, Germany and Scandinavia but by the end of the century the centre of gravity had shifted south: Italy alone was sending 400,000 a year by 1910.

Intra-European movements were also under way before 1940. In the nineteenth century, industrializing nations drew workers from neighbouring countries, the Irish moving to Britain, Italians to France and central Europeans to the German Empire. Later, Kosiński (1970) estimated that 7.7 million Europeans were involved in 'political migrations' associated with the First World War. The German census of 1925 recorded that 770 000 persons had moved into Germany from German-conquered territories such as Alsace-Lorraine, northern Schleswig and Eupen-Malmédy when they reverted to their original countries (France, Denmark and Belgium, respectively). Regional wars provoked large transfers of population between Greece and Turkey, and an estimated 300 000 Spaniards moved to France during and after the Spanish Civil War. Political migrations continued in the late 1930s with the *Volksdeutsche*, the 'gathering in' of ethnic Germans attracted by pan-German propaganda and economic prosperity in pre-war Germany.

The Second World War was accompanied by further massive shifts of population across national frontiers which were themselves changing. One estimate gives 25 million people moving (Kosiński 1970), the majority of them (13 million) Germans. Most of the remaining displacements were in Eastern Europe. Germans drafted for military service were replaced by foreign workers recruited from conquered lands. Amounting to 7 million, of whom 5 million were civilians and 2 million were prisoners of war, they accounted for 20 per cent of the German workforce by the end of the war.

At the end of the war, further *Volksdeutsche* took place from the Baltic states, Ukraine, Romania, Bulgaria and from different parts of dismembered Yugoslavia. As a result of an agreement with Italy, Germans from the Bolzano region were given a chance to emigrate to Germany. About one-third of them, around 100 000, actually moved. Six million Germans emigrated from Poland, Czechoslovakia and Hungary during 1945–49, 485 000 Finns were repatriated in two waves in 1939 and 1944 during the Soviet–Finnish wars, and about 200 000 Italians left Istria when the Italo-Yugoslav border was adjusted at the end of the war.

This description of pre-1945 migrations, admittedly skeletal, is necessary to show that the post-1945 'economic' migration of migrant labourers did not arise as a sudden and totally new phenomenon – an impression which is sometimes given. Italy, the major sending country of the early post-war period, already had 900 000 members of its population living in France in 1931. In West Germany there were 9.8 million *Heimatsvertriebene* ('expelled from homes') recorded in 1964; because of their young age structure (2.5 million of them below 20 years of age in 1964), they contributed greatly to the labour supply, holding down wages and acting as a key factor in the West German economic miracle before the subsequent influxes of non-German migrants from southern Europe (Kosiński 1970).

Later, politically stimulated migrations resulted from decolonization processes in the Maghreb and elsewhere. The most dramatic among this group were the 1 million *pieds noirs* from Algeria who were repatriated to France after the Treaty of Evian in 1962. Most of the *pieds noirs* settled in the Midi or in Paris; by 1973, 80 per cent were living in towns, mostly large ones, and 60 per cent had settled in a belt stretching between Bordeaux and Nice. The settlers opting for southern France were mostly the poorer classes among the repatriates; the numerically fewer civil servants and professionals tended to head for Paris and the north (Baillet 1975).

General character of post-war labour migration

The post-war labour migrations from southern European and North African villages to northern European towns and cities are the world's most massive in recent decades. Estimates vary as to the numbers involved because of the considerable quantity of clandestine movers and the large amount of fluid temporary movement which goes partially unrecorded. The various countries have different ways of recording migrants based partly on different definitions of exactly what constitutes a migrant. Switzerland differentiates 'permanent', 'annual', 'seasonal' and 'frontier' migrants, whereas France has a much more lax recording system. Official statistics indicate that at the eve of the recession in 1973 there were about 7 million migrant workers in the EC Nine, accompanied by a further 5 million dependants. Enlarging the figures to account for unrecorded moves and for non-EC countries like Sweden and Switzerland yields 'guesstimates' of around 10 million for workers and perhaps 15 million for the total migrant population at this time.

What distinguished this migration from others in the past was its essential *temporary* character. At least to start with, neither the migrant himself nor the reception country had any ambitions for permanent settlement. The migrants came basically to sell their labour where there was a labour shortage. The jobs they filled had certain common characteristics: poorly paid, insecure, unpleasant, boring and heavy. Until recently, non-EC migrants had few rights or claims outside the reality of their jobs. Berger and Mohr (1975) wrote that it was not men who migrated but machineminders, sweepers, diggers, cement mixers, etc.

There is no doubt that the recruitment of foreign workers became a central plank for the growth and prosperity of the members of the EC and of other European industrial countries between the 1950s and the early 1970s. What was initially regarded as a temporary expedient to ease labour shortages has become something close to a necessity, a permanent structural feature of the EC space economy. The 'self-feeding' process of labour migration has seen the penetration of migrants into other sectors of the economy (Böhning 1972). To a certain extent this is because of a natural process of upward mobility that some migrants enjoy through their drive and perhaps their ability to adapt and to learn a new language quickly. In other cases the existence of a migrant community provides its own opportunities for entrepreneurship to flourish within the 'ethnic economy'. A good example of this is the Cypriot community in London which is almost completely self-sufficient with its own housing market, eating places, lawyers, travel agents, shops and other services (Constantinides 1977).

Expansion of migrants into other job fields also takes place as rising standards of living induce a flight on the part of the native population from employment types that are perceived as undesirable. Labour migration thus has two interrelated but very different effects: it ensures a plentiful supply of cheap, low-grade labour; and it acts as an agent for the social promotion – the *embourgeoisement* – of the indigenous workforce.

In very general terms, the post-war labour migration into Western Europe's industrial heartland had well-defined characteristics (Salt 1976). Most migrants were male, unmarried and unskilled, departing in their twenties. The temporary nature of the movement was partially governed by the nature of the work contracts, which rarely lasted for more than a year or two, though most were renewable. This 'rotation' of migrant workers found its most overt expression in West Germany in the 1960s where the policy of *Konjunkturpuffer* was a clear attempt to use migrant labourers as an industrial 'reserve army' to be hired and dismissed according to the dictates of the economy's fluctuations, thereby protecting the host country's own workers from the unemployment effects of cyclical recession (Castles and Kosack 1985). Berger and Mohr (1975) exposed the exploitative character of migration and the way industrialized countries benefited from it. As far as the economy of the industrialized country was concerned it was as if migrant workers were 'immortal' since they could be changed continually. It seemed as if they were not born or brought up, nor did they age, get tired or die. As far as the receiving economy was concerned their single function was to work.

Nevertheless, the fundamental ambiguity of the European migrant worker situation, still today, must be recognized. It is a paradox that millions of people are living in foreign countries in Europe but are not designated as immigrants. The concept of 'immigrant' has been cryptically paraphrased by the use of such terms as 'guestworker' (*Gastarbeiter*), 'foreign employee' and 'migrant labourer'. Yugoslavian emigrants were termed 'Yugoslav residents in temporary employment abroad' by the sending country: a clear if clumsy refusal by the Yugoslav government to acknowledge that the emigrants might not return. A consequence of this collective denial of the objective existence of emigrants and immigrants and of emigration and immigration countries is that no clearly defined migration policies have been developed, even in those countries which have been receiving post-war migrants for forty years. The migrants' status remains permanently provisional – a situation which is as absurd as it is inhumane (Hoffmann-Nowotny 1978).

Institutional background

Articles 48 and 49 of the Treaty of Rome provide for the free movement of labour within the EC and for the abolition of discrimination between Community nationals with regard to employment, pay and other rights of work and civic life. The free movement provisions were ushered in between 1961 and 1968. The articles have been described as little more than a liberal–capitalist prescription for bringing manpower shortages and surpluses into balance within a given trade area (Böhning 1972). They were inserted at the

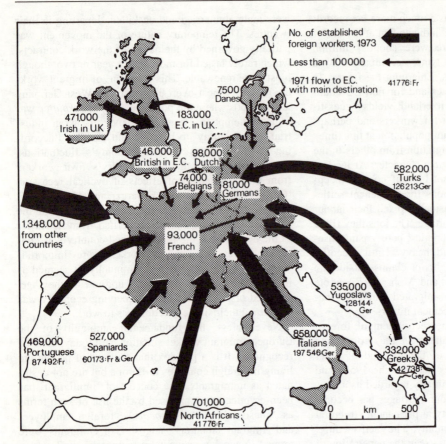

No. of established foreign workers, 1973

Less than 100000

1971 flow to E.C. with main destination   41776:Fr

7500 Danes

471,000 Irish in U.K.

183,000 E.C. in U.K.

46,000 British in E.C.

98,000 Dutch

74,000 Belgians

81,000 Germans

582,000 Turks 126 213 Ger

1,348.000 from other Countries

93,000 French

535,000 Yugoslavs 128144: Ger

858,000 Italians 197 546 Ger

332 000 Greeks 42 738 Ger

469,000 Portuguese 87 492 Fr

527,000 Spaniards 60173 Fr & Ger

701,000 North Africans 41 776: Fr

0   km   500

Fig. 4.1
Labour migration in the European Community in the early 1970s.

prompting of Italy, the Common Market country then most affected by labour surpluses, but the provisions were also actively supported by industrial countries whose governments and large employers wanted a flexible and non-permanent labour supply to perform the least desirable job functions during periods of rapid economic growth and labour shortage.

In practice, the 'free movement' policy had negligible effects. Migration between EC countries actually declined over the period of free movement, a paradoxical situation which is partly explained by the growth of the Italian economy in the 1960s and the diversion of southern Italian migration from Europe to northern Italy. This, however, is not the whole story, for Italian emigration to Switzerland continued to grow and in most years during the 1960s more Italians migrated to Switzerland than to the entire EC. In fact, Articles 48 and 49 failed to halt a trend that was already established before 1961 – the decline of intra-EC movement as a proportion of all migration into the EC (65 per cent in 1958, 51 per cent in 1961, 28 per cent in 1964). In the same year as the free movement provisions came into force, West Germany concluded bilateral

agreements with Spain, Greece and Turkey for labour importation under German controls. An agreement with Yugoslavia followed soon after. As Italian migration to West Germany tailed off (141,000 in 1960; 30 000 in 1970), the Yugoslavian and Turkish streams swelled to account for over half of West Germany's annual intake of *Gastarbeiter* by 1970.

In short, the 'liberal–capitalist prescription' had failed: labour movements had not equalized income and prosperity differentials within the Community – if anything, regional disparities became accentuated. Labour migration had shown itself to be one of the most powerful links binding the economic core of Europe to an ever-widening periphery (King 1982).

Geographical pattern of flows

Figure 4.1 shows the spatial pattern of international migration into the major receiving countries of the EC at the eve of the mid-1970s recession. The pattern has a certain spatial symmetry, the magnet of the 'core' attracting inflows from the periphery. France drew most of its migrants from the south and south-west

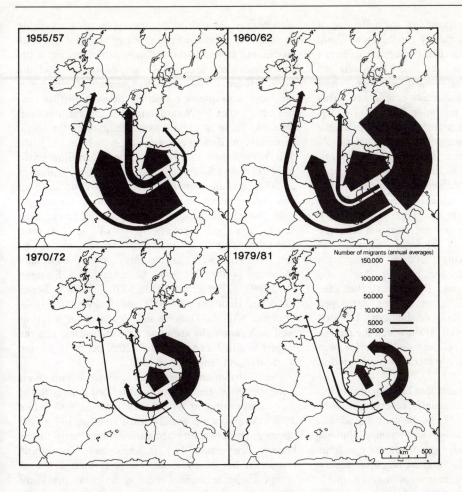

Fig. 4.2
Emigration from Italy
to European countries.

(Spain, Portugal and Algeria), while West Germany's hinterland lay to the south-east in Yugoslavia, Greece and Turkey. Italy has sent large numbers of migrants to both France and West Germany, as well as to Switzerland. Britain has drawn mainly from the Irish periphery as well as from New Commonwealth territories, while in the far north (not shown on Fig. 4.1) the Nordic Common Labour Market, dating from 1954, has enabled many Finns to migrate across the Gulf of Bothnia to Sweden.

In fact, the pattern evolved in a series of stages, with different sending and receiving countries dominant at different times. During the late 1940s and the 1950s, France, Belgium, Switzerland, Sweden and the UK were the principal destination countries; Italy, the Republic of Ireland and Finland were the main suppliers. West Germany then still drew labour from territories to the east, but this changed in 1961 when the East German source was shut off. The period 1961–73 saw the rise of West Germany as the major destination for southern European migrants. At the same time, the

Italian source began to dry up, to be replaced by Greeks, Yugoslavs and Turks in West Germany and by Spaniards, Portuguese and North Africans in France. Details of the Italian post-war flow to Europe can be seen in Fig. 4.2. The four triennia were chosen for their representativeness in charting the changing pattern and volume of flow. In the 1950s, France and Switzerland dominated the exodus, although over 20 000 were going each year to the Benelux countries (mostly to Belgium) and over 10 000 to Britain. Five years later, the West German and Swiss flows had risen to dominance; the British flow was maintained but the Belgian and French were halved. The early 1970s and the triennium 1979–81 reveal contraction across all major destination countries, especially marked being the fall-off to France. Now, in the 1980s and 1990s, Italy is a country of immigration.

Although the sending countries shared common features of relative overpopulation, underemployment and underdevelopment, their individual migration profiles varied. Some were 'old' emigration countries whose post-war flows to Europe represented the deflection of streams which had headed mainly for the New World (Greeks and Italians to America, Portuguese to Brazil and Angola). For other countries, like Yugoslavia and Turkey, emigration was an almost completely new phenomenon, lacking significant precedents. This notion of the 'maturity' of migration flows is also relevant in helping to explain regional differences in the intensity of outmigration within the individual supply countries. In Portugal, Spain, Italy and Greece, emigration during the 1960s was heaviest from overpopulated, rural, upland districts like Tras-os-Montes, Galicia, Andalusia, the southern Apennines, Epirus and Thrace. For Yugoslavia and Turkey, the pattern was the opposite, with heaviest outmigration from the more urbanized regions of Slovenia, Croatia and Istanbul (see the map in King 1976: 71). The contrast corresponds to the difference between 'old' and 'young' emigration countries. The Yugoslavian case also demonstrates how different regions developed specialized links with certain destinations. Detailed maps presented by Baučić (1973) show that the pattern of outmigration to France, concentrated in Serbia, was the complete opposite to the pattern for outmigration to West Germany.

The flows of labour migrants that built up during the 1960s were matched by invisible, reverse flows of money. Remittances are held to be the principal advantage emigration brings to the sending country. Hume's (1973) estimate of an annual flow of remittances from migrants in Europe of $2500 million has been widely quoted but is certainly an undervaluation for much money finds its way back not through banks and official channels but via letters and friends. Thousands of southern European villages have come to rely on remittances as their main source of income, in some cases accounting for over half or even three-quarters of the communities' cash resources. Rarely, however, is this money used to promote the development of the regions concerned: most of it goes on maintaining members of the family left behind, improving housing and importing consumer durables. Since most consumer durables are imported from industrial Europe, the money goes straight back whence it came.

Economic and social characteristics of migrants in European cities

The distribution of migrants exhibits even more distinct spatial characteristics at the other end of the migratory chain. With one or two exceptions, such as Spanish labourers working in French vineyards and Yugoslavs working in Austrian ski resorts, migrants have clustered in urban and industrial regions, their distribution determined by the types of work made available to them. In France, 40 per cent of migrants live in and around Paris, and most of the rest are in the major industrial areas of Lyons, Marseilles and the north (Ogden 1985a). In West Germany, *Gastarbeiter* are more widespread, reflecting the more diffuse spatial structure of the West German urban and industrial network. There is a relative concentration in the south, especially in Baden-Württemburg which in 1973 had 570 000 migrants in a population of 3.5 million. Also important were Südbayern and Hessen, both with over 250 000, and the populous Nordrhein-Westfalen with 680 000. Between them these four out of Germany's nine *Landesarbeitsamtsbezirken* (regional labour markets) contained three-quarters of the Federal Republic's migrants. The largest single clusters of foreign workers were in the main cities, especially Munich (with 156 900 *Gastarbeiter*), Stuttgart (130 200), Frankfurt (122 300), West Berlin (82 500), Hamburg (69 500) and Cologne (63 000). Stuttgart, with the Daimler and Mercedes car plants as well as a range of expanding light industries, had the largest relative concentration: 26.5 per cent of the workforce were migrants in 1973.

The distribution of migrant groups within West European cities has always been related closely to the types of housing to which they have access (White 1984). Migrants are usually constrained to live in the cheaper, poorer-quality accommodation. Partly this is because they have low incomes and few capital resources, but an additional factor is their desire to accept cheap accommodation in order to maximize their savings and hasten their dream of returning home. As a result, migrants tend to be forced into quasi-ghettos which take three main forms.

Firstly, they may be concentrated in the low-quality, older type of tenement buildings found in the inner city. A 1970 housing survey in the Ruhr found that migrants, even though they paid 30 per cent more than Germans for rented accommodation, were twice as likely to be in pre-1945 housing and three times as likely to lack their own bath and lavatory. Also analogous to inner-city flats are the squalid *hôtels meublés* of the twilight zones of French cities; usually these are crammed full of North Africans. Second, and prevalent mainly in France, are *bidonvilles* or peripheral shantytowns. These are now very much reduced in extent but in the 1960s existed as festering and insanitary sores around most major French cities. The third type of quasi-ghetto was the worker hostel built for *Gastarbeiter* in Germany and Switzerland. These single-sex dormitories, many of which were built within factory compounds, represented the ultimate in social segregation; their atmosphere has been well captured in the photographs of Jean Mohr (Berger and Mohr 1975).

Patterns of migrant employment naturally differ from country to country, which makes a full account impossible, but most migrants entered low-status, low-paying unattractive jobs shunned by the native workforce. The pattern of employment has also changed somewhat over time (Salt 1976). During the 1950s, most migrants were in agriculture, mining, construction and the service sector. Subsequently, there was a massive expansion into the industrial sector. By 1973, migrants made up 20 per cent of the industrial workforce in France, 12 per cent in Germany and 40 per cent in Switzerland. Within the industrial sector, migrants were especially prevalent in the most disagreeable and hazardous jobs, on boring production lines and in health-sapping industries like asbestos, rubber and bricks. In Europe's big car plants – Ford in Cologne, Volvo in Gothenburg and the Renault factories in France – 40–45 per cent of the labour force were migrants by 1973. As foreign workers have increasingly penetrated into other sectors of the economy, some stratification has developed with regard to the status of jobs taken by different nationalities. By the early 1970s, Italians no longer filled the more menial and unpleasant jobs. They had moved on to the lighter and more pleasant factory and service-sector jobs, while jobs like refuse collecting and street sweeping were left largely to Turks in West Germany and to Algerians in France.

Although migrants form a kind of sub-proletariat or *Unterschichtung* beneath and unrelated to the social structure of the receiving country (Hoffmann-Nowotny 1978), it is clear that the migrant stratum itself is differentiated with respect to job types and housing. Italians do not sweep the streets, neither do they live in *bidonvilles*; they are the aristocrats of the brawn drain. The Spaniards are not far behind. At the bottom of the migrant hierarchy are the Turks and the Maghrebins; the Portuguese, Yugoslavs and Greeks form a middle layer. This idea can be developed further by examining interrelationships between socio-economic status, cultural background and degree of 'acceptability'. In France, for instance, the more acceptable Italians and Spaniards come from a cultural, religious and linguistic background that is not dissimilar to the French; not so the Turks and North Africans with their totally different languages and religion.

Much has been written of the social and cultural problems of migrant workers (Castles and Kosack 1985; Hammar 1985). Such issues may not be very 'geographical' but they are vitally, shamedly important. Space permits only a crude listing of some of these problems: they include general lack of integration, linguistic isolation, culture conflict, sexual frustration, separation from kin, discrimination, poor access to social services, a high rate of industrial accidents, poor health and many more.

International migration after the oil crisis

The oil crisis is widely thought to have marked a turning point in the history of post-war European migratory movements (Lebon and Falchi 1980). Although this statement is broadly true, it needs to be qualified in two senses.

First, it can be demonstrated that moves aimed at stemming if not inverting migration flows had been taken well before late 1973. The UK, France and the Netherlands had taken steps to limit inflows from ex-colonial territories as early as the 1950s and 1960s. Switzerland attempted to stabilize its foreign workforce in 1963 and, in the 1970s, in the light of vigorous political debate associated with referenda on immigration controls, immigration legislation was considerably tightened. Denmark temporarily suspended immigration of foreign workers in 1970 and Norway narrowed the availability of its work permits in 1971. In 1972, West Germany strengthened its legislation against illegal immigrants, including the abolition of worker entry on tourist visas. The West Germans were also concerned about the build-up of ethnic concentrations in certain regions and attempted to prevent further immigrant settlement in areas where the foreign workforce was above 12 per cent (Salt 1981).

The second qualification concerns the assumption that economic factors were the overriding motive behind the ban on further recruitment. The racial factor was very important too. Looking back, it is possible to discern that anti-immigrant tension and the increasing power and ambitions of the migrants themselves were leading to a crisis (Slater 1979). The tension exploded in a series of ugly racial incidents in France in 1973, mostly directed against Algerians. Other serious race riots occurred in the Netherlands, as a result of which a ceiling was imposed on the number of foreigners allowed to work in Rotterdam. West Germany's stop to the recruitment of *Gastarbeiter* occurred in November 1973: the fuel crisis was the official explanation, but it was as much to do with a seething undercurrent of racial friction and an anxiety over growing numbers of foreign labourers.

From the point of view of the statistical monitoring of post-recession migratory movements, the year 1973 also saw the establishment of the SOPEMI continuous reporting service on migration (*Système d'Observation Permanente des Migrations*). Although SOPEMI acknowledges that its national suppliers of migration

Table 4.1   Annual labour migration for selected European countries, 1973–85 (all figures in thousands)

	1973	1974	1975	1976	1977	1978	1979	1980	1981	1982	1983	1984	1985
(A) Inflows to receiving countries													
Workers:													
West Germany (1)	319.1	46.3	21.9	24.1	29.7	19.5	37.9	82.6	43.9	25.9	24.4	27.5	33.4
France (2)	143.5	53.4	15.8	17.3	14.2	10.0	9.2	9.5	25.7	89.2	9.0	5.9	5.5
Belgium (3)	5.9	6.1	4.1	4.2	4.7	3.9	3.4	3.8	3.5	2.3	1.8	1.7	1.9
Switzerland (4)	49.7	31.3	18.0	15.6	19.7	21.9	25.5	32.3	35.3	33.0	24.4	27.5	33.4
Austria (5)	263.4	189.8	115.7	97.2	116.9	81.1	83.1	95.4	81.9	57.2	52.7	55.2	60.3
Gross immigration:													
Netherlands (6)	73.7	95.1	55.2	48.9	49.9	55.6	72.2	79.8	50.4	40.9	36.4	37.3	46.3
Sweden (7)	24.9	31.9	30.1	39.7	38.7	31.7	32.4	34.4	27.4	25.1	22.3	26.1	27.9
(B) Outflows from sending countries													
Workers:													
Spain (8)	96.1	50.7	20.6	12.1	11.3	12.0	12.5	14.5	15.9	18.5	22.0	21.4	18.2
Greece (9)	12.0	11.0	10.0	3.0	8.0	6.5							
Yugoslavia (10)	73.5	20.0	15.0	10.0	4.9	3.5	35.0	30.0	30.0	22.0	16.0	15.0	19.5
Turkey (11)	136.0	20.0	4.0	11.0	19.1	18.9	23.6	28.4	58.8	49.4	52.5	45.8	47.4
Gross emigration:													
Finland (12)	6.7	9.6	12.1	15.7	18.2	11.9	12.5	12.0	6.9	4.6	4.6	5.1	5.5
Italy (13)	123.3	112.0	92.7	97.2	87.7	85.4	89.0	84.9	80.5	98.2	85.1	77.3	66.7
Portugal (14)	120.0	70.3	44.9	33.2	28.8	30.3	20.6	18.0	16.5	10.2	6.9	6.6	7.2

Source: *SOPEMI Annual Reports*, 1973–86, OECD Directorate for Social Affairs, Manpower and Education, Paris.

Remarks:

(1) Permanent workers excluding EC nationals;
(2) Permanent workers excluding EC nationals; 1981 and 1978;
    1982 figures comprise exceptional regularizations;
(3) Wage-earners excluding EC nationals;
(4) Workers employed permanently or yearly;
(5) Work permits issued;
(6) Based on aliens' files at *communes'* registers;
(7) Estimates for all immigrants;

(8) Assisted emigration to Europe;
(9) All destinations; Greek emigration statistics ceased in 1978;
(10) Based on data from Yugoslav Employment Bureau; after
    1979 modified by host country data;
(11) All destinations;
(12) To Nordic countries only;
(13) All destinations;
(14) Emigrants registered by Secretary for State for Emigration.

data lack consistency and that its sources on return flows are especially fragmentary, the post-1973 picture is clear in its essential features. With the exception of Italy, protected by EC free movement laws, all southern European countries experienced a drastic cut in emigration across the 1974 threshold. Return movements accelerated, although massive waves of return migration have not occurred.

Table 4.1 presents a synthesis of the available SOPEMI data on annual migration from sending countries and to host countries for the period 1973–85. The lack of comparability of the national data sets should again be stressed; see the 'Remarks' section in Table 4.1. While all sending countries except Finland experi-

enced a reduction of flows during the mid-1970s, the host countries can be divided into two groups: those in which the foreign workforce diminished rather sharply between 1974 and December 1978 – Austria (–20.5 per cent), West Germany (–19.8 per cent), Switzerland (–17.5 per cent) and France (–9.2 per cent); and those in which some continued increase is discernible – the Netherlands (+20.2 per cent), Sweden (+12.2 per cent) and Belgium (+10.2 per cent). Böhning (1979) calculated a total return to Mediterranean countries of about 1.5 million migrants during 1974–78, the bulk of whom returned during the first two post-recession years.

By the early and mid-1980s the pattern of flows had

Table 4.2 Migration into and out of West Germany, 1967–88 (all figures in thousands

Year	Inflow	Outflow	Balance
1967	330.3	527.9	−197.6
1968	589.6	332.6	+257.0
1969	909.6	368.7	+540.9
1970	976.2	434.7	+541.5
1971	870.7	500.3	+370.4
1972	787.2	514.5	+272.7
1973	869.1	526.8	+342.3
1974	538.6	580.4	−41.8
1975	366.1	600.1	−234.0
1976	387.3	515.4	−128.1
1977	422.8	452.1	−29.3
1978	456.1	405.8	+50.3
1979	545.2	366.0	+179.2
1980	631.4	385.8	+245.6
1981	501.1	415.5	+85.6
1982	312.7	433.3	−120.6
1983	273.2	424.9	−151.7
1984	331.1	545.1	−214.0
1985	324.4	366.7	−42.3
1986	378.8	347.8	+30.9
1987	414.9	334.0	+80.9
1988	545.4	359.1	+186.3

Sources: Statistisches Jahrbuch für die Bundesrepublik Deutschland, various issues; recent data from SOPEMI.

become generally more stable. After the severe post-1973 recession cutbacks many countries, both sending and receiving, saw their respective out and inflows rise somewhat to reach a new 'steady state', albeit at a much lower level than the 1960s and early 1970s. Where exceptional departures from this new equilibrium pattern occur, specific circumstances have operated, as in the 'regularization' of illegal immigrants' status in France in the census year of 1982.

Changes in the national recording of migrants make it difficult to update Table 4.1 beyond 1985. However, it is worth paying special attention to the West German figures for immigration and returns, not only because they are more tenable than most other countries' statistics, but also because they show interesting trends in very recent years. Table 4.2 records immigration and return migration of foreigners from 1967 to 1988. After the temporary negative balance of the recession year 1967, immigration grew rapidly to nearly a million a year during 1969–70, coupled with low levels of return. The next inversion of the trend came in 1974 and lasted for four years. This was a period when returning migrants leaving West Germany exceeded new entrants, the stimuli being the laying-off of many

migrant workers and the reduced job opportunities for new entrants. In 1975 the annual return flow reached its post-war peak of just over 600 000. During 1978–81 a new four-year cycle emerged, one of renewed net inflow. Declining return over these years could be interpreted partly as a function of the ending of the chaotic mass return effects of the recession. In the slightly longer term, many immigrants decided to stay rather than return. With increasing restrictions on free entry, especially for Turks (the majority immigrant group in West Germany), the decision to go home appeared irrevocable and so a growing section of labour migrants have preferred to stay on, even in times of hardship and unemployment. Secondly, the rallying of the West German economy in the late 1970s permitted the resumption of a healthier migratory balance. A third reason for the resumption of a positive balance during 1978–81 was the relaxation of restrictions on family members joining male workers: hence an increasing proportion of the inflow during the late 1970s and early 1980s is made up of wives and young children. Fourthly, West Germany's liberal asylum policy encouraged many political refugees, both genuine and spurious, to seek a safe haven in the Federal Republic. Most came from Asia and entered via the Berlin 'loophole'. The lax control over such asylum-seekers has since been tightened somewhat. Finally, the growth in the migration of high-level professional managerial and technical manpower contributed to the expanded inflow. This is also borne out by the disaggregated country of origin data which show increasing numbers of Belgians, French and Dutch moving into West Germany. Then in 1982 the balance turned negative again, with escalating net and gross outflows resulting from renewed job cuts consequent upon the second oil crisis. Finally, the late 1980s saw a resumption of net immigration in response to economic growth and German demographic stagnation: increased quotas of Turks and Yugoslavs were allowed in, but this was but a prelude to the more dramatic political and population events of 1989–90, which will be considered later in this chapter.

Impact of return migration on sending countries

During the 1960s, one of the chief benefits of labour migration to industrial Europe was thought to be the formation of returning cadres of skilled workers who could aid the development of countries like Greece and Italy. Twenty years and a lot of empirical research later, the debate is put in very different terms. Firstly, southern European 'temporary' migrants are now universally recognized as having among them a much larger proportion of permanent emigrants than was thought likely in the 1950s and 1960s. Although most emigrants abroad hold fast to the ideal of return, for many this becomes a myth: as the return is continuously postponed, they are steadily turning themselves into permanent expatriates. This is probably even more true of Britain's immigrants, as Anwar (1979) has shown in a study of Pakistanis in Rochdale. Second, it is clear that while emigration involves a positive selection of the young and the vigorous, return migration entails a negative selection from among the emigrant community, sending back those who have not adapted, who are not upwardly mobile and who suffer problems of physical and mental health. Cerase (1974), writing of returns to Italy, has called this the 'return of failure' or the 'return of conservatism'.

At the village level, the behaviour of the returning migrant appears to follow a consistent pattern right across southern Europe from Portugal to Turkey (King 1979). The first ambition, indeed for many the raison d'être of the whole migration process, is to improve the housing environment. This usually implies building a new house rather than improving the old. In rural Portugal, the casa francesa is the symbol of success of returning migrants; in appearance, the new dwellings resemble the French suburban houses built by the same migrants during their 'other existence' as emigrant construction workers (Bretell 1979).

The second main characteristic of return migration, at least in rural areas, is the mushrooming of small, remittance-financed businesses. Typical enterprises are shops, bars, hairdressers, cobblers, carpenters, car-repair workshops and taxis. Being one's own boss appears to confer prestige, but these self-employed enterprises rarely flourish in villages continually denuded by depopulation. Opportunities for larger-scale industrial investment are limited, as are restaurants and hotels since most migrant villages are well off the tourist beat. Few returnees go back to farming or invest productively in agriculture. They may buy small plots of land for security, speculation or just to potter about in, but their impact on the local agricultural economy – except to inflate land prices – is nil. As former farmhands or general rural labourers, their experience of farm management is limited and their emigration will not have furthered this at all. In any case, a return to working the land is not the migrants' ambition: they emigrated in order to escape being peasants and farm labourers.

Much less is known of returnees who go back to live in urban areas. In Greece it is clear that many emigrants who originated from rural districts do not return there but resettle in regional towns or in Athens. After their experiences abroad, their preference is for an urban life style and they perceive better returns for

their accumulated capital in urban property than they could get by investing it in the rural economy (Unger 1986). Migrants who are fortunate enough to be able to return to an area undergoing some tourist development have even better chances, for here there are many possibilities to invest their savings in small hotels, restaurants, snack bars and other services. Evidence from many parts of southern Europe and from western Ireland confirms the leading role that returnees often play in the evolving tourist economy of coastal and island regions (King, Mortimer and Strachan 1984; Manganara 1977; McGrath 1991; Mendosa 1982).

In one country, return migration in the 1970s had an added dimension. Following Portugal's revolution of 1974 and the decolonization of its African territories, an estimated 800 000 retornados or colonial settlers returned (Lewis and Williams 1985). Most arrived from Angola in late 1975. Their absorption posed special problems for the Portuguese economy, both because of their number, equivalent to nearly a tenth of the Portuguese population, and because most arrived with few possessions and little or no money. Most of the retornados seem to have been successfully integrated. They tended to settle in the north-east, where many had come from originally, and around Lisbon, their arrival point from Africa. As a fairly entrepreneurial group, many have made good use of government funds made available to them. They have some parallels in this respect with other European ex-colonial refugees such as the French pieds noirs retreating from Algeria in the early 1960s, the Belgian refugees from the Congo and British settlers (including Asians) driven out of East Africa.

The migrant labour issue is now seen in terms of stocks rather than flows. No longer are migrants overtly viewed as a kind of cyclical shock-absorber, although this does not mean that some renewed migration might not occur if Western Europe's economic horizons were to brighten. Turkey had 1 million workers on the waiting list for recruitment to West Germany when the 1970s recession first occurred, and the pent-up potential has probably increased since then.

Return migration continues to play a significant part in the European international migration scene. Numbers of returnees have fallen somewhat since the mass returns of the 1970s, but a steady stream is maintained by the essentially temporary character of much post-war labour migration and by those who wish to return home upon retirement. France and West Germany have tried to hasten immigrants' departure by financially aiding the return home. The French scheme of 'assisted voluntary repatriation' was introduced in 1975, initially for unemployed migrants resident in France for less than five years, but later extended to employed and longer-established immigrants. The incentive was a free ticket to their home country and a grant of 10 000 francs – the famous 'million centimes' (Poinard 1979). By 1981, when incoming President François Mitterrand rescinded the return policy, 93 000 individuals had taken advantage of the return bonus, a small reduction in France's immigrant population of 4.4 million (Ogden 1985a). The German return policy is more recent, although it should be remembered that the *Konjunkturpuffer* ideology always implied a rapid return and denied, at least in theory, the possibility of much permanent settlement. It was not until 1983, following urban riots, escalating racial tension and rising unemployment, that the West German parliament approved legislation granting lump sums to those immigrants who became unemployed through company closure or bankruptcy. The 'buy them out' policy offered each prospective repatriate 10 500 DM (£2500 approximately), together with 1500 DM for each child, plus refund of social security payments. Following the French experience, most of those who take the money and go are merely those who were planning to return anyway.

Towards a European melting pot?

More important than a declining trend in return migration is what is happening to the migrant stock. The main feature of change here is the appearance on the labour market of wives and children of migrant workers, thus considerably altering the structure of the minority labour supply. Processes of integration are beginning, helped by a very definite shift in host country policy towards assimilation, respect for ethnic background and family reunion. However, progress is patchy. Salt (1985a) asks whether the difficult-to-integrate and racially abused second-generation Turkish children in West Germany are 'young Turks or

little Germans'. He maintains (1985b) that the immigrant children are a social time bomb ticking away. In 1981, France and West Germany each had around 1 million foreign children under the age of 16. Generally poorly educated by the host country's standards, these young people are being poured on to a labour market already burdened by high rates of unemployment; their chances of finding satisfying jobs are extremely poor. On the other hand, in some of the earlier-established migrant communities, such as the Italians in France and Belgium, the third generation is now being born, a generation that generally has few ties to its ancestral homeland and may not even speak the language. Among these second- and third-generation migrants from other European countries, rates of intermarriage with the host population are quite high. International labour migration, for all its negative aspects, is beginning to play a role in European integration and in ethnic melting.

However, among other ethnic groups, especially those originating from outside Europe, the 'cultural gulf' with the host society is still wide. This statement applies especially to Western Europe's estimated 5.5 million Muslims whose presence received strong media attention during the Gulf War. Even before that, the proliferation of mosques and Islamic schools for the teaching of Koranic principles had focused attention on religious and cultural differences. While these clashes of belief and life style are very real, insufficient attention is given to the children of Islamic immigrants who witness these conflicts – over dress, diet, arranged marriages, Western permissiveness, etc. – at the personal, individual level. It also needs to be stressed that the Islamic presence in Western Europe is highly heterogeneous, both in terms of country of origin (mainly Turks in Germany, Algerians in France, Pakistanis in Britain) and in terms of religious rite and degree of fundamentalism (Peach 1992). Table 4.3 gives estimates of the Muslim population of selected

Table 4.3 Distribution and origin of Muslim population in Western Europe, *c.*1990 (in thousands)

Country of origin	France	Germany	UK	Netherlands	Belgium	Switzerland
Algeria	820.9	5.1		0.7	10.6	2.0
Morocco	516.4	52.1		139.2	135.5	1.8
Tunisia	202.6	21.6		2.7	6.2	1.6
Turkey	146.1	1523.7		176.5	79.5	56.8
Pakistan			376.1			
Bangladesh			111.4			
India			121.8			
Other	57.9	56.3	190.7	76.9		10.9
Estimated total	1743.9	1660.0	800.0	405.9	237.2	73.1

Source: after Peach (1992).

West European countries by country of origin. The number of Muslims in France may be a considerable underestimate because of the large number of second and third generation who are not reflected in the statistics on foreign birthplace. The growing number of Islamic migrants in Europe, including many in recent countries of immigration such as Spain and Italy, poses a great challenge to policy-makers of whether to encourage integration or preserve cultural separatism.

New forms of international migration

After a period of relative quiescence, the late 1980s and early 1990s have seen international migration pushed right back on the European political agenda. Enlargement of the EC, moves towards the Single Market, the extraordinary events in Eastern Europe, the growth in asylum migration and continuing 'push pressures' from the southern Mediterranean and beyond were the main contextual circumstances which moved migration centre-stage once again.

Some of the new trends are the enlargement of existing flows. As noted earlier, recruitment of foreign labour by the richer economies of Western Europe started to pick up after about 1986. The data on total migration flows into West Germany (Table 4.2) confirm this, and similar trends can be observed from the SOPEMI data for France, Austria, Switzerland and Sweden; in Belgium and the Netherlands the post-1986 upturn is only slight. Some of these increasing migration flows come from traditional emigration countries such as Turkey and Yugoslavia and involve continuing processes of family reunion migration, but others originate from new sources of labour migration in Africa and Asia such as Egypt and the Philippines.

A second continuing but enlarging trend is the migration of high-level technical and professional manpower among the countries of Western Europe. Higher salaries, the desire for international experience and career advancement appear to be some of the motives for this kind of migration (Salt 1984), but few detailed studies have been made. With the exception of some net outflow or 'brain drain' from peripheral countries like Ireland, Scotland and Greece, much of this migration consists of two-way 'brain exchange' between the more highly developed regions of Europe, especially those with significant clusters of high-tech industry, banking and financial services, corporate headquarters and international organizations. Given the extent of multinational enterprise in Western Europe, a good deal of this skilled international migration involves movement within the same firm or organization.

The newer forms of international migration, which came to the fore during the late 1980s and for which West European policy-makers (both international organizations such as the EC and national and regional authorities) were not really prepared, include asylum migration, the threatening tide from Eastern Europe, and the growing numbers of Third World immigrants who have managed to gain entry, mainly in southern Europe. We will consider each of these in turn, before concluding with some remarks on the Single Market and future migration policy.

Asylum migration

After the cessation of guestworker migration the only option for non-EC migrants unable to use the family reunion scheme has been asylum. First, migrants from Turkey, Yugoslavia and the Maghreb countries used this procedure, but they were soon joined by numerous others from more far-flung parts of the world such as Central America, Iran, Sri Lanka and Zaïre. Many of them were victims of generalized repression and violence as well as severe poverty; however, most do not reach the strict definition of refugees as laid down by the 1951 United Nations convention. Increasingly, therefore, asylum-seekers have sought ways of bypassing the orderly admission procedures (which quickly became overwhelmed), merging with the vast army of irregular migrants which has been swelling continuously since the 1970s.

Unlike the earlier refugee flows from Hungary after 1956 and Czechoslovakia after 1968, which were largely composed of intellectuals, most of the recent influxes of asylum migrants from Africa and Asia have been absorbed only as highly marginal reserve workers in the lowest occupational classes. Even so, many of these Third World immigrants do have good educations: most of the Egyptian newspaper-sellers who patrol the streets of Vienna, for instance, are graduates or university students.

Table 4.4 documents the enormous increase in asylum migration in Western Europe since 1983. The numbers have grown from less than 100 000 per year in the early 1980s to around half a million per year in 1990–91. Just under three-quarters of asylum-seekers entering Europe apply to EC countries, 42 per cent to Germany alone. Standardizing for population size, however, reveals that Sweden and Switzerland take proportionately more refugees than other European countries. It must be stressed that the figures in Table 4.4 refer to asylum-seekers entering the various countries, not 'successful' refugees. Given the lengthy procedures required to settle claims, and the variability of these procedures from one country to another, comparable annual data on persons who acquire definitive

Table 4.4 Countries receiving asylum migrants, 1983–91 (all figures in thousands)

	1983	1984	1985	1986	1987	1988	1989	1990	1991	% 1983–91
Germany	20	35	74	100	57	103	121	193	256	41.9
France	14	16	26	23	25	32	59	56	46	13.0
Sweden	3	12	14	15	18	20	30	29	27	7.3
Switzerland	8	7	10	9	11	17	24	36	42	7.2
UK	4	3	5	5	5	5	16	30	58	5.7
Austria	6	7	7	9	11	16	22	23	27	5.5
Netherlands	2	3	6	6	13	7	14	21	22	4.3
Belgium	3	4	5	8	6	5	8	13	15	2.9
All EC Countries	47	68	125	155	127	166	226	333	432	73.4
All Western Europe	67	102	168	201	182	234	346	453	537	100.0

Source: United Nations High Commissioner for Refugees. Only countries receiving more than 50 000 or more than 2 per cent of the total for 1983–91 are included in the above list.

refugee status are hard to assemble. Moreover, many apply for asylum only after entry, and the entry point is not always the country of intended settlement.

The United Nations High Commissioner for Refugees estimates that half the claims for asylum in Europe are false and are made by economic migrants pretending to be political refugees. The pressure of increasing numbers of bogus claims for political asylum has forced receiving governments into a more unsympathetic stance: only 4 per cent of asylum claims submitted to Germany in 1990 were accepted as genuine, while some countries, including the UK, have decided to impose heavy fines on airlines carrying incoming passengers who do not possess the full documents necessary for entry into the country (Salt and Kitching 1991).

The tightening situation for asylum-seekers, genuine or otherwise, has fuelled illegal immigration, both from the newly liberated countries of Eastern Europe and from the less developed countries of the South. Arriving by various clandestine routes overland and by sea, or flying in on tourist visas and then staying on, these illegal and semi-legal immigrants have generally had to resort to informal occupations in the black economies of Europe. They become street-traders, domestic maids, construction workers, hotel and hospital cleaners: all jobs which are poorly paid, insecure and devoid of pension or welfare rights. By definition, their numbers are hard to estimate but there are thought to be about 3 million undocumented immigrants in the EC alone (Ghosh 1991). Attempts to regularize the position of clandestine migrants through amnesties encouraging them to register and legalize their status have had mixed results. The French amnesty of 1981 drew out 132 000 'hidden' immigrants, but the Spanish scheme of 1985 attracted only 44 000. The Italian amnesty of 1987 registered

119 000 'illegals', only about a quarter of the estimated total. A second Italian amnesty in 1990 was more successful, with 300 000 (Ghosh 1991).

Three final points about refugees need to be made. First, attention needs to be paid to effective policies for settling and integrating refugees. The British policy of dispersing Vietnamese refugees was a failure. Sweden on the other hand tries to preserve a measure of ethnic identity and community solidarity by settling different refugee nationalities in different towns. Second, the restrictive concept of a refugee as a person seeking justified political asylum needs to be questioned and perhaps broadened, for are not poverty and environmental disaster not equally repressive and unfortunate conditions provoking human flight? The third point involves setting the European experience of refugees in its world context. While it is true that growing numbers of asylum migrants seeking entry into Europe pose a problem of coordination and policy formation among the European countries, it is a mistake to view the phenomenon as some threatening, unrestrainable tide of humanity fleeing from the South and pouring into the North. Such a fear is at least partly due to parochial self-interest. The vast majority of refugee movements take place within the developing world and involve some of its poorest countries such as Sudan, Somalia, Zaïre and Malawi (Kliot 1987b).

Migration from Eastern Europe

Flows of migrants from Eastern to Western Europe have increased dramatically in recent years. In 1989 alone more than 1.3 million people emigrated from the countries of Eastern Europe. Of these 80 000 were asylum-seekers (mainly from Poland, Romania and Yugoslavia), 150 000 left the former USSR (most of these were Jews), and over 720 000 were Germans –

either migrants from the former East Germany (345 000 so-called *Übersiedler*) or ethnic Germans (the 377 000 *Aussiedler*) from Poland, Romania and other East European countries (Widgren 1990). While the unification of Germany in October 1990 means of course that migration from former East to former West Germany is now treated as *internal* migration, the immigration of ethnic Germans will continue. This movement offers particular interest because of its irony. In a country in which Turkish immigrants of twenty or thirty years' standing, perhaps with German-born and German-speaking children, cannot claim German citizenship or full civic rights, the immigration and granting of full citizenship to ethnic Germans, who left Germany generations ago and whose descendants often cannot speak German, proceed apace since the German constitution requires that citizenship be granted, without limit, to all persons proving German origin. In reality, Germany is finding it increasingly difficult to cope with the new arrivals, less than a tenth of whom have technical and educational skills appropriate to their new lives. Consequently a third of the *Aussiedler* remain unemployed (Treasure 1991).

Continuing ethnic conflict in Eastern Europe and the painful but probably temporary problems of economic readjustment will lead to further migration flows to the West. In 1990 real output in Eastern Europe shrank by 8 per cent and consumer prices rose by 34 per cent; on average, incomes are less than a third of Western European levels. People who have lived all their lives under command economies and state paternalism are now exposed to unemployment, market prices, inflation and changing social security arrangements (see Ch. 13). Such disorienting changes may sharpen their desire to move to the West. The real unknown is the scale of emigration from the former Soviet Union: perhaps between 4 and 6 million may emigrate in the coming years, after a new emigration law comes into force in 1993 (Ghosh 1991).

However, the emigration potential of countries in Eastern Europe will probably only be temporary. As their economies become more integrated with those of the West and as the demographic increase stabilizes close to zero, economic migration could become much lower too. At that point, perhaps in the late 1990s, migration pressures from the South will probably take over.

Push-pressures from the South

Up until twenty years ago Europe's 'migration frontier'. separating areas of net immigration from those of net emigration, ran along the Pyrenees, followed the south coast of France and the Alpine crest between

Italy and Switzerland, and then passed to the north of Yugoslavia and Greece to the threshold of Turkey. Now it has shifted south and runs east–west through the Mediterranean from Gibraltar to the Bosporus, passing south of Sicily (King 1991). Many of the countries which were the main exporters of labour up to 1973 – Spain, Portugal, Italy and Greece – are now countries of immigration. Now the immigrants come from North Africa and beyond.

Three reasons lie behind this migration reversal in southern Europe. First, these countries have become prosperous. Italy is now one of Europe's richest countries and Spain and, to a lesser extent, Portugal and Greece have modernized rapidly since joining the EC (King 1992; Naylon 1992). Second, the demographic regimes of these countries have changed: they now have low birth rates (that in northern Italy is reputedly the lowest in the world) and therefore less youth unemployment than previously. Third, immigrants coming in from North Africa, Eastern Europe and other parts of the world have found entry easy, partly because of long coastlines for clandestine landings, and partly because they have simply filtered in with tourist arrivals. Italy may have successfully repulsed the Albanian 'boat people' in 1991 but it did not stop the steady infiltration of Moroccans and Tunisians arriving by boat along the south and west coasts of Sicily.

Demographic 'push-pressures' from the southern and eastern Mediterranean Basin are probably the biggest challenge facing Western Europe's population planners over the coming decades (Salt and Kitching 1991). On the southern shore of the Mediterranean the average woman has five or six children; in Western Europe she has one or two. While most West European countries will experience little or no population growth between 1990 and 2000, Algeria's population will increase by 31 per cent, Sudan's by 33 per cent and Libya's by 43 per cent. Currently 46 per cent of the population of Algeria are under 15 years of age, with figures of 42 per cent in Morocco and 40 per cent in Egypt. In most West European countries the figure is only 20 per cent and in Italy it is barely 10 per cent (Ghosh 1991). This 'nightmare demographic scenario' is reinforced by political instability, by economic contrasts (average GDP per capita in the EC and EFTA countries is nearly five times the southern Mediterranean countries), and by environmental deterioration in North Africa as global warming pushes the Sahara northwards, exhausts water resources and erodes soil. As Salt and Kitching (1991) point out, this combination of sharp demographic gradient, poverty, political instability and ecological crisis provides a heady cocktail of push factors for migration.

Labour market restructuring in Europe offers some

possibilities for Third World migrants through the growth in demand for casual labour in the informal sector but the pressures will far exceed the number of jobs available. Even for those who do enter, life is far from ideal: 'their family life is limited or prohibited, their housing is inferior, their rights as employees are markedly worse than indigenous workers, they are disenfranchised and unorganized in Trade Unions . . . they are subordinate . . . and defenceless' (Cohen 1987: 111).

The Single Market and the future of international migration

The Single Market – which is also a single market for labour – will probably have less effect on population movements than on other aspects of EC economic life, for two reasons. First, as was noted earlier, the free movement provisions of the 1960s had little direct effect of stimulating intra-EC migration. Second, the movement of people tends to respond to the behaviour of capital and in particular its concentration in employment-generating activities in new and expanding industries and services. Therefore the Single Market's effect will probably be indirect, and limited to the higher-skill components of the international division of labour. Its effect on periphery-to-core migrations, for example from Ireland or Portugal, will depend on the balance of capital concentration in the core versus structural-fund economic growth in the periphery: at present it is difficult to see which will be the dominant process. At least in the short term, the cultural barriers (language, national customs, etc.) to mass migration of low-skill labour within the Community will remain. Regional economic convergence may make such movements less likely anyway.

The Single European Act reflects the humanistic spirit of building a Europe of freer movement of goods, capital, ideas and people – with the proviso that the movement in question is only for Europeans. The Berlin Wall comes down but the 'ring fence' of 'fortress Europe' goes up. For those immigrants already inside, complications arise depending on country of birth, nationality, parental history and country of residence (Marie 1991). The Schengen Agreement of 1991 favours free internal movement within the EC Six, but not for non-EC nationals already established there for decades or even since birth. Such international accords on migration will become increasingly important over the next few years as Western Europe struggles to achieve a coherent policy on immigration, particularly in the face of growing numbers of illegal immigrants and asylum-seekers (Salt and Kitching 1991).

For the immediate future – to 1995 – the signs are for rather minimal change. The modest revival of migration flows since the late 1980s and the pressure from some employers for a relaxation of controls will probably be countered by the political unacceptability of a return to open migration from outside the EC. Structural changes in the European economy and the pursuit of more labour market flexibility point to the further development of more casual, temporary types of immigrant labour. These new migrant groups, perceived both as outsiders and inferior, are just as readily assigned to economically marginal roles and to an incomplete citizenship as were the *Gastarbeiter* of the 1960s (Gordon 1991).

The longer-term future opens up the possibility of a more radical role for international migration in Western Europe. Without immigration and with a continuation of the present low fertility levels, the populations of many European countries will begin to shrink early in the next century, following the trend of the national population of former West Germany which started to decline as early as 1972. Very high net immigration into Western Europe would be needed – about 1.7–1.8 million per year – to sustain populations at their present levels. This level is considerably greater than the influx during the height of the labour migration boom of the 1960s and early 1970s (Wils 1991). Yet no European country admits an enhanced role – or in most cases *any* role – for immigration in its medium-term population projections.

Trends in internal migration

The international migrations considered so far in this chapter must be set alongside the more profound if less dramatic internal movements that have occurred within individual countries. At least in the West European context, internal migration is a much more complex phenomenon than international labour migration for it encompasses a whole range of types and scales of movement. Identification of these patterns by the geographer or the demographer depends very much on the data frameworks used. Inter-regional tabulations based on data available at province or county level conceal more local-scale moves between villages and towns or within conurbations. While rural–urban migration has been historically the most important form of migration in Western Europe, we should not ignore urban–rural movements, of considerable importance in the 1970s and 1980s, as well as inter-urban, intra-urban and rural–rural migration. All these patterns exist in Western Europe; the problem is one of finding the requisite data and, if that can be done, of devising appropriate statistical and cartographic modes for their proper portrayal.

Rural–urban migration

As the most universal form of spatial relocation in Western Europe, at least until recently, rural–urban migration deserves special attention. It is, however, as Clout (1976) points out, an elusive phenomenon, for it depends on definitions of 'rural' and 'urban' and is frequently confused with rural depopulation and with the decline of the agricultural workforce, processes considered in more detail in Chapter 9. For many countries, rural–urban flows are impossible to disaggregate from available inter-regional migration statistics which incorporate rural–urban movements to varying degrees depending on the nature of the regions involved. In highly urbanized countries such as the UK and Germany, the majority of inter-regional movement is composed of inter-urban flows. In rapidly urbanizing countries like Portugal and Greece, on the other hand, most inter-regional movement consists of rural migrants moving to expanding towns, notably capital cities. It should also be remembered that much of Europe's post-war *international* migration has been rural to urban.

The rural–urban drift in Western Europe is centuries old, accelerating and then decelerating according to various economic and cultural trends. One of the earliest waves of rural–urban migration took place in medieval Italy as Renaissance urban and industrial growth drew in workers from the surrounding countryside to powerful city-states like Venice and Florence. Later it was the Industrial Revolution which sucked in population to north European coalfields and to growing towns specializing in products like iron, textiles and pottery.

Since 1945 virtually all rural regions of Western Europe have experienced outmigration. In areas of particularly severe migratory loss, more than half the population has departed. The myriad of migratory filaments which tie together villages and towns, near and far, can never be properly mapped for they are so complex. At a macro-regional scale, outmigration has been most marked from the agricultural periphery of Western Europe. But local studies show that young people living in countryside areas which are close to towns depart more readily than their counterparts in remote areas where facilities may actually be poorer and employment possibilities more narrow. At the same time, there is also a clear indication that the smaller the settlement, the greater its chance of rapid loss through migration, for larger villages, with their wider range of social facilities and economic opportunities, have a better base for retaining their populations (Clout 1976).

Migration from rural areas has important impacts on both the departure and reception areas. These are somewhat analogous to the effects of international migration. The dramatic increase in the pace of rural–urban migration in southern European countries since the war has provoked exceptional strains on urban areas like Barcelona, Milan and Athens. Rural migrants to cities like these not only undergo a difficult adjustment to city life, they also suffer many objective difficulties such as cramped and insanitary housing, discrimination and poor access to public services (see Ch. 8). Nowhere is this seen more clearly than in Italy where the post-war Italian 'miracle' brought millions of rural people pouring into northern industrial cities that infrastructurally and socially were quite unprepared (Bethemont and Pelletier 1983). If the mayor of Turin complained that his had become a 'southern' city, others profited from the influx of south Italian workers, not least the Fiat empire which dominates the city's economy and unscrupulous landlords who packed the migrants into inner-city garrets (Fofi 1970). In other southern European cities, illegal squatter settlements appeared: *bidonvilles* for the internal migrants. These peripheral slums, not unlike the shantytowns of Third World cities, have been studied in Istanbul and Athens (Karpat 1976; Leontidou 1990).

Main patterns of inter-regional migration

Fielding (1975) has accomplished pioneering work in constructing maps of internal migration rates for the 1960s on a common scale, minimizing the distorting effects of the absence of a uniform data base. The 1960s was the boom period for internal migration, as it was for international migration. The overall picture (Fig. 4.3) largely speaks for itself so only a few descriptive and interpretative remarks will be made. There is a major, if somewhat attenuated, area of migration balance and gain in the central industrial belt of Western Europe; further areas of net gain, some of them quite dramatic, are to be found around the Mediterranean coasts of eastern Spain, southern France and northern Italy. These are surrounded by a dominantly rural periphery of net migration loss. The areas with highest net gains include favoured retirement areas such as south-west England and the Côte d'Azur and some rapidly expanding capital cities like Copenhagen, Oslo and Madrid. Other areas of positive migration are newer industrial areas such as the southern Paris Basin, Munich, western Switzerland, Barcelona and Valencia–Alicante. The regions of biggest migration loss are mostly remote upland areas of forestry, pasture and poor-quality farming: interior Spain and Portugal, southern Italy, northern Scandinavia, Scotland and western Ireland. Less

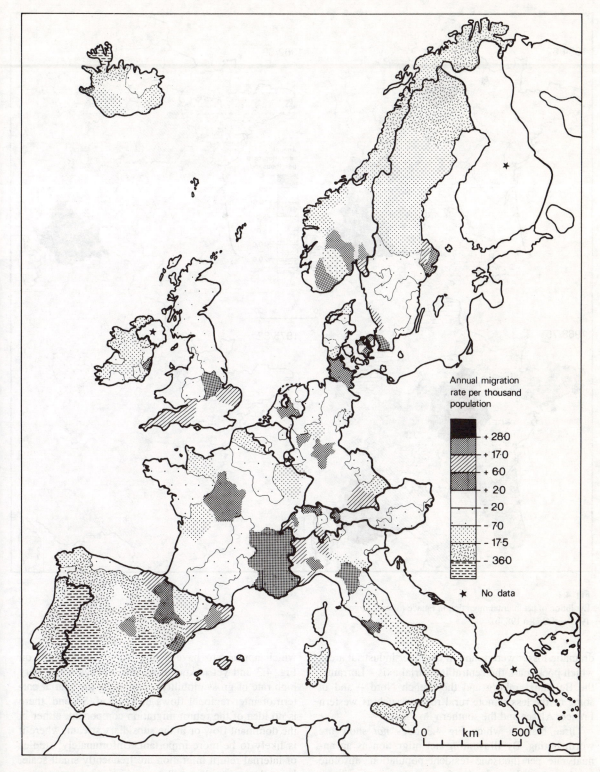

Fig. 4.3
Net internal migration in Western Europe during the 1960s (after Fielding 1975).

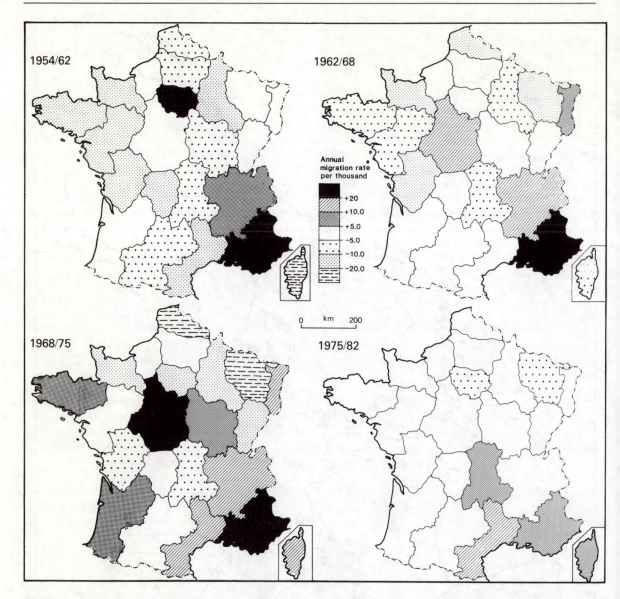

Fig. 4.4
Evolution of net internal migration in France (after Courgeau
1978 and Ogden 1985b).

dramatic losses were found in the older industrial areas
which pock-mark the central industrial axis – Lorraine,
the Belgian Limburg and the French Nord – and in
some of the less remote rural regions such as western
France, Alsace and the southern Po Plain.

Three features which Fig 4.3 does *not* show are
worth noting. By illustrating net migration as an annual rate per thousand resident population, absolute
numbers and the volume of gross flows are missed.
This is crucial in the case of London and the south-east

which appears to have a migratory 'low profile' on
Fig. 4.3 and yet is known to have always had a very
high rate of gross mobility, dominating the spatial pattern of inter-regional flows in England. Second, there
is no idea of the return migration component, either in
the dominant flow or in the subsidiary stream, where it
is likely to be more important. Unfortunately, studies
of internal return migration are frequently small scale,
inconclusive and unable to demonstrate dramatic
effects on sending areas (Nicholson 1975; Townsend

1980a). Third, there is no indication of the regional patterning of the flows themselves in terms of significant pairs or clusters of origins and destinations. Net flow maps do exist for some of the larger European countries and very clear trends emerge, particularly for the 1960s (Federici and Golini 1972; Wood 1976). For France, the flows for this period are to Paris and the Mediterranean littoral from the centre and from the eastern and western peripheries. For Italy, the streams are from the south and the islands of Sicily and Sardinia to Rome and the north. British flows focus on the London region. Spanish flows are more complicated, with four main areas of attraction: Madrid, Barcelona, Valencia–Alicante and the Basque country. During the 1950s, Madrid and Barcelona accounted for 80 per cent of the net immigrants, a proportion which fell to 66 per cent in the 1960s, due largely to the relative decline in the attractive power of Madrid. During the 1970s and 1980s, the Mediterranean provinces have been the main focus of growth. Valencia attracts migrants mainly from adjacent provinces but Barcelona has been able to attract from a wider area. Barcelona's ability to 'capture' most of the Andalusian migrants dates from the 1920s when large numbers of southerners worked on the Barcelona Fair and the Metro (Bradshaw 1972, 1985).

The case of France

France is perhaps the most representative European country to study from the point of view of internal migration: it still has a large residuum of rural population; it has experienced massive metropolitan and resort-oriented growth; and there has been some decentralization of population to regional nodes and to ex-urban rural hinterlands. With France's strong traditions in demography and population geography, many detailed and rigorous studies are available which trace the evolution of internal migration in the period since 1945.

The early post-war period was dominated by migration from rural areas to the dominant city-region of Paris. This was a trend traceable at least to the mid-nineteenth century since when France has lost 12 million rural dwellers, many *départements* shedding over 50 per cent of their rural populations between 1851 and 1968. Paris, meantime, more than tripled its population, fed by a steady stream of migrants from all parts of the provinces, from all age groups and from all social classes. During the 1950s and early 1960s, an average of 100 000 migrants came to Paris each year; migration ties then were especially strong with the belt extending west to Brittany. The internal migration process was not, however, a simple link between rural periphery and urban metropolis. Only 41 per cent arrived in Paris direct from their village of origin; most had 'step migrated' with one or two stops en route to the capital.

A 1961 survey of more than 6000 migrants in Paris (Pourcher 1970) showed that 60 per cent came to the French capital because of work reasons: to find a job, to improve prospects for promotion or to obtain a higher income. These results contradict the oft-repeated opinion that rural migrants move to big cities to find better living conditions and social facilities – the 'bright lights' argument. For manual workers, farmers and service personnel, Paris offers, through its vast range of urban jobs, the decent standard of income that cannot be guaranteed in rural areas. For higher-status managers, professionals and civil servants, coming to Paris sets the seal on a successful career. Not only does the move to Paris enable the migrant immediate advancement, it also creates the conditions for the furtherance of upward mobility in the future. Up to the early 1960s it could be said that Paris was the only really major pole of attraction for 'economic migrants' in France (Pourcher 1970). But the disadvantages of such a monolithic magnet became all too well known: they included housing shortages, pressure on services, choking traffic, social problems, psychological tension. . .

The problems were equally serious – though of a very different type – for the rural areas of outmigration. Ogden (1980) has described what happens to rural *communes* sapped by prolonged migration. A sample of *communes* in the Ardèche (eastern Massif Central) recorded a collective loss of population of nearly 60 per cent between 1861 and 1975. The selectivity of the out-movement causes imbalance in the residual population which becomes disproportionately old and female, deprived of innovators and social leaders. Schools, shops and other services close down. This downward rural spiral is not, of course, peculiar to France; it can be found in most countries of Western Europe. Traditional peasant society, with its local dialects, customs, crafts, and also hardships, collapses.

During the late 1960s the predominantly rural–urban pattern described above changed. The 1968 and 1975 censuses showed that the pace of internal mobility was maintained with 10 per cent of the French population changing its residence each year. But the big cities, especially Paris, no longer attracted as many immigrants, and rural losses lessened. A process of population redistribution had started in which the two-way flows between origins and destinations were more balanced. The rural–urban imbalances created in the past were being redressed and in many areas rural repopulation was in evidence. This trend was confirmed by the 1982 census which showed more definite urban and metropolitan decline counterbalanced by rural in-migration (Ogden 1985b).

The evidence for the reversal in internal migration trends can be presented in two forms: by region and by size of settlement. The regional picture is given in Fig. 4.4. This shows the pattern of losses and gains for the four intercensal periods 1954–62, 1962–68, 1968–75 and 1975–82; in each case the net migration rate is standardized by the population of the region concerned. It is clear that the general high rate of internal migration noted above is highly differentiated regionally. The main region of gain initially, Paris, became a region of migratory loss by 1968–75. Conversely,

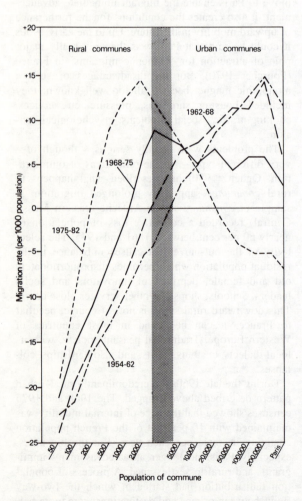

Fig. 4.5
France: net intercensal migration rates, by size of *commune* (partly after Bourgeois-Pichat 1981).

many regions of initial migratory loss started to gain, some as early as 1962–68, many more by 1968–75 and 1975–82. In the cases of Brittany and Corsica, the switch from heavy losses to quite strong gains has been dramatic (Dean 1986). Overall, it is Paris and the north-east where migratory attractiveness declines most; and the south, south-east and west where the pulling power increases most. The notable thing about the 1975–82 intercensal period is the general decline in inter-regional mobility.

The second mode of analysis is by settlement size. Figure 4.5 shows the net annual migration rates for the three intercensal periods according to the various size categories of *communes*. Moving from 1954–62 to 1968–75, the curves pivot clockwise around a point near the small towns (5000 to 20 000). Net immigration to urban *communes* declines through time; the more urbanized the *commune,* the more marked the decline. In fact, there has been a decline in immigration since 1962 in 95 per cent of French towns with over 50 000 inhabitants. At the same time, net emigration rates have declined for rural *communes,* and in the larger rural *commune* classes net outmigration has given way to net immigration. Now it is the rural *communes* that are growing more quickly than their urban counterparts – the first time in more than a century that this has happened (Ogden 1985b).

In terms of time, the balance was tipped for the rural *communes* around the mid-1960s. Initially, the communes most affected were those found on the outskirts of major towns. Later other *communes* were affected, deep in the countryside. Here the net in-migration was often relayed out by the intermediary functions of small towns attracting industries. In the case of remote *communes,* the initiative of local entrepreneurs and *commune* mayors often played a crucial role (Limouzin 1980). Increasing numbers of elderly people move to the country on retirement. For some, this involves a return to the area of origin; for others, it involves a move dictated by scenic and climatic attractiveness, perhaps influenced by holiday experiences.

The recent publication of provisional and summary results of the most recent (1990) French census gives a further twist to the above account. These results indicate a reaffirmation of earlier trends of urban and metropolitan growth and rural decline: in other words the end of counter-urbanization. During 1982–90, nineteen of the twenty-six French agglomerations with more than 200 000 inhabitants gained population. Paris grew by 353 000 or 4 per cent, Lyon by 3.3 per cent, Toulouse by 12.4 per cent and Bordeaux, Montpellier, Nice and Toulon each by 6–7 per cent. A breakdown of the size distribution of French towns shows a U-shaped relationship with fast growth and in-migration into both large urban areas (over 200 000) and small towns (below 20 000), but stagnation in middle-sized towns which in previous intercensal periods had been favoured by high rates of increase. Regionally there was a strong drift of the French population to the south (Jones 1991; Scargill 1991).

The extent to which the French experience in internal migration is replicated in other West European countries has been examined by Fielding (1986) for the counter-urbanization phase and by Champion and Illeris (1990) for the urban re-growth phase. It seems that the counter-urbanization process diffused southwards from Scandinavia and northern Europe as far as northern Italy. By the end of the 1970s only Spain, Portugal and Greece failed to show clear symptoms of the end of urbanization. Even more interest attaches to trends in the 1980s. Here judgements are limited by the crude size of the regional units in the analysis and by other data problems. Some countries, such as Germany and Italy, show evidence of continuing counter-urbanization in the 1980s; others show reversion to rural depopulation and renewed urban growth (Britain, the Netherlands, some Nordic states); yet others show no relationship between settlement size and net migration. Just as the overall pattern varies, so do the explanations. While economic restructuring tends to produce new agglomeration economies for the advanced producer services which characterize the most recent phase of the post-industrial scene, it is also true that new information technologies enable many people and activities to locate anywhere, including remote and environmentally attractive districts. Perhaps all that is clear is that there are many population redistribution processes in action which operate in an overlapping fashion at different scales and in different regions of the continent.

Inter-urban migration

The urbanization versus counter-urbanization debate tends to obscure the fact that inter-regional migration, particularly in the more urbanized, industrialized countries of northern Europe, is increasingly dominated by inter-urban movements. Net flows are irrelevant and misleading in this type of migration which is characterized by high rates of *exchange* between urban areas (Wood 1976). Also important, of course, is intra-urban residential relocation, the greater part of which takes place in and around the larger conurbations; it has very different social and economic significance to inter-regional migration, not least because it is normally independent of employment change.

A theoretical rationale for the growth in importance of inter-urban migration has been provided in a visionary paper by Richmond (1969), who placed changing types of internal migration in the context of three broad stages of socio-economic evolution: the traditional, the industrial and the post-industrial. Traditional society was agrarian, quasi-feudal, with non-mechanical modes of transportation and a tight network of communities (*Gemeinschaft*); the form of population movement was rural–rural with coercive push factors dominant. The move to industrial society, which was a more associative, class-based, mechanical *Gesellschaft*, was accompanied by rural–urban migration operated by dominantly pull factors. Finally, the second half of the twentieth century, in north-west Europe and North America, is called by Richmond *Verbindungsnetzschaft* – a social system in which interaction takes place rapidly through a multitude of communications systems which increasingly make distance irrelevant. In this post-industrial phase, migration is predominantly inter-urban, two-way and, for certain categories of personnel, *is the norm*. Like other forms of occupational and social mobility, spatial mobility has become a functional imperative in advanced post-industrial societies; its role is not only to allocate human resources in a more economically productive way, but also to provide choices and opportunities for individuals.

The most promising explanations of inter-urban movements involve life-cycle stages and career patterns. Once more, there is a clear parallel with the international migration of high-level manpower between urban–industrial regions of Europe noted earlier. In general, mobility is greatest in the early years of single adult and married life. In England and Wales, mobility is often initiated by the move of the educationally better qualified away from home to institutes of higher education. Educational background is also related to career paths. The available evidence clearly indicates that long-distance inter-urban migration tends to take place most intensively among higher-income professional and managerial workers. The mechanism of 'spiralism' is classically invoked to explain this feature, since the motivations of these groups are towards promotion and higher salaries (Wood 1976). Manual workers, on the other hand, are much more sedentary: stronger home ties, poorly developed career structures, relatively low incomes and the virtually ubiquitous existence of unskilled and semi-skilled manual work provide little incentive to move. Much less clear, however, are the migratory propensities of intermediate status occupations such as skilled manual workers, technical workers and the junior white-collar grades.

Further reading

Berger J. Mohr J 1975 *A Seventh Man: the story of a migrant worker in Europe*. Penguin, London

Champion A G (ed.) 1990 *Counterurbanization*. Edward Arnold. London

Hammar T (ed.) 1985 *European Immigration Policy: a comparative study*. Cambridge University Press, Cambridge

King R (ed.) 1993 *Mass Migration in Europe: the legacy and the future*. Belhaven, London

Ogden P E, White P E (eds) 1989 *Migrants in Modern France*. Unwin Hyman, London

Salt J, Clout H (eds) 1976 *Migration in Post-war Europe: geographical essays*. Oxford University Press, London

5

Energy

The evolution of energy demand

The importance of energy in the economies of Western Europe was abruptly demonstrated by the 1973/74 and 1979/80 oil crises, and it is appropriate that any examination of this sector should give prominence to the post-crisis era. Yet it would be wrong to view this period in isolation. Understanding the impact of the OPEC-induced upheavals requires a knowledge of major trends, and the causal factors behind them, in the preceding decades. In addition, our interpretation of post-crisis developments must recognize that, in a number of important instances, they have been founded on initiatives taken before the crises struck. Although the crises were momentous, they were essentially turning points rather than starting points. The present chapter therefore examines energy issues in this broad temporal context, focusing particularly on attempts to contribute to the economic development of Western European nations by planning their fuel industries. In doing so it concentrates on energy supply questions, but at the outset it is necessary to sketch the other side of the equation, the evolving scale and geography of energy demand.

Energy consumption in Western Europe has risen beyond all expectations in the post-war period. In the 1950s the proportional increase was one-third; in the 1960s it rose to two-thirds; and only under the influence of the crisis did it fall back to little more than one-tenth in the 1970s. Some of this upswing can be attributed to population growth, but by no means all: per capita energy consumption in the early 1990s was more than twice as great as in 1955. The primary impetus behind the growth came from the industrial sector, which at its peak – in the early 1970s – accounted for almost half Western Europe's energy consumption. But industry's dominance has been undermined since the 1973 oil crisis as major energy-consuming industries have contracted, as surviving

Table 5.1 Structure of European Community energy demand

	1979	1981	1986	1987–89 average
Industry	35.1	33.8	30.3	31.0
Transport	22.5	24.6	27.7	29.7
Residential and other consumers	42.4	41.6	42.0	39.3
	100	100	100	100

Source: Commission of the European Communities (1988b, 1992e).

activities have used energy more efficiently and as non-industrial consumers have prospered. By 1990, manufacturing in the EC accounted for less than a third of total demand, compared with 39 per cent absorbed by private and service-sector customers (Table 5.1). This sectoral division of demand is, of course, the summation of consumption patterns at the national level, and the precise balance between industrial and other consumers varies considerably throughout Western Europe. Above all, this reflects the diversity of development paths followed by individual countries. For example, Luxembourg's industrial consumption is exceptionally high simply because of the steel industry's unusually prominent position in the economy. Denmark and the Republic of Ireland, in contrast, share common ground in their relatively recent movement from agriculture to light industrialization, a trend betrayed by industry's low share of total energy consumption.

Chiefly because national energy consumption is dictated by population size and the scale and nature of the economies – variables which evolve only slowly – the geography of total energy demand (Fig. 5.1) is likewise slow to change. Even so, long-term shifts can be identified, as can be demonstrated by correlating national

Table 5.2 Estimated per capita energy consumption, 1991 (tonnes of oil equivalent)

Luxembourg	8.9	Switzerland	3.6
Netherlands	5.3	Denmark	3.4
Sweden	4.8	Ireland (Rep.)	3.0
Belgium	4.7	Italy	2.8
Finland	4.5	Greece	2.4
Norway	4.2	Spain	2.2
Germany	3.9*	Portugal	1.2
France	3.8		

* Estimate for unified Germany.
Source: Commission of the European Communities (1992e).

energy consumption figures for 1951 with those for 1961, 1971, 1981 and 1991. The resulting coefficients decline at an accelerating rate from +1.0 to little more than +0.8. The geography of this adjustment in the pre-crisis era is identified when demand growth is expressed in per capita terms. In the West European core, two major countries – France and West Germany – experienced demand increases corresponding closely with the West European average. The UK, Belgium and Luxembourg, countries which had unusually high per capita consumption figures at the outset, recorded growth rates well below the average. And on the northern and southern peripheries the typical pattern was for growth to exceed the average, especially in southern Europe. The net effect of these contrasting trends, which continued with modifications into the 1980s, was to reduce – but by no means eliminate – the peaks and troughs of the energy demand 'surface' (Table 5.2).

These, then, are three major features of post-war energy demand: impressive growth up to the oil crises, perceptible change in the geography of energy consumption and, coupled with this change, the emergence of a less accidented demand surface. On their own, these features might well have made the problem of adjusting energy supply to demand a relatively easy one to solve. But efforts to achieve this adjustment have been handicapped by a fourth demand characteristic: major shifts in consumers' fuel preferences. In addition, OPEC's use of oil for political ends has complicated the supply problem by exposing the dangers of external dependence. In seeking to reduce these dangers, a number of countries have discovered that inexperience can be a substantial obstacle to the satisfactory transition to new energy sources. As we shall see, these factors have enveloped the supply side of the energy sector in uncertainty, strengthening the argument for rational planning yet simultaneously acting as major constraints on effective management.

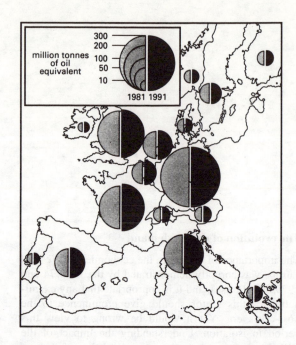

Fig. 5.1
Total energy demand, 1981 and 1991.

Oil versus coal in the cheap energy era

The first phase of coal industry planning has been thoroughly documented and need not be reconsidered in detail (Gordon 1970). Briefly, Western Europe was still 85 per cent dependent on coal in the early 1950s and, almost without exception, the industry's planners saw coal as the unchallenged fuel of the future. Energy demand was strong and planning for maximum production became the norm. Only the newly established European Coal and Steel Community (ECSC), which had no direct control over production, warned that the era of European energy self-sufficiency was over and that coal's future depended on ensuring that it could face rigorous competition. The experience of countries without coal industries bore out the ECSC's conclusion – in at least eight energy-deficient countries, imported liquid fuels had already captured a third or more of the market – and the reality of the threat from oil was forced on coal producers in the late 1950s and early 1960s when coal demand failed to recover from a minor economic recession. Thereafter, almost every market shrank rapidly as oil displaced coal and captured the lion's share of demand growth (Fig. 5.2). By 1970, coal's share of EC demand was down to 27 per cent and, instead of leading the industry into a retrenched yet defensible position, the ECSC was

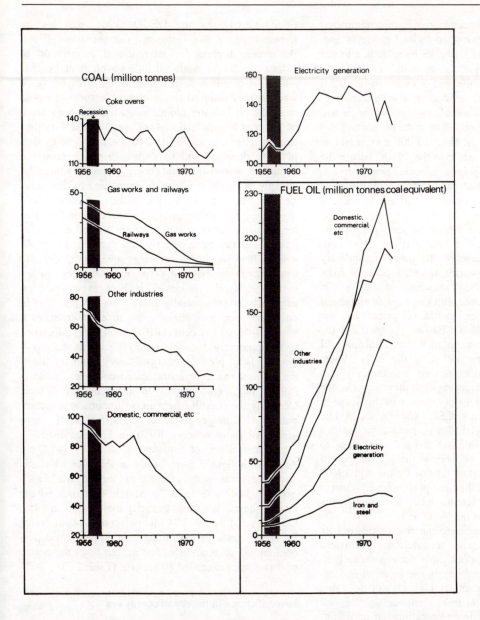

COAL (million tonnes)

Coke ovens

Recession

140

110

1956   1960   1970

Gas works and railways

50

Railways   Gas works

0

1956   1960   1970

Other industries

80

60

40

20

1956   1960   1970

Domestic, commercial, etc

100

80

60

40

20

1956   1960   1970

Electricity generation

160

140

120

100

1956   1960   1970

FUEL OIL (million tonnes coal equivalent)

230

Domestic,
commercial,
etc

200

Other
industries

150

100

Electricity
generation

50

Iron and
steel

0

1956   1960   1970

Fig. 5.2
Use of coal and fuel
oil in Western Europe
up to the first oil crisis.

obliged to assist governments by acting as a fire brigade dealing with the regional and social consequences of collapse (Collins 1975). Moreover, the situation would have been even worse if one substantial coal market, electricity generation, had not been artificially expanded in coal-producing countries by the development of preferential policies – between 1959 and 1964, power-station coal consumption in Western Europe rose from 110 to 150 million tonnes. Even so, from then on the growth of even this relatively sheltered market was captured by oil.

Why were coal industry planners so conspicuously unsuccessful in combating the advance of oil? As is well known, at its simplest the problem was that, as the 1960s progressed, refiners were increasingly able to market fuel oil at prices well below the cost of coal (ECSC 1964; Ray 1976). In Hamburg, for example, the price advantage of 43 DM per tonne in 1960 had risen to 162 DM in 1972. But this was only the basic arithmetic, behind which lay a battery of factors working to oil's advantage.

One was the oil industry's low labour requirements; these enabled it to contain the effects of rising wages while wage inflation forced up the cost of coal. A second was the rapid expansion of the European refining industry, from 50 million tonnes capacity in 1950 to

863 million in 1973. Simultaneously, this minimized reliance on expensive imported refined products and, because larger and larger refineries were built, allowed Western Europe to enjoy the benefits of increasing economies of scale (Molle and Wever 1984a; Odell 1986). These economies, of course, also became significant in the transportation of crude oil by larger and faster tankers – by the mid-1960s costs per tonne kilometre were twice as great for a 20 000 dwt vessel as for one of 70 000 dwt; while in the 1950s, before the scale factor became highly significant, heavy investment in new tankers had produced a world surplus which, through the large charter market, also drove down transport costs.

Other factors included the oil policies pursued by the USA and the USSR. The American strategy of controlling imports to safeguard the domestic industry diverted oil on to the world market, especially from Venezuela, while the USSR sold more to the West in exchange for hard currency. But the most fundamental supply factor contributing to low oil prices was the dominant position of Middle Eastern producers in the European system – 65 per cent of Western Europe's oil supplies came from this region by the early 1960s. Middle East production costs were substantially less than those of other producer regions: in Venezuela, the Middle East's most serious competitor in this respect, they were twice as great (ECSC 1964). Also, Middle East supplies became plentiful as producer countries and companies multiplied, and this led companies to discount selling prices well below the official posted levels. This process went so far that by the mid-1960s the OECD was arguing that low profits posed a threat to exploration in the Middle East, a warning now familiar in the context of the North Sea.

Cost factors were crucial to the rise of oil, but its qualitative appeal was also considerable. Compared with coal, fuel oil was much more convenient to handle, allowed greater flexibility in the location of storage areas and, because of its higher calorific value tonne for tonne, was a far more compact fuel. Beyond this, the precision of refining operations ensured consistent fuel quality, while advances in the design of oil-fired equipment led to increasingly efficient consumption coupled with precise temperature control. Additionally, efficient combustion meant lower pollution levels, a significant factor in countries seeking to control air pollution by legislation. (At that time, of course, the wider climatic implications of large-scale oil consumption were not appreciated.)

Governmental response

At the international level the strategic dangers of heavy reliance on the Middle East were stressed by the ECSC (1964), by the OECD (1964) and by the Commission of the European Communities (1968). But repeated pleas for international cooperation to reduce the danger produced little governmental enthusiasm for common defensive policies, irrespective of whether they referred to the energy sector as a whole or to the oil industry alone. Instead, throughout much of Western Europe national policies adopted very limited objectives: to avoid shortfalls between energy supply and demand, and to ensure in the interests of economic expansion that energy was available as cheaply as possible. Given the prevailing buyers' market, oil was the fuel most capable of meeting these objectives, and many policy initiatives therefore fostered an open-door policy to crude oil and its derivatives. This was made easier by the fact that cheap oil, even when purchased in large quantities, did not impose a major balance of payments burden. In the early 1960s oil's share of national expenditure on imports was substantially more than 10 per cent in only one country – Spain – and in most instances it was less than 5 per cent (OECD 1964). Yet adherence to an apparently desirable policy – such as the provision of cheap, plentiful energy – may eventually generate unforeseen and undesirable consequences, as Lucas (1979) has demonstrated in the case of France. Certainly the pursuit of cheap oil imports was incompatible with the geographical diversification of source areas. African producers, it is true, had established a 29 per cent share of European markets by 1972, but Arab states figured prominently in this rise which, in any case, was not achieved at the Middle East's expense. In fact in 1972 the Middle East took 64 per cent of the market, almost exactly the same as in 1959. On the eve of the 1973/74 oil crisis, no less than twelve European countries were more than 60 per cent dependent on this source region, and in four cases the degree of dependence exceeded 80 per cent (Table 5.3).

Diversification in the cheap energy era

Although the main theme before 1973 was imported oil's advance at the expense of coal, the argument that governments were indifferent to the potential problems of dependence on the Middle East should not be overstated. Well before the stimulus of the oil crisis, attention was given to a wide range of possibilities for developing more secure resources.

Dutch gas and the North Sea

Many European countries already possessed small oilfields, a fact which encouraged governments to license more widespread exploration. The significance of this

Table 5.3 Oil consumption and sources, 1972

Source regions for West European consumption

	(million tonnes)	(%)
Africa	190.8	29
North America	0.0	0
Caribbean	17.1	3
Middle East	418.7	64
Communist bloc	24.0	4
Others	1.4	0
	652.0	100

National dependence on Middle East

	Total oil consumption (million tonnes)	Middle East dependence (%)	Oil as % of total energy consumption
Belgium	36.2	78	55
Denmark	9.7	81	94
France	117.8	66	68
Germany (West)	104.6	41	53
Ireland (Rep)	3.4	62	65
Italy	118.7	67	76
Netherlands	67.6	73	34
UK	107.2	68	45
Austria	5.7	65	54
Finland	9.2	32	75
Norway	6.4	82	51
Portugal	4.3	88	83
Sweden	11.2	44	81
Switzerland	4.8	35	82
Greece	7.2	85	68
Spain	38.7	67	65

Source: United Nations (1976).

work lies not in its immediate impact on the oil market (indigenous production in the late 1960s was a mere 20 million tonnes) but in the chance discovery of one of the world's largest natural gas deposits at Slochteren in the northern Netherlands in 1959. With initial proven reserves put at 1100 billion m^3, the consequences of this discovery could not be confined to the Netherlands. Wider exploration received a powerful boost and, in particular, the location and geological characteristics of the Slochteren deposit turned attention in a new direction – towards the North Sea.

Investigations here quickly confirmed the existence of two potentially productive zones, the northern and southern basins. Because at this stage countries encircling the North Sea had not all ratified the 1958 Geneva Convention on the Continental Shelf, precise territorial claims could not be established immediately; but ratification was completed in 1964, and exploration rapidly gathered momentum. Initially attention focused on the

southern basin, between the UK's Yorkshire–Lincolnshire coast and the Dutch Wadden Islands, where exploration and exploitation were helped by relatively shallow water. Then in 1969 the discovery of the Ekofisk field in Norwegian territory stimulated interest in the northern basin, and by the 1973 oil crisis a sharp contrast had been identified between the resources available in the two areas (Bailey 1977; Ion 1977). Natural gas predominated in the south, but in the north oil deposits were typically associated with greater or lesser volumes of gas. What was also apparent was that the benefits of this new diverse resource region were very unevenly distributed among countries with North Sea rights. The northern basin, extending along a narrow north–south axis for 700 km, came almost entirely under the jurisdiction of the UK and Norway, while sovereignty over productive parts of the southern basin lay with the UK and, to a lesser extent, the Netherlands. Fortunate though they were, however, the countries which benefited significantly from this major wave of exploration also faced the challenge of devising policies for rational exploitation. We shall return to their varied responses in a later section.

Alternatives for electricity generation

Electricity generation was a major growth industry in this period, with output increasing more than fivefold between 1950 and 1973. By the early 1970s, oil accounted for half Western Europe's conventional thermal power production and was poised to lever coal out of its last major stable market. Yet the choice of energy inputs was not simply that of coal or oil, and many countries were able to pursue alternative strategies which would retard industry's rapidly growing dependence on oil.

One such strategy was increased use of lignite or brown coal, a fuel of low calorific value, but an attractive one when capital-intensive opencast mining is possible. The Republic of Ireland and Greece, two countries particularly poor in energy resources, both expanded their output of this fuel in the 1960s, but the country best placed to exploit it was West Germany. Here output rose from 96 million tonnes in 1960 to 118 million in 1973. In terms of import substitution, this was equivalent to approximately 23 million tonnes of oil, compared with actual oil imports of 113 million tonnes in 1973. Within West Germany the dominant source area was, and remains, the Rhenish brown coalfield, where electricity generation is vertically integrated with mining, and where advanced opencast technology may eventually allow exploitation up to 1000 m deep (Ray and Uhlmann 1979; Ion 1977).

Hydroelectric power was an option open to many

Table 5.4 Hydroelectric power development, 1950–74

	Output (billion kWh)			
	1950	1960	1970	1974
Western Europe	110.4	229.5	337.3	385.1
EC	49.5	105.8	125.9	123.5

HEP as % of all electricity generation

	1950	1974
Western Europe	43	27
EC	26	12
France	49	32
Germany (West)	19	6
Italy	92	28
Austria	79	67
Finland	88	47
Norway	100	100
Sweden	95	76
Portugal	46	74
Switzerland	98	77
Spain	74	38

Source: United Nations (1976).

Table 5.5 Early development of the nuclear power industry

	Initial production	Nuclear power as % of total electricity generation, 1974
Belgium	1962	0.4
Finland	1977	0.0
France	1958	7.8
Germany (West)	1961	3.9
Italy	1963	2.3
Netherlands	1968	5.9
UK	1956	12.3
Sweden	1964	2.7
Switzerland	1969	18.0
Spain	1968	8.9
EC		6.4
Total Western Europe		5.7

Sources: United Nations (1976, 1981b).

more countries and, from Scandinavia to Iberia, was exploited on a rapidly increasing scale. Thus Western Europe's production more than doubled in the 1950s and rose again by two-thirds before the first oil crisis. This development must be placed in perspective: despite the absolute expansion, HEP's share of total electricity generation slipped from 35 per cent in 1960 to 27 per cent in 1974. Yet in the early 1970s Norway, Sweden, Switzerland, Austria and Portugal all obtained the large majority of their electricity from this source, while in Finland, Spain, France and Italy it remained far from insignificant (Table 5.4). For these countries, past investment was shortly to provide its most valuable economic return as inflation and balance-of-payments problems intensified under the pressure of escalating oil prices.

The final major alternative for electricity generation lay in the development of nuclear power. As with oil, this strategy necessitated external dependence with respect to supplies, but the dangers were perceived to be far less, primarily because relationships with the main producer countries – Canada, the USA, South Africa and Australia – were stable. In terms of economic advantages, nuclear power appeared similar to HEP. High investment costs would be offset, it was predicted, by low operating costs – ECSC calculations indicated that nuclear power stations would produce electricity for roughly half the price that coal-fired plant would achieve (ECSC 1964). This encouraged widespread interest in nuclear investment, but by the oil crisis nuclear energy's contribution to Western Europe's total electricity consumption was less than 6 per cent. Only two countries –

the UK and Switzerland – generated more than a tenth of their electricity in this way (Table 5.5). These achievements fell well short of the protagonists' hopes.

At this stage, delays in the expansion of nuclear power were not primarily a reflection of environmentalists' hostility to the industry (Collingridge 1984a). Opposition was gaining momentum, but political and technical factors were much more influential. Most countries were unwilling to cooperate on reactor development and construction, even in the EC, which had established Euratom for these and similar purposes (Pearson 1981). Within the industry itself, power struggles were significant, particularly over the problem of which reactor system should be adopted. In France, for example, Electricité de France fought a protracted battle with the Commission à l'Energie Atomique and was eventually successful in ending the gas graphite reactor programme imposed by the Commission (Lucas 1979). More generally, many delays were caused by the difficulties of transposing high technology from the research environment to the commercial world, and these difficulties were compounded when – as in the UK and France – decisions were made to abandon one system in favour of another. Finally, increasing economies of scale in conventional generators raised average thermal efficiency to approximately 30 per cent and, although this was far from ideal, the savings made contrasted sharply with the growing financial burden caused by policy arguments and technical delays in the nuclear branch of the industry (Reid, Allen and Harris 1973).

Oil crisis and the European response

The history of oil price dynamics is well known and need not be covered in great detail (Bailey 1977; Odell

1986). As oil prices sank in the 1960s, the trend was increasingly interpreted in centre–periphery terms by the unindustrialized producers. Within OPEC, which had been formed in 1960, a more united attitude emerged and prices were roughly doubled in 1970–73. But the first real shock came in 1973–74 when Arab unity was hardened by the Arab–Israeli Yom Kippur war: an oil embargo was placed on the Netherlands in retaliation for her pro-Israeli stance; supplies to the West in general seemed suddenly insecure; and the price of crude oil was virtually tripled. A period of quiescence followed but was abruptly ended by the start of the Iran–Iraq conflict, with OPEC prices doubling once more in 1979–80. In real terms, this raised the price of Saudi Arabian crude to six times the 1973 level (Odell 1981). The effect of this second oil crisis was to give a major boost to economic restructuring in Western Europe which lasted throughout the 1980s.

What is particularly interesting is that this impact was sustained even though OPEC proved unable to maintain prices at early 1980s levels. This failure partly reflected a decline in demand in industrialized countries, but was also a consequence of dissent within OPEC. Dissension led several producers to raise output above agreed quota levels, and even major production cutbacks by Saudi Arabia could not prevent the emergence of a crude-oil surplus. Oil prices at first fell steadily. Oil imported to Europe in 1981 averaged $36.5 a barrel, compared with $29 a barrel in 1984, but the value of oil then crashed in 1985/86 as Saudi Arabia rejected its balancing role and OPEC cooperation broke down. In what has sometimes been called the third oil crisis – one which primarily hit LDC (Less Developed Country) producers – Arabian light oil prices fell to approximately $13 a barrel, recovered to $17 in 1987 and subsequently have shown no sustained rise above that level.

It is important to appreciate that the severity of this decline was closely connected with a geopolitical miscalculation by OPEC countries. These argued that if their output rose, non-OPEC producers would compensate by reducing production to sustain prices and, therefore, tax revenues. This did not happen, and in Europe the UK was particularly resolute in maintaining output. Industrialized countries in general welcomed the anti-inflationary effects of cheaper oil, and even some producer countries were prepared to trade lower-revenue income for these benefits.

Reactions

While one immediate effect of the 1973 upheaval was to provoke numerous national energy reviews, it also prompted many OECD countries to take the first significant step towards international cooperation in energy sector development – the formation of the International Energy Agency (IEA) in 1974. This organization quickly produced agreement on an oil-sharing scheme to be activated in future crises (IEA 1980a), but more importantly it has subsequently been deeply involved in monitoring and in attempting to guide the energy sector. As the only major Western European country that has not joined the IEA is France, these activities are highly relevant to the present discussion.

The attempt at guidance has taken place within the framework of twelve principles for energy policy and a further fourteen principles for action on coal (IEA 1980a). The central element in the design is reduced dependence on oil imports, with substitution by coal, natural gas and nuclear power providing the chief medium-term insurance against energy shortfalls. In addition, member countries are strongly encouraged firstly to improve the sector's efficiency by pursuing wide-ranging conservation policies and, secondly, to concentrate on the long-term development of renewable energy sources, such as solar, wind, biomass and geothermal power. Emphasis is placed on the need to price energy realistically to promote interest in conservation and renewable power schemes, while intensive research, development and demonstration (RD and D) are considered essential for the emergence and adoption of new technologies capable of effective competition with oil (IEA 1980b; UN Economic Commission for Europe 1983). Although some of this RD and D expenditure can be expected to come from industry, governments have been left in no doubt that their leadership may be crucial to success.

Several indicators of progress towards these goals can be identified. As a later section demonstrates with respect to oil and gas resources, unrealistically cheap energy has been abandoned as a policy platform. In addition, a number of outstanding conservation programmes have been devised, those in Denmark and Sweden being particularly advanced (Schipper 1983, 1987), while almost all European IEA countries have increased in real terms their RD and D expenditure. This exceeded $2 billion a year by the turn of the decade, with relatively equal attention being given to conservation projects, new coal technologies, alternative energy sources and support technologies for the electricity industry.

There has also been a substantial reduction in oil imports and in their contribution to primary energy consumption (Table 5.6). Indeed, this reduction in imported oil dependency has progressed much further than seemed possible during the second oil crisis, even though oil consumption has now started to creep up once more. In 1980, for example, the IEA predicted

Table 5.6 Oil import dependency in Western Europe

	Oil imports (mtoe)	Oil imports as % of primary energy consumption
1973	716.2	59.4
1979	598.7	46.5
1980	514.9	41.5
1981	441.8	36.6
1982	383.0	32.4
1983	342.2	29.0
1984	337.7	28.1
1985	311.6	25.2
1986	346.9	27.9
1991	413.0	29.8

Sources: British Petroleum (annually) *Statistical Review of the World Energy Economy*. British Petroleum, London.

that European oil imports would be approximately 500 million tonnes in 1985 and would account for almost 40 per cent of primary energy. In reality, the figures were 311 million tonnes and 25 per cent, and by 1991 they remained well below the pre-crisis levels.

Despite these favourable signs the IEA's success to date must be considered partial. Outstanding conservation policies in Denmark and Sweden only highlight the inadequacy of steps taken in much of the remainder of Western Europe. Reduced oil imports are partly a consequence of restructuring efforts, but they also reflect recessional influences and increased European oil production. There are indications that cheap oil has recently stimulated imports (Table 5.6), and declining prices have not encouraged governments to maintain the expansion of RD and D expenditure into alternative energy sources. The IEA's (1980a) judgement that in the RD and D context 'There is no country that cannot and should not do more' remains valid more than a decade later. Previously this conclusion reflected the fact that the lion's share of diversification was absorbed by a single controversial energy source – nuclear power. Between 1982 and 1986, the European Investment Bank (EIB), which primarily responds to investment demand in EC countries, allocated 56 per cent of its power production loans to nuclear projects (Table 5.7). In contrast, geothermal energy and alternative energy sources accounted for only 3 per cent of investment, while combined heat and power projects absorbed even less. Even when nuclear power was excluded from the calculations on the grounds that its capital intensity ensured that it dominated investment, the proportions allocated to new forms of energy generation remained modest relative to the remainder of the sector. More recently, nuclear power investment has waned rapidly in importance as safety fears and environmental opposition have grown. But this has not encouraged national governments to seek EIB money to invest in new and alternative energy sources. Instead the emphasis has switched to conventional power stations and oil and gas deposits, which in the last few years have absorbed three-quarters of EIB finance for energy production.

Although progress in the development of entirely new energy sources has been unimpressive, the fact that more mainstream forms of energy production have received high levels of investment indicates that energy planning activity has been intense at the national level and within the individual energy industries (IEA 1984). Reappraisals have to some extent led to the reversal of past strategies – for example, as we shall see in a later section, with respect to consumption. But recent years have also been marked by a strengthened commitment to the process of diversification which was already in progress before the oil crisis. Above all, this has meant increased determination to derive maximum benefit from newly discovered oil and gas resources. As the following discussion of these major themes demonstrates, however, widespread appreciation of the economic and strategic value of secure oil and gas supplies has not brought about a golden age for energy planning. Indeed, in many respects governments and planners have experienced a rougher ride since 1973 than at any time in the post-war period.

New indigenous resources – the exploitation challenge

Before considering the problems of oil and gas exploitation, the background to their development must be sketched. The evolving geography of their extraction has been examined in detail by Odell (1986) and is summarized by Fig. 5.3, in which they are shown according to a common measure – million tonnes of oil equivalent (mtoe).

By the time of the first oil crisis, Dutch gas production was already well advanced. Output had reached 60 mtoe a year, 70 per cent of the peak it would eventually achieve towards the end of the decade. Gas from the British sector of the southern North Sea was also flowing strongly by this time – 1973 production was 25 mtoe – and, as in the Netherlands, it was tracing a smooth upward output curve. In Norwegian waters, and in the British sector of the northern basin, later exploration and longer lead times meant that product flows were minimal before the mid-1970s. But northern basin yields then rose markedly, although with different results on either side of the median line. Norwegian production was 118 mtoe in 1991, almost a

Table 5.7 European Investment Bank allocations to energy production projects, 1982–91

	1982–86 (million ECUs)	(%)	1987–91 (million ECUs)	(%)
Nuclear power	3249.4	56.4	60.0	1.2
Conventional thermal power stations	500.6	8.7	1358.3	27.4
Hydroelectric power stations	673.5	11.7	656.5	13.2
Geothermal energy and alternative energy sources	175.1	3.0	100.1	2.0
Heat and power generating plant	112.8	2.0	394.6	7.9
Development of oil and gas deposits	972.5	16.9	2325.4	46.8
Solid fuel extraction	75.9	1.3	73.1	1.5
Total	5759.8	100	4968.0	100

Sources: European Investment Bank, *Annual Reports.* EIB, Luxembourg.

quarter of this being accounted for by gas. British northern basin production consists primarily of oil and, since peaking at 120 mtoe in 1986, has been cut back to little more than 80 mtoe a year. This decrease partly stemmed from the closure of the Piper field from July 1988 to mid-1992, following a disastrous fire on the production platform. However, this factor accounted for only a fifth of the downturn, most of which reflected field depletion and company perceptions of optimum production rates.

Rapid exploitation of the North Sea since 1975 has depended partly on the discovery of giant fields, such as Statfjord, Brent and Forties. The operators' initial estimates of recoverable oil reserves in these three fields alone totalled 900 million tonnes. But these giants are not typical. In the British sector, for example, more than half the oilfields currently in production had initial reserve estimates of less than 50 million tonnes, and 85 per cent were thought to contain less than 100 million tonnes. In the difficult operating conditions found in the North Sea, the diseconomies associated with exploiting relatively small fields can be substantial. For this reason, it might be anticipated that falling prices have recently hindered exploitation of, and exploration for, additional deposits. At first sight the evidence suggests that this has in fact occurred throughout most of the 1980s. In the British sector, for example, company investment in field development was virtually halved between 1981 and 1987 (Department of Energy 1987). However, it will be shown later in this chapter that a distinction should be drawn between exploration and exploitation activity before and after the 1985/86 oil price crash.

For those fields which are in production, size has also influenced the mode of transport to land, although this should not be overstressed. Some small fields, such as Argyll (peak production 1.1 million tonnes)

and Montrose (1.4 million tonnes), load directly into tankers, but so do the Beryl A and Statfjord fields (5.8 and 37.5 million tonnes respectively). The majority of active fields are, however, linked with the Shetlands and the mainland by the now extensive pipeline system, one major advantage of which is its ability to operate irrespective of weather conditions (Fig. 5.3).

A search for additional reserves has been pursued outside the main producing areas, although recently this activity has been affected by the downward price trend. In UK waters, little has been discovered outside the main northern basin, although the Morecambe Bay deposit off the Lancashire coast, which contains 114 mtoe of gas condensate, is an exception to this rule. Norwegian waters are more promising – exploration to date has been less comprehensive than in the British sector and as yet has been largely limited to the area south of 62° N. The Danish sector, too, is under-explored, one reason being that until recently all rights were vested with a single consortium which felt no urgency to invest. Evidence from nearby sectors, however, suggests that there may be nothing of outstanding interest in Danish waters. West German discoveries have been modest and, although numerous Dutch gasfields have been proven on land and north-west of Den Helder, the large majority of the country's reserves still lie in and around the Slochteren deposit. Meanwhile, the exploration wave has spread to the Mediterranean and is concentrating initially on Italy where, before the advent of Dutch gas, several of Europe's more important gasfields were to be found. Most Italian discoveries have been in the Adriatic, but the Ionian Sea off Calabria is also promising, as is the Sicilian offshore zone. Under the National Energy Plan of 1988, the aim is to cut dependency on oil and gas imports from 81 to 24 per cent. As yet, however, indigenous production accounts for only 5 per cent of oil consumption and (more encouragingly) 36 per cent

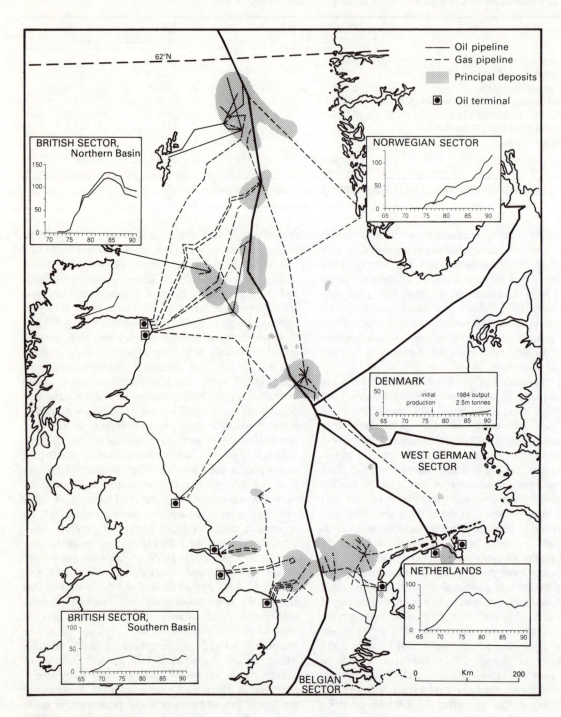

Fig. 5.3
North Sea oil and gas. Output is measured in millions of
tonnes of oil equivalent. For the northern basin in the British
sector, and for the Norwegian sector, the lower curve represents oil; the upper curve represents oil and gas.

of gas supplies (Cranfield 1992). In 1991 Italy borrowed over 90 per cent of the finance made available by the EIB for the development of oil- and gasfields, the large majority of the loans being employed in offshore projects. At present the most productive oilfields are Vega, off the coast of south-east Sicily, and Rospo Mare, off mainland Italy close to Pescara. These fields account for half the annual output (Estrada *et al.* 1988).

The vast programme of exploration and production that has been undertaken to date has necessarily involved the oil companies in an intensive learning process. This is particularly true of technological aspects of the industry, and the experience of operating in the North Sea's extreme weather and sea conditions is already proving beneficial in other parts of the world. But the governments and energy ministries of the countries concerned have also had to learn rapidly in order to erode the oil companies' natural advantage of past experience, to establish rational exploitation programmes and to ensure that society derives an adequate financial return from the new resources. The production curves in Fig. 5.3 may suggest that this governmental learning process has proceeded smoothly but, if so, they are highly deceptive. In reality, the obstacles to, and the disagreements about, effective control have been too numerous to consider individually. Thus we must focus on a selection of major themes.

Dutch natural gas – exploitation, pricing and income

Partly because of state participation with the original discoverers, Shell and Esso, and partly because of the tax structure imposed on the industry, since the late 1970s natural gas has provided more than a tenth of the Dutch government's income. In 1985, before the energy price collapse, the proportion was almost a fifth. Although this may imply the pursuit of a well-implemented exploitation strategy, much of the evidence suggests that the policy adopted in the era of cheap oil severely impeded the maximization of returns in the high-cost energy era. Moreover, the strategy eventually adopted to solve this problem hit government income hard when the energy price collapse occurred.

The initial policy was one of rapid development to exploit the resource before nuclear energy came into its own. From the mid-1960s a price structure which undercut competing fuels, coupled with the exceptional cleanliness and efficiency of natural gas, ensured that the consumption curve rose steeply. By the early 1970s the fuel had captured 60 per cent of the domestic heating market and was supplying approximately half the energy inputs to industry, electricity generation and horticulture. Naturally, a great deal of this remarkable

progress was made at the expense of coal and, indeed, of oil: between 1965 and 1972, industrial oil consumption fell by 45 per cent and oil's share of the total energy market declined from two-thirds to a half (Pinder 1976). This trend was officially encouraged because of its import-substitution effect on the balance of payments, but in this connection a further major impact was anticipated from gas exports, which were in fact a central element in the policy of obtaining swift returns. Although they only began in 1963, by 1970 36 per cent of production was sold abroad.

Gas demand continued its remarkable growth in the early 1970s, but the first oil crisis and the Arab oil embargo caused an urgent review of energy policy, the outcome being an almost complete reversal of the exploitation strategy. According to guidelines adopted in 1976, consumption was to be reduced to the minimum level consistent with contract obligations, no new export contracts were to be concluded and existing contracts were not to be renewed. If domestic demand ran ahead of production, as was anticipated by 1985, the shortfall was to be covered by oil and gas imports rather than by higher output, despite the balance of payments penalty. After a second policy review in 1979, the emphasis on conservation was reinforced by declaring the Slochteren deposit a strategic reserve, the depletion of which should be retarded (OECD 1980b).

Price rises aimed at conservation, coupled with intensifying recession, produced a 15 per cent fall in home demand between 1975 and 1980. But exports behaved differently, achieving a fivefold increase during the decade. West German consumption rose particularly quickly, to account for half the exports, while Belgium and France each purchased a fifth and Italy a tenth. Altogether, 60 per cent of the gas extracted in 1980 was exported. Although this rapid increase occurred partly because foreign consumers used their contractual rights to obtain secure supplies following the oil crisis, an additional major factor was that export prices became increasingly attractive. This was chiefly because, even though contract terms had been reviewed in the aftermath of the first oil crisis, they remained too inflexible to allow the Netherlands to control foreign demand through the price mechanism. In fact the contracts were particularly deficient in two respects: gas price increases triggered by rising oil prices could take up to a year to come into effect, and price indexation was inadequately linked to the cost of oil. By 1980, therefore, exported gas was underpriced by 30 per cent. Even without new contracts, the Netherlands was committed to supplying 550 billion m^3 of natural gas to her neighbours by the mid-1990s, an amount equivalent to 447 million tonnes of oil. Thus the most immediate problem became the need for contract revision to minimize foreign-exchange losses.

Discussions opened in 1979, but the fourteen major purchasing companies understandably showed little enthusiasm for renegotiation. Lack of progress quickly made the problem a major political issue, the most trenchant view being that quick results should be achieved by invoking a law of 1974 and halting supplies. Ultimately, however, the issue was tackled more subtly by the appointment of a retired diplomat, well known throughout the EC, as the country's official negotiator. Armed with the threat of cessation of supplies, by late 1980 this emissary had successfully concluded substantial contract revisions. These included the rapid upward adjustment of gas prices to rises in the price of oil, and a phased increase which raised the average price of gas at the Dutch border by 20 per cent. In this way, export earnings were improved by up to 2 billion guilders a year, compared with an original target of 3 billion guilders, and total demand for Dutch gas was reduced by almost a quarter between 1981 and 1988. But the formula which allowed gas prices to rise in line with those of oil also required gas charges to be lowered if the value of oil fell. When the oil prices collapsed, therefore, the consequences for the national exchequer were dramatic. Income from exports is estimated to have fallen by 67 per cent between 1985 and 1987, while government revenue from domestic and export sales fell by 54 per cent. Moreover, although rising gas demand since 1988 has partially offset this trend, production in 1991 remained 8 per cent lower than in 1981.

The decline in natural gas revenues has necessitated urgent reappraisal of government expenditure, one of the main targets being economies in the extensive Dutch welfare state. (In the mid-1980s, total social security costs accounted for a third of the national income.) There have, it is true, been benefits. Disposable income in private households has risen as gas prices have fallen, and production cost reductions have helped export industries. Yet in an economy as open as that of the Netherlands, higher disposable incomes primarily stimulate imports, and low gas prices make it increasingly difficult to sustain energy conservation as a significant national policy.

Controlling oil and gas development – the British experience

The relationship between private enterprise and the state has aroused more political controversy in the UK than in any other part of the North Sea petroleum province. As in the remainder of the province, private enterprise was essential to initiate exploration and development, and more than sixty companies now have a financial stake in British sector resources. The private enterprise view has been that this diversity is in itself a protection against capitalist manipulation. More than half the companies have an interest in only one field, and half the financial holdings in individual fields are of 20 per cent or less. The policy initiated by the Labour Party, however, was to counterbalance private interests – particularly multinational capital – by extensive state participation (Department of Energy 1974). Thus the British National Oil Company (BNOC) was created in 1976 to negotiate a majority (51 per cent) holding in existing private sector production and to exercise the option of majority participation in all future commercial developments. Subsequently the production arm of BNOC became Britoil, which was then sold in 1992 by the Conservative government. One justification for this step was that Britoil had become the target of heavy criticism from the remainder of the industry. This argued that the nationalized company curtailed competition, reduced investment by creating uncertainty in the private sector and occupied a favoured position by being both a quasi-independent producer and a government agency. But, in addition, the Britoil sale was devised to further two major government policies, privatization of nationalized activities and the realization of assets to reduce the public sector borrowing requirement (Webb 1985). The main issue raised by the sale was whether the proceeds would in the long run offset the loss of earnings from a nationalized Britoil. Light was shed on this issue towards the end of 1987 when British Petroleum made a takeover bid which valued the company at £2.27 billion, almost four times the £548 million raised for the Exchequer by the original sale five years earlier. This high valuation was made despite the fact that oil prices had almost halved over this five-year period.

Contrasted political attitudes have also been intertwined with the pricing strategy for British gas. Anti-inflation policies pursued by the Labour governments in the 1970s prevented household gas prices rising as rapidly as those of oil, with the result that by 1980 gas was being sold to domestic consumers for less than half its true value. The consequent rapid increase in consumption led to winter supply shortages, discounts were offered to companies willing to sign interruptible contracts and new customers were only accepted if they had a legal right to supplies. Revenue losses and the disincentive to conservation caused by low prices led the Conservative government elected in 1979 to impose closely phased price increases, but consumption proved so insensitive to price that little was done to bring demand in line with supply. Here we may note that the strong demand for gas assisted the sale of the British Gas monopoly in 1986 as part of the government's privatization programme.

One policy response to sustained demand has been to increase supplies by imposing strict controls on the flaring of 'waste' gas considered by companies to be an uneconomic by-product of oil production in the northern basin. As late as 1982, the amount flared off was equal to 11 per cent of British gas production, but by then the new controls were operating and the practice was rapidly diminishing. By 1991 the amount flared was equal to only 4 per cent of total production. The gas saved is either reinjected for later extraction or is piped ashore via extensions to the pipeline network. Gas landed from the northern basin amounted to almost 13 mtoe in 1991. The second policy response has been to encourage increased production from the main gas-producing area, the southern basin. Several new fields have been brought into production and improved extraction methods have been employed on older gasfields. This has reversed a production decline which developed in the late 1970s as output from the established fields fell through depletion. Output in 1991 was 31 mtoe, compared with 23 mtoe in 1986.

Gas demand has been so buoyant, however, that these measures have done no more than prevent a deterioration of the demand/supply gap. The outcome has been that this gap has had to be bridged by substantial imports from Norway. These have fluctuated from year to year but for some time accounted for between 20 and 25 per cent of total gas consumption (Stern 1986). Increased production has, however, reduced this proportion to 11 per cent in 1991.

Governments of both persuasions have been careful to ensure that the selling price of North Sea oil reflects its world market value. Two factors have prevented a repetition of the Dutch and British mistakes of selling gas cheaply. Firstly, exploitation began in the era of high prices so that there was no policy legacy from the cheap energy years to be overcome. Secondly, the nature of the oil, which has a low sulphur content and is best suited to the production of light distillates, meant that in the early years of production it was not well matched with British demand and with British refinery technologies. At the outset, therefore, the efficient use of the resource demanded that large quantities of oil should be exported at market prices to pay for more suitable imports from traditional suppliers. Almost two-thirds of production was exported in the 1980s, mainly to Western European markets.

Successive governments have also agreed on the necessity for a strong tax regime, the result being a long-running conflict between industry and the state over the relationship between taxation policy and resource development. The regime adopted after the 1973 price rises was intended to divert 70 per cent of the companies' net revenues to the Exchequer, and the industry's view is that numerous subsequent adjustments reduced the developers' share of marginal income to a mere 10 per cent by 1982. After allowing for oil price rises and general inflation, this meant a 50 per cent reduction in real income per barrel. This, it is argued, contributed substantially to a 20 or 30 per cent shortfall between production forecasts and achievement because production became decreasingly attractive. It is also claimed that the harsher treatment encouraged a rapid decline in investment, which gave rise to widespread talk of an investment strike (Department of Energy 1976, 1982).

Recent taxation changes and incentives intended to stimulate investment are detailed in the Department of Trade and Industry's annual publication on the *Development of Oil and Gas Resources in the UK*. Indicators of company attitude provided by this report suggest that North Sea investment did become more attractive in the 1980s, even though oil prices began to decline very early in the decade (Table 5.8). Although the number of oil and gas development wells drilled showed no clear upward trend, exploration and appraisal activity was significantly higher in the early 1990s than ten years previously. Similarly, the volume of mobile drilling rig activity tripled over this period. As Kemp, Rose and Dandie (1992) have emphasized, however, investment in these activities continues to respond to external political factors, and not simply to the UK tax regime.

Norwegian strategy and the national economy

In theory, several factors reduce the argument for rapid development of Norway's oil and gas: HEP is attractively priced; the small-scale economy has a very low demand for oil – only 8 or 9 million tonnes a year; and it has long been argued that the swift build-up of the industry and the injection of large revenues into the economy pose a threat to society and the manufacturing sector. Moreover, constraints of physical geography cannot be ignored. The Norwegian Trough, running parallel to the coast, has hindered pipeline construction to the mainland, while topography on the mainland continues to pose problems for pipeline development. Against this background, Norway has long professed a commitment to a 'moderate' pace of development, which apparently distinguishes her depletion policy from the early rapid exploitation strategies pursued by both the Netherlands and the UK (Ion 1977; Noreng 1980).

Moderation, however, is a term requiring definition, and the intention to curtail production must not be overemphasized. Indeed, the cynical might conclude that, with respect to the presentation of policy, Norway has attempted to have her cake and eat it. By 1980, oil

Table 5.8 Drilling activity in UK waters

	Exploration wells started	Appraisal wells started	Development wells started	Mobile rig activity (rig years)
1978	37	25	96	18.1
1979	34	15	102	16.1
1980	32	22	122	20.6
1981	48	26	137	24.6
1982	68	43	118	30.1
1983	77	51	95	34.2
1984	106	76	108	49.1
1985	93	64	133	51.7
1986	73	40	85	35.3
1987	69	63	124	35.2
1988	90	70	161	50.7
1989	92	75	143	49.5
1990	152	54	108	58.3
1991	106	60	125	59.1

Sources: Department of Energy (1987); Department of Trade and Industry (1992).

and gas output was six times greater than domestic oil consumption, the equivalent figures for 1985 and 1991 being nine and thirteen times, respectively. In reality, a strategy of establishing the industry as Norway's primary exporter has been pursued and, with the construction of pipelines to the UK and Germany (Fig. 5.3), has been achieved with ease. Between 1981 and 1991, oil and gas accounted for more than 40 per cent of all Norwegian visible exports. During this period the industry's share of exports did not fall below a third, and in 1985 it reached 56 per cent. In this way, Norway became the largest net exporter of oil and gas in the OECD.

Table 5.9 Government subsidization in the Norwegian economy

	1982–91 average	1989	1990	1991
	(billion krone)			
Total subsidies*	21.9	19.1	21.1	19.5
	(%)			
Agriculture	60	64	60	64
Fishing	7	6	7	6
Manufacturing	30	23	27	23
Shipbuilding	6	5	5	6
State companies	8	4	5	2
Private services	1	3	3	3
Other activities	2	4	3	3
	100	100	100	100

* 1991 prices.
Source: OECD (1992).

Successful though Norway's experience may appear to have been, it has become clear that the general economic policy that has been built on oil-sector expansion has created major structural problems (OECD 1987a). In the late 1970s the country embarked on a wide-ranging and costly anti-recession policy which included, among other things, heavy subsidization of employment in non-oil activities. The annual bill for these subsidies has typically been 20 billion krone, with agriculture and manufacturing being the dominant beneficiaries of the policy (Table 5.9). One apparently advantageous effect of this policy was to reduce unemployment to remarkably low levels. Between 1984 and 1987, unemployment fell from only 3 per cent to less than 2 per cent. But the very welcome social benefit of low unemployment was in itself an indicator that necessary labour rationalization was being discouraged by the subsidization policy. Productivity failed to rise and low unemployment caused inflationary wage pressures. The result was that Norwegian producers rapidly became less competitive and, therefore, more addicted to government policy support. Figures for the manufacturing sector in the period 1981–90 illustrate this weakness. During this period, manufacturing output fell in four out of ten years, and rose by less than 1 per cent in two others. As was indicated above, despite such problems successive governments have maintained substantial subsidies to the business sector, at rates approaching twice the OECD average. Despite this high level of support, however, economic pressures have recently overcome the power of subsidies to control unemployment. Total unemployment rose from 2 to 5.2 per cent between 1986 and 1990, while youth unemployment soared from 5.0 to 11.8 per cent (OECD 1992).

Table 5.10 Nuclear power in the post-crisis period

	Nuclear power as % of total electricity generation				Nuclear output (million kWh)	
	1974	1980	1985	1989	1985	1989
Belgium	0.4	23.3	60.3	61.1	32 692	39 045
Finland	0.0	17.3	48.0	35.4	17 782	17 976
France	7.8	23.6	64.9	74.5	213 087	288 715
Germany (West)	3.9	11.3	31.1	34.3	119 461	251 310
Italy	2.3	1.2	3.8	0	6 717	60
Netherlands	5.9	6.5	6.1	5.4	3 674	3 785
UK	12.3	13.0	19.3	21.7	53 767	63 602
Sweden	2.7	22.0	44.0*	45.0	50 995	62 708
Switzerland	18.0	28.3	39.6*	40.6	17 396	21 543
Spain	8.9	4.9	22.1	38.6	26 768	53 748
EC	6.4	12.6	30.6	35.6	456 166	590 090
Total Western Europe	5.7	11.6	26.5*	33.5	542 339	692 317

* 1984 data

Sources: United Nations (1976, 1981b); Commission of the European Communities (1987d, 1992e).

A further problem has been that, as the competitive prospects of the non-oil sector have declined, support in the form of orders from the oil industry has increased. The majority of the industry's orders are placed with Norwegian producers; the large metal, machinery and equipment companies are heavily dependent on the oil sector; and shipbuilding is strongly reliant on oil-related business. At first sight this support appears essentially advantageous. But its effect, coupled with stagnation in many non-oil activities, has been to make the oil industry an increasingly dominant growth pole in the Norwegian economy. Given the instability of oil prices, the desirability of this is highly questionable. In the mid-1980s it appeared that falling oil prices would undermine the growth pole, but this has been avoided as a result of industrial orders related to the rapid expansion of oil production since 1988 (Fig. 5.3). However, it is clear that this small nation cannot continue to expand its oil industry indefinitely, and the consequences of a halt to growth would be serious for the Norwegian economy. So far as policy is concerned, therefore, revitalization of the non-oil sector has become an urgent necessity in order to reduce over-reliance on an energy industry that is essentially controlled by global, rather than national, forces.

Nuclear power in the post-crisis era

The nuclear power industry's growth record has been far from insignificant (Table 5.10). France generates three-quarters of her electricity in nuclear power stations; Belgium almost two-thirds; Sweden and Switzerland almost half. Overall, the nuclear power industry has increased its share of European electricity production from 6 to 33 per cent. Placing nuclear power's contribution in a wider context, its availability reduces oil usage by at least a quarter, and nuclear plants generate an eighth of Western Europe's primary energy.

It is necessary to appreciate, however, that much of this growth has resulted from the commissioning of stations which were already under construction or planned when the first oil crisis struck and, as Maull (1980) has stressed, the revised capacity targets set in the immediate post-crisis period have not been met. This is well illustrated by the industry's output figures, which almost doubled between 1981 and 1985, but then grew by only a quarter up to the early 1990s. One reason for the failure to meet construction targets is that the revised programmes adopted in the mid-1970s were wildly optimistic. In the EC, for example, a five-fold capacity increase was proposed by 1980 and a further tripling by 1985. Relatively slow progress also reflects the continuing influence of the technical arguments and delays so familiar in the 1960s; Pearson (1981) has demonstrated the importance of this factor in the British case. In most of Western Europe, however, the argument was not about the merits and demerits of competing nuclear technologies, but about the desirability of any form of nuclear power.

The environmental confrontation

The nature of this sometimes violent debate is well known and need not be examined here (Lucas 1979; Maull 1980; Pilat 1980; OECD 1979a). What is significant is that, while the protest movement's precise impact cannot be established, there is no doubt that it has exerted a major influence on the industry's international growth performance.

In several countries anti-nuclear campaigns have prevented governments from embarking on nuclear programmes or substantially expanding existing toeholds in the industry. Both Denmark and the Republic of Ireland have drawn back from earlier proposals; in Austria, the Zwentendorf station on the Danube stands complete but uncommissioned; it is probable that widespread opposition to the Netherlands' two nuclear stations will at least prevent continuation of the expansion programme (de Vries and Dijk 1985); and Belgium is unlikely to embark on additional nuclear plants. Furthermore, the anti-nuclear movement has achieved considerable success by inflicting delays on the more ambitious programmes of other countries, particularly West Germany and Italy. By 1990, West Germany's nuclear capacity was only a third of the original target for that year, while Italy had only 5 per cent of its planned capacity. These delays have worked in the opposition's favour by forcing higher safety standards and substantial increases in capital costs. In order to speed progress by reassuring the public that fears of nuclear power are unfounded, numerous safety modifications have been made to basic designs. This has involved additional material costs, but the main penalty has been longer construction periods, causing increased labour costs and escalating interest charges on unproductive capital. These cost factors are now so substantial that more than a third of the capital cost of the UK's pressurized water reactor at Sizewell can be attributed to safety measures.

Four other facets of the nuclear debate must be noted. Firstly, despite British emphasis on remote locations, the opposition movement has been strengthened because many nuclear power stations are close to major population centres (Openshaw 1982). For example, in Belgium the Tihange station on the Meuse lies only 20 km upstream and upwind from Liège; Spain's Lemoniz development is adjacent to Bilbao; and in France the siting policy has frequently preferred locations near the load to reduce transmission costs and losses (Lucas 1979). Similarly, the fact that at many sites an accident could lead to transfrontier pollution has been a strengthening factor – dense populations could be threatened by events only kilometres away yet beyond their country's jurisdiction. This argument was powerfully reinforced in 1986, when a major escape of radioactive material from the Chernobyl power station near Kiev spread within days to affect the whole of Western Europe. We shall return to this disaster, and the general problem of transfrontier pollution, in Chapter 6.

Second, although opponents may be in a majority at the local level, this is not invariably the case, and at the national level it is not normally so. The Austrian nuclear programme, it is true, was halted by a referendum: and, after Chernobyl, the Italians closed down their single operational station, while the Swedes voted for eventual decommissioning of existing nuclear plants. But in Switzerland a referendum has rejected abandoning this source of power, and in national elections around Europe anti-nuclear candidates have rarely polled more than 5 per cent of the total vote. The movement is, therefore, essentially a minority force which has chiefly operated outside parliamentary systems yet has exerted considerable influence on the evolution of energy policy.

Third, the opposition has not always been successful, especially in the case of France. Here the attack has been particularly powerful and prolonged, yet there has been little sign of compromise on the basic policy (Lucas 1979, 1985; Collingridge 1984b). It is true that in the early 1980s the programme was scaled down, but this primarily reflected slower demand growth, rather than effectiveness on the part of the environmental lobby. More than forty French nuclear power stations were operating in the early 1990s and, in addition to supplying three-quarters of the nation's electricity, the industry satisfied more than 30 per cent of all French energy consumption. One reason for this relatively smooth progress is that Electricité de France and the monopoly construction firm, Fromatome, have made the building programme a highly standardized affair, but its implementation is dependent upon resolute government support and, indeed, a willingness to use force to put down opposition. Despite recent socialist hesitation, this strong government stance is in turn dependent on the official commitment of all political parties to the transition to nuclear power, and it is assisted by a licensing system which severely restricts opportunities for effective public participation in the decision-making process. Moreover, as the nuclear construction industry has grown, its contribution to the economy as a whole has become widely appreciated – all the major trade unions have joined the political parties in stressing the importance of an industry employing more than 125 000 people.

Fourth, while the nuclear debate is primarily conducted at the national level, it also has an important international dimension. The most widely recognized aspect of this is the international protest movement, and the fact that there has been supranational pressure for its development – chiefly from the European Commission – has been largely overlooked. In practical terms, the EIB has been the primary vehicle for implementing the Commission's policy of promoting nuclear power. As Table 5.7 reveals, between 1982 and 1986 the construction of nuclear power stations absorbed more than half the Bank's energy loans, the principal beneficiary being the UK. This impressive investment programme was pursued without either the Commission or the Bank coming under heavy attack from the protest movement, most probably because their policies were operated through national governments. Thus neither of these powerful organizations was in the front line.

Since the mid-1980s, however, the stance adopted by the Commission and the Bank on nuclear power has changed substantially. Initially, the Commission argued that 'The pursuit of vigorous [nuclear power] programmes is an essential element in an economic policy for Europe,' but this attitude was later tempered to the view that nuclear power development 'is clearly only possible if the public is reassured as regards safety' (Commission of the European Communities 1981a, 1987a). Most recently, EC documents have placed almost no emphasis on nuclear power (Commission of the European Communities 1992a). Similarly, EIB loans to the industry have fallen sharply (Table 5.7) and references to nuclear power are virtually absent from documents such as annual reports. This quiet revolution undoubtedly reflects the influence of the 1986 Chernobyl disaster which, in much of Western Europe, immediately called into question the feasibility of gaining acceptance for new nuclear power stations. Evidence that this is the case comes from the fact that recent years have also witnessed the Commission's rapidly growing concern for the impact of fossil fuels on the environment. Nuclear power has the capacity to lessen that impact by reducing the need for fossil fuel consumption. As we have seen, the industry already reduces European oil consumption by a quarter. But large-scale nuclear programmes are no longer pursued because it is not believed that the public can be persuaded to accept that the immediate environmental gains outweigh the long-term risk of accidents.

Coal – a glowing future?

One popular assumption in the immediate post-crisis period was that the new circumstances heralded the resurrection of the coal industry, but attitudes varied in the industry itself. In France the wave of optimism found little support, the programme which emerged being one of continued contraction, admittedly at a slower rate, to a production plateau of 10 million tonnes in the late 1980s. Similarly, Belgium saw no scope for re-expansion, but adopted the goal of halting the production decline at the 7 million tonne level. Greece and Spain, on the other hand, embarked almost

immediately on more intensive exploitation of lignite and low-grade coal deposits, while Western Europe's major coal producers also adopted optimistic scenarios for the remainder of the century. West Germany planned to halt contraction and then re-expand capacity by up to a quarter, while the British coal industry proposed an ambitious two-phase investment programme to increase output by between 25 and 50 per cent (Greene and Gallagher 1980; Manners 1981; North and Spooner 1977, 1978; Spooner 1981). Initially, little difficulty was experienced in expanding lignite production, but problems arose in West Germany in the 1980s as demand for this inefficient and environmentally unfriendly fuel declined. Failure to develop a growing market then spread to other countries, so that by the early 1990s only Greece was producing substantially more than in the mid-1980s (Table 5.11). This fuel's ability to substitute for oil has therefore proved to be modest: by 1991, production in 'traditional' Western Europe was reducing oil consumption by no more than 60 million tonnes. Here, however, we must note that the reunification of Germany has more than trebled West Germany's lignite production capacity and has more than doubled that of Western Europe as a whole. Thus total lignite output in the 'new' Western Europe in 1991 was equivalent to nearly 120 million tonnes of oil, or almost a fifth of actual oil consumption. Whether lignite will be able to maintain this position through the 1990s is, however, doubtful. As has been indicated, it is no longer gaining ground in most producer countries and, although precise figures are not yet available, there is evidence to suggest that economic problems caused production in the former East Germany to fall by a third in 1990–91.

So far as hard coal is concerned, consumption in the 1980s has fluctuated, and this at first sight suggests that producers have at last stemmed the tide of decline that characterized the pre-1973 era (Table 5.12). As Table 5.11 demonstrates, however, the end of dramatically declining demand has not led to stabilization of the supply side of the industry. Since 1985 coal output has continued to decrease in all producer countries except Spain and the UK (where the data are distorted by the effects of the 1984/85 miners' strike). In Western Europe as a whole, output contracted by a fifth in the 1980s. The reality is, therefore, that more stable demand has benefited low-cost overseas producers rather than Western Europe's indigenous coal industry. Imports have fluctuated in sympathy with consumption and have doubled since the early 1970s (Table 5.12). Before the first oil crisis, imports supplied less than a fifth of the coal burned; by the 1990s the proportion was approaching 40 per cent.

Table 5.11   Post-1973 lignite and coal production trends

Lignite	Production (million tonnes)				Change (%)		
	1973	1980	1985	1991	1973–80	1980–85	1985–91
Austria	3.6	2.9	3.1	2.4	− 19	+ 7	−23
France	2.8	2.6	1.8	0.4	− 7	−31	−78
Germany (West)	118.7	129.9	120.7	110.0*	+ 9	− 7	− 9
Greece	13.3	23.2	35.9	53.2	+ 74	+55	+48
Italy	1.2	1.4	1.9	1.0	+ 17	+36	−47
Spain	3.0	15.5	23.6	19.7	+417	+52	−17
Total	142.6	175.5	187.0	186.7*†	23	+ 7	0

Coal	Production (million tonnes)				Change (%)		
	1973	1980	1985	1989	1973–80	1980–85	1985–89
UK	130.2	128.2	90.8	98.3	− 2	−29‡	+ 8§
Germany (West)	103.7	94.5	88.8	77.5	− 9	− 6	−13
France	25.7	18.1	15.1	11.5	−30	− 17	−24
Belgium	8.8	6.3	6.2	1.9	−28	− 2	−69
Spain	10.0	12.6	16.1	19.2	+26	+28	+19
Total	278.4	259.7	217.0	202.4	− 7	−16	− 7

* Estimate.
† Former East Germany 169.0.
‡Output affected by industrial dispute.
§ Recovery from industrial dispute.
Sources: Commission of the European Communities (1982b); OECD (1987d); British Petroleum (1992).

Bleak though the picture is for the indigenous coal industry, the situation would be much worse but for the fact that one branch of the coal market – electricity generation – has been protected as governments have renewed earlier efforts to reduce the power industry's oil consumption (Schaller and Motamen 1985). In the UK, for example, the oil-fired power stations were loaded very lightly after the 1979/80 oil price rise until a prolonged miners' strike temporarily forced them into full production. Welcome though the growth of power station demand has been, however, the consequence is that 80 per cent of Western Europe's indigenous coal industry is now reliant on a single market. Moreover, for four reasons, the future of this market is uncertain. Firstly, the cost of fuel oil – the main conventional fuel which competes with coal for electricity generation – has fallen substantially in the wake of the 1985/86 oil price crash. Secondly – as has been shown above – many nuclear power stations were commissioned in the 1980s, and others are due for completion. If there is surplus capacity, these can be operated at the expense of coal-fired plant. Thirdly, although the pace of nuclear power expansion has lessened in the aftermath of Chernobyl, it is by no means certain that substitute power stations will burn indigenous coal. For example, natural gas is increasingly seen as an eco-

nomic and environmentally friendly fuel for electricity generation (Odell 1990). Fourthly, political willingness to maintain the indigenous coal industry is now far less firm than in the 1970s and early 1980s. Nowhere is this more evident than in the UK, where Western Europe's largest coal industry is due to be privatized by 1994, and where the already privatized electricity generating industry has been freed from the obligation to buy British coal.

These fundamental changes will lead to further sharp contraction in the British coal industry in the 1990s, although the actual pace of the decline is as yet unclear. Late in 1992, in response to clear evidence that the electricity producers were moving towards increased use of imported coal and gas-fired power stations, British Coal announced the closure of 31 of its 50 pits. Ten were to close at once, and the remainder within a matter of months, reducing the total workforce from 44 000 to 14 000.

Compared to earlier announcements of colliery closures, public reaction to this precipitate rundown was dramatic, forcing two immediate reviews of the future market for coal – one to be conducted by a parliamentary Select Committee and the other by the industry itself. In addition, British Coal's unilateral announcement of the closures was declared illegal by the courts

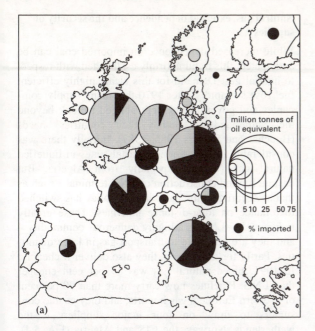

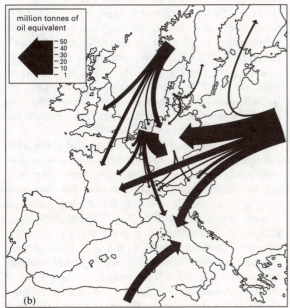

Fig. 5.4
Natural gas, 1991: (a) consumption and percentage imported (b) exports

Table 5.12 Post-1973 coal consumption and import trends

	1973	1980	1985	1989
Consumption (million tonnes)				
European Community	324.4	330.0	319.1	313.7
Western Europe	330.0	338.1	328.6	334.3
Imports (million tonnes)				
European Community	52.8	99.0	110.7	110.7
Western Europe	58.2	106.7	121.0	128.4
Imports as % of consumption				
European Community	16	30	35	35
Western Europe	18	32	37	34

Source: OECD (1987d).

unless there is a fundamental change in the principle that the privatized electricity producers should be able to use the energy sources they prefer, in the long run it seems unlikely that the storm of protest which followed the initial closure announcement will save a substantial number of pits. Now that electricity generation is privatized, the imposition of strategic constraints on the industry, of the type which would be needed to sustain British coal production, has become extremely difficult. Moreover, if an attempt were made to introduce such constraints, it is highly likely that the EC would intervene on the grounds that government policy would obstruct free competition. In both Spain and Germany the EC is scrutinizing policies designed to support the coal industry by ensuring its market among electricity generators.

Coal for Europe or European coal?

Although the European coal industry's prospects have not been revolutionized, some sections of it hope to recapture lost ground as a result of world-wide efforts to convince consumers that the fuel's reputation for being dirty, inefficient and inconvenient is false. These efforts entail extensive research into new consumption technologies, coupled with an aggressive attitude to advertising coal's attractions. Some lines of investigation, such as fluidized-bed combustion, which is likely to improve efficiency and reduce pollution, are highly promising and should eventually appeal to many industrial users as well as the electricity industry. Others, including coal gasification and liquefaction, are in relatively early development stages and, if they are economically viable at all, are unlikely to exert a substantial influence on the market before the end of this century. Here it may be noted that the return to lower oil prices has handicapped research and

because it ignored an agreed review procedure which required the industry's management to consult widely on proposals to close pits. However, this illegality can be made good by observing the correct procedure and, development of gasification and liquefaction processes because consumers' potential savings from oil substitution have been greatly reduced.

Even if these research and marketing efforts prove highly successful, European producers will still be faced with the stark reality of price competition (Gordon 1987). The UK coal industry has become profitable but cost analyses produced by the EC show that the imported product can now be marketed in European ports for no more than half or three-quarters of the cost of most European coal. Given this continuing large price differential, it is widely accepted that a high proportion of any demand increase which occurs up to the end of the century will be met by overseas suppliers. This has long been the view of the Commission of the European Community (1980), which has predicted that imports will satisfy 80 per cent of additional demand by the year 2000. What must be emphasized, however, is that existing – as well as new – demand is a potential target for imported coal. The evidence in Tables 5.11 and 5.12 is that indigenous coal producers are poorly placed to defend their current markets. Although the growth of coal imports is not as yet dramatic (Pinder 1992), all the signs are that output from most producer countries will continue to decline, to the benefit of the imported product.

A final point is that, because the industry is usually far from benign at the point of production, the future balance between European production and imports may be influenced by environmental, as well as cost, considerations. Opencast mining destroys landscapes, flora and fauna; and it may also cause social dislocation, as in the Rhenish brown coalfield, where whole villages have been compulsorily relocated. Deep mining is less devastating, yet totally new collieries almost always entail the colonization of previously rural areas. In the past the impact of mining on the countryside was usually accepted, but surface workings, colliery spoil, dust emission, industrial traffic and incoming population are now commonly regarded as threats to the environment and the local community (Department of the Environment 1981a). Against this background it is increasingly argued that future demand estimates are insufficiently reliable to justify new large-scale developments when, if required, coal could be purchased from stable and politically sympathetic countries. This was the issue raised during the public inquiry into major new colliery proposals for the UK's Vale of Belvoir and, although the opposition's victory was not total, it is clear that the industry failed to establish that national benefits would decisively outweigh local environmental and social costs (Department of the Environment 1981b; Manners 1981).

Natural gas, external dependence and the security of supplies

While increased dependence on imported coal can be anticipated, this trend is firmly established with respect to natural gas. Demand for this clean, highly efficient fuel rose so rapidly after 1970 that many supply companies and governments were obliged to look beyond the Netherlands to satisfy their countries' needs (Boucher and Smeers 1985, 1987). Initially there was considerable interest among coastal states in liquefied natural gas (LNG), especially from Algeria. But despite limited construction of LNG terminals, such as that at Zeebrugge, the basic import policy has switched to reliance on long-distance pipelines. These greatly reduce risk – LNG is a highly dangerous commodity – and they are now able to transport gas in large quantities. Partly for this reason, they also deliver it cheaply. In 1991, piped natural gas was 17 per cent cheaper than LNG. Pipelines now carry more than 90 per cent of Western Europe's gas imports, and they enable consumers to draw on four major suppliers – the Netherlands, Norway, the CIS and Algeria (Fig. 5.4). Of these, the most important single source is the CIS which, despite political upheaval and poor pipeline maintenance, supplied 44 per cent of Western Europe's gas imports in 1991. Maintaining these exports is, of course, a priority for the CIS, which has an urgent need for hard Western currency. In contrast, the Netherlands chose to reduce exports in the 1980s, not least for conservation reasons. Thus Dutch exports fell by 31 per cent between 1985 and 1988.

Gas usage has been actively encouraged at the international level on the grounds that it will substantially reduce oil demand and will therefore benefit the security of energy supplies. Imports from the four main suppliers account for well over half the gas consumed in most importing countries (Fig. 5.4) and, in Western Europe as a whole, gas imports are equivalent to almost 130 million tonnes of oil a year. Without these supplies, oil imports would therefore be substantially higher. The positive official attitude to the growth of the international natural gas trade is again seen most clearly in the stance of the European Commission and in the activities of the EIB. Between 1973 and 1981 the Bank loaned almost 1600 million ECU for Community oil and gas development projects, almost half of this being for import diversification through projects such as the Trans-Austria (TAG) link with the USSR and, most important of all, the Algeria–Italy pipeline. In the period 1984–86 the Bank invested 1400 million ECU in oil and gas development and – although the gas supply network was much nearer completion than in the 1970s – the proportion allocated to pipelines was still almost a third. Since the mid-1980s the Bank has placed more emphasis on the development of indigenous oil and gas resources, yet in the early 1990s almost a quarter of its investment in these forms of energy was for projects aiming to improve access to imported gas (EIB Annual Reports 1980, 1982, 1987, 1991).

In the 1980s this strategy caused little controversy in Western Europe but highly adverse reaction in the USA because of the relationship between energy planning and geopolitics (Adamson 1985; Manne, Roland and Stephan 1986). The American view was that because almost half the gas came from potentially hostile sources – the USSR and Algeria – there was an unnecessary threat to security. With the demise of Communism this scenario has changed, but it may still be argued that Western Europe faces continuing uncertainties. These relate chiefly to the possibility that supplies from the CIS could still be disrupted, either by internal political conflict or by breakdowns in the poorly maintained pipeline system. Given these dangers, it is important to evaluate the threats posed and how supply disruption might be combated.

Several factors point to the conclusion that the apparent threat to the security of supplies is exaggerated. One powerful disincentive for the CIS to halt exports is that gas sales provide badly needed hard currency for the former Soviet economy. Also, although the popularity of this fuel has rapidly increased in Western Europe, its share of total energy consumption is still limited. Except for the Netherlands, no country is more than 30 per cent reliant on gas, and in many cases the proportion is 20 per cent or less. A third point is that Western Europe's internal pipeline system is now so highly developed that in times of crisis indigenous gas could rapidly be substituted for curtailed imports from the CIS or North Africa. And, finally, Norway and the Netherlands have acknowledged their ability to guarantee European supplies in times of crisis. It is highly unlikely that in an emergency these countries would not come to the assistance of their neighbours, and indigenous gas resources can therefore be seen as an international asset providing much-needed insurance against import disruption in this major branch of the energy sector.

Retrospect and prospect

While the geography of energy demand has evolved at a gradually decelerating pace, the geography of supply has changed beyond recognition through market forces and, to a degree, the planning process. From a state of heavy self-reliance in the early post-war period, Western Europe became fully integrated into the world energy system in the 1960s. At the same time a num-

ber of strategies for indigenous resource development were adopted, but it required the 1973 and 1979 oil crises to achieve general acknowledgement of the need to reshape the geography of supply to improve national and – to a lesser extent – international security.

Impressive progress in this direction has been made, firstly, through a sharp reduction in oil imports and Middle Eastern dependence (Table 5.6) and, secondly, through the compensating development of Western Europe's oil and gas resources (Fig. 5.3). But these departures should not be allowed to create the impression that the days of external dependence are over. Instead, this phenomenon is assuming a modified form through large-scale purchases of natural gas and, to a lesser extent, coal imports. As has been indicated, it is arguable that these imports are much less strategically dangerous for Western Europe than was the earlier dominance by the Middle East. Coal is readily available from politically friendly countries and European gas reserves provide important insurance against the interruption of supplies from either the CIS or Algeria. Yet it is still the case that, in Western Europe in general, the energy gap is better described as a gulf. Despite higher output, energy production in 1991 was equivalent to only 58 per cent of consumption (Table 5.13). Moreover, looking to the future, there is little to suggest that the gap will be narrowed substantially (Guilmot et al. 1986). Indigenous coal output is likely to fall, while North Sea output is officially expected to dwindle rapidly in the early decades of the next century. If the North Sea and adjacent areas follow the pattern set by petroliferous areas in other parts of the world, where initial resource predictions have proved to be substantial underestimates, the timing of this decline will be delayed. But its eventual onset will re-emphasize the significance of the energy gap.

This highlights the potential importance of new, renewable energy sources, which in the years following the oil crises were expected to reduce significantly Western Europe's external energy dependence. Despite this optimism, renewable energy industries have gained very little impetus, to the extent that they have not justified separate consideration in this chapter. For example, the total contribution of wind, solar, geothermal and tidal energy to total electricity generation is less than 0.2 per cent and is rising extremely slowly. Even in Denmark, where emphasis is placed on energy conservation and the exploitation of alternative and renewable resources, these forms of power account for no more than 0.6 per cent of the electricity consumed. Only in countries unusually endowed with conditions favourable for geothermal energy does the proportion rise significantly higher than this, to 1 per cent in Italy and 4.8 per cent in Iceland. This failure partly reflects

Table 5.13 Western Europe's energy, 1980–91

	Production		Consumption		Balance	
	(1980)	(1991)*	(1980)	(1991)*	(1980)	(1991)*
Total (mtoe)	545	790	1072	1357	−527	−567

	1980	1991	Change
Consumption by			
energy type (%): Solid fuels	25	18	−7
Liquid fuels	52	46	−6
Gases	18	21	+3
Primary electricity	5	15	+10
	100	100	

	1991 (% of total oil imports)
Sources of imported crude oil:	
USA and Canada	3
Latin America	7
Middle East	38
North Africa	21
West Africa	11
Former USSR	14
Others	6
	100

* 1991 data include the former East Germany.
Sources: OECD (1987d); British Petroleum (1992).

the strength of well-entrenched lobbies in the energy sector. The nuclear power industry, for example, established itself firmly as the chief recipient of investment in alternative energy sources in the 1950s and 1960s, making it difficult for fledgling competitors to establish themselves in later years. But slow progress is also a consequence of the long lead times required to develop new energy sources, and of the fact that energy price rises caused by the oil crises have not been sustained. The fact that energy has not remained expensive has significantly reduced the pressure to accelerate the exploitation of new energy sources.

If a significant increase in the development and adoption of alternative energy technologies is to be achieved, it may well be that the main driving force will prove to be the European Commission. Since the Chernobyl disaster, the reduction in the Commission's commitment to nuclear power has been balanced by a new emphasis on technological progress in the energy sector. This has rapidly gained ground as realization of the environmental consequences of energy consumption has grown. The broad concepts of the Commission's new stance are set out in its European Energy Charter (de Bauw 1992) and practical progress is being sought through initiatives such as the *Thermie*

programme. Under the fourth round of this programme, launched in 1992, the Commission is promoting technologies leading to the wider use of renewable energy resources, improved efficiency in the use of all energy sources, the cleaner use of solid fuel and optimum exploitation of oil and gas resources (Commission of the European Communities 1992a). Moreover, while support is naturally being provided for innovatory projects, funding is also being made available to disseminate the progress that has been made by completed research. Even though this programme is potentially of great significance, however, the scope for rapid change should not be overestimated. Projects receiving Commission support will mature over a substantial timespan, and it will be the next century before many effects are felt.

With respect to energy planning, as much of the discussion in this chapter shows, the record is littered with examples of governments and agencies struggling, and often failing, to meet goals. Evidence underlining the difficulties of guiding this vital, yet volatile, sector is abundant. To a great extent, this reflects the continuing power of oil prices, the control of which lies firmly outside Europe. Inexperience of managing new energy sources has also been important, as the Norwegian

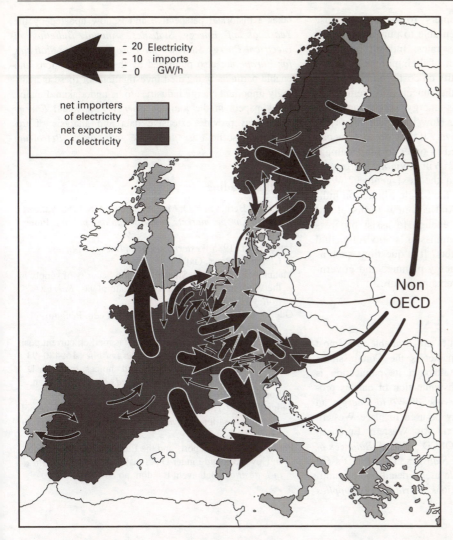

20 Electricity
10 imports
0 GW/h

net importers
of electricity

net exporters
of electricity

Non
OECD

Fig. 5.5
International electricity
exchanges, 1989.

energy saga graphically demonstrates. But two other neglected factors that have increased the problems of planning are the political fragmentation of Western Europe and the multi-faceted character of the energy sector itself. Political fragmentation has militated against international coordination, even in the EC. Programmes such as *Thermie* are therefore designed to stimulate change at the national level, and in no sense do they amount to supranational plans. This has left the planning process in the hands of a multiplicity of agencies, many of which have been responsible for relatively narrow sections of the energy sector within individual countries. Although these agencies are increasingly subject to the overall control of government departments, each has its own viewpoint and interests to protect, with the result that their conflicting claims inevitably impede progress towards rational planning.

Once again, however, change in this context may

come about as a result of EC policy. The completion of the Single Market in January 1993 applies to energy as much as to any other sector, and it must be anticipated that national energy industries will be exposed to considerably greater international investment and competition in coming years (Odell 1990; Padgett 1992). This may well bring to the energy sector a significantly higher level of efficiency than exists today. The potential benefits of energy sector integration are demonstrated by the international electricity market which is already functioning. In Western Europe as a whole, and not simply the EC, linkages between the various national grid systems allow sales to be made in neighbouring countries on a significant scale. Italy, for example, imported 14 per cent of her electricity requirements in 1989, while France exported 12 per cent of her output (Fig. 5.5). In this way, the efficiency of Western Europe's power industry is markedly improved, allowing consumers to benefit because importing countries are often able to purchase surplus production at advantageous rates. Indeed, importers frequently adjust the pattern of their purchases on an hour-by-hour basis, essentially on cost grounds.

Finally, we must note that uncertainty remains a central feature of Europe's energy sector, and that major questions are legion. How long will oil prices remain at a relatively low level? If they do so, how can energy demand be contained in order to limit damaging environmental impacts? Will a viable indigenous coal industry be feasible in the next century? What will be the precise effect of the Single Market on individual energy industries? And can new forms of energy be developed rapidly enough to assist progress towards the European Commission's environmental and energy-dependence targets? It is questions such as these which confront the energy planners and governments of today. The future remains unsure.

Note

Many series of energy statistics are available. The most useful for the EC energy market is the Commission's *Energy Statistics Yearbook*, although its publication is inclined to be slow. Rapid dissemination of data is provided by BP's *Statistical Review of World Energy*, an annual publication which may be used to place Western Europe in its world context. All Western European countries are covered by the OECD's *Energy Balances of OECD Countries* and *Energy Statistics*. The OECD's *Oil and Gas Statistics* also provide useful detail on consumption and trade. The United Nations' *World Energy Supplies* takes a historical perspective and is now updated by the *Yearbook of Energy Statistics*. *Annual Bulletins of Electrical Energy Statistics for Europe* and *Gas Statistics for Europe* are also published by the United Nations and enable particularly detailed investigations of these increasingly important energy industries to be undertaken. Finally, many papers in the journals *Energy Policy* and *Energy Economics* provide extensive data and analysis of the European scene, as does the periodical *Petroleum Economist*.

Further reading

Estrada J, Bergesen H O, Moe A, Sydnes A K 1988 *Natural Gas in Europe: markets, organisation and politics*. Pinter, London

Lucas N J D 1985 *Western European Energy Policies*. Clarendon Press, Oxford

Manne A S, Roland K, Stephan G 1986 Security of supply in the Western European market for natural gas. *Energy Policy* **14**: 52-64

Odell P R 1986 *Oil and World Power* 8th edn. Penguin, Harmondsworth

Odell P R 1988 The West European gas market: current position and alternative prospects. *Energy Policy* **16**: 480–93

Odell P R 1990 Energy: resources and choices. In Pinder D A (ed.) *Western Europe: challenge and change*. Belhaven, London, pp 19–36

Padgett S 1992 The Single European Energy Market: the politics of realisation. *Journal of Common Market Studies* **30**: 53–76

Pinder D A 1992 Seaports and the European energy system. In Hoyle B S and Pinder D A (eds) *European Port Cities in Transition*. Belhaven, London, pp 20–39

6

Industry

Long-term development

It is increasingly easy to overlook the contribution of the post-war industrial upswing to the present-day scale, structure and geography of Western Europe's industrial systems. By the mid-1950s the recovery from wartime devastation was complete, and in most countries the growth process was clearly established. Between 1955 and 1963 output in Western Europe as a whole rose by 54 per cent, and on the eve of the oil crisis production was virtually two and a half times greater than in the mid-1950s. Moreover, in most countries the growth curve after the early 1950s was remarkably smooth, the most marked hesitations prior to the first oil crisis being episodes of relatively slow growth in the late 1950s, around 1967 and in the early 1970s.

Naturally enough, individual industries contributed to this expansionary impulse in varying degrees (Table 6.1). The most typical growth industries were, however, those which roughly doubled their production over the fifteen-year period. These included a broad cross-section of the manufacturing sector: from metal products (a catch-all category including many producer and consumer goods) to furniture manufacturing, paper products and publishing

Manufacturing's performance as a whole was far better than many governments – still haunted by the spectre of the 1930s Depression – had dared to anticipate. So great was the upward thrust that, at least in central and northern parts of Western Europe, labour shortages began to afflict the manufacturing sector, even though population growth and the contraction of agriculture simultaneously increased labour supplies. It is true that these shortages were also a reflection of the rapid expansion of the service sector, but the latter's progress was heavily dependent on industry as a creator of wealth and, therefore, of demand.

One important indicator of the strength of labour

Table 6.1 Industrial output, 1960–75

	Growth index, 1974 (1960 = 100)	Share of Western European output, 1975 (%)
Chemicals, petroleum products, rubber and coal products	325	14.8
Wood, furniture, etc.	215	4.4
Metal products	207	38.2
Non-metalllic mineral products	203	4.6
Paper, printing and publishing	192	6.8
Basic metals	177	7.6
Food, drink and tobacco products	177	12.9
Clothing industries	139	4.5
Textiles	138	5.1
Other industries	203	1.1

Sources: United Nations (1977, 1981a).

demand in the more successful economies was the attraction of migrant workers in large numbers to satisfy both the manufacturing and service sectors. This process operated at a variety of scales, its earliest expression being the flow of Italian workers into the West German, French, Belgian and Swiss economies (King 1976). But as Italy expanded economically and Italian labour became less readily available, other countries on the edge of and beyond Western Europe's southern periphery were quickly integrated into the international labour supply system. 'Guestworkers' from Spain, Portugal, Greece, Algeria, Morocco, Yugoslavia and Turkey flowed into the main industrial economies, except the UK, in impressive numbers (see Ch. 4).

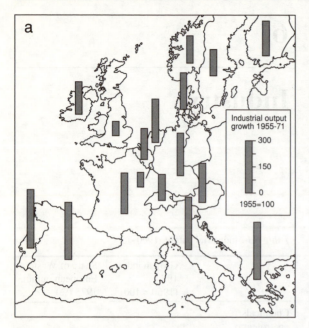

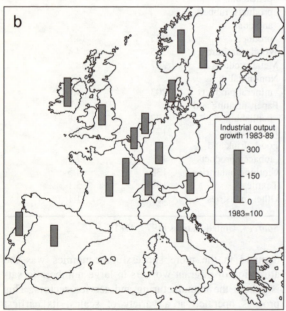

Fig. 6.1
Industrial output trends, 1955–71 and 1983–89.

Despite the polarization of labour from south to north, the most important differences in national manufacturing growth rates before the mid-1970s were not simply those distinguishing an advantaged core from a disadvantaged southern periphery. As Fig. 6.1 reveals, in reality the distribution of growth was much more complex and simultaneously resulted in divergence from and convergence towards a more even industrial development 'surface'. So far as divergence was concerned, one major feature of the geography of industrial expansion was the average and below-average performance of the economies in the northern periphery. But divergence was also increased because countries in Western Europe's industrial core recorded growth rates that were far from identical. While substantial successes were achieved in the Netherlands, West Germany and France – where growth was significantly higher than the norm – Belgium's performance was not outstanding, and in the UK and Luxembourg industrial output rose by little more than 50 per cent. Convergence, on the other hand, was brought about by unusually rapid expansion in the southern periphery as Portugal, Spain, Greece and Italy all more than trebled their industrial production. In absolute terms this trend was most significant in Italy and Spain, which came to rank fourth and fifth in Europe's manufacturing league table and account for 85 per cent of the southern periphery's industrial employment. Impressive though it appears, however, the relative advance of the Mediterranean zone must be placed in perspective. As we have seen, at least in Portugal, Spain and Greece, industrial expansion was insufficient to stem the flow of guestworkers to stronger European economies. Moreover, in the mid-1970s seven out of ten industrial workers were still to be found in Western Europe's traditional industrial core, compared with one-fifth in the southern periphery. Indeed, although subsequent industrial restructuring has severely affected the industrial core countries, by the early 1990s they still accounted for almost 60 per cent of all industrial employment (Table 6.2). In addition, recent data reveal that industrial growth rates in these core countries have not been consistently overshadowed by those in the southern periphery (Fig. 6.1b).

Multinational firms and Western Europe's industrial system

Many factors contributed to the buoyancy of industrialization before the oil crisis but, despite intensive investigation, their individual significance remains a matter of contention (Aldcroft 1978; Boltho 1982; Postan 1977). One factor which proved particularly interesting for the geographer, however, was the speed with which multinational firms developed and penetrated Western Europe's industrial system (Table 6.3).

Although multinationals grew most rapidly in the 1960s, the foundations were laid decades earlier. By the inter-war period, two basic processes for multinational development were taking root in Western

Table 6.2 Manufacturing employment in Western Europe's core and periphery (%)

	1975	1989
Southern periphery (Italy, Spain, Portugal, Greece)	19.9	28.3
Industrial core	68.9	58.9
of which: Netherlands, West Germany, France	*43.6*	*40.1*
Belgium, Luxembourg, UK	*25.3*	*18.9*
North-western periphery (Finland, Norway, Sweden, Denmark, Ireland)	6.4	7.6
Other countries (Austria, Switzerland)	4.8	5.2
	100	100

Source: International Labour Office (1989).

Table 6.3 The distribution of foreign investment in the EC (%)

	1970–79	1980–89
Belgium	9.4	8.6
France	16.4	15.3
Germany (West)	16.7	6.2
Italy	6.9	7.7
Netherlands	8.2	8.4
Spain	4.3	8.3
UK	31.9	40.0
Other EC	6.1	5.5
Total	100	100

Source: Commission of the European Communities (1992c)

Europe. The first involved the transatlantic transfer of capital by US companies; largely because of language and cultural affinities, this investment flow was strongly biased towards Britain, although this predisposition must not be exaggerated. In the car industry, for example, Ford established factories at Manchester (1911), Berlin (1925), Dagenham (1931) and Cologne (1932). Similarly, General Motors' acquisition of the British Vauxhall company in 1925 was followed in 1928 by the takeover of the German firm Opel. The second process entailed the growth of indigenous European firms to multinational status (Hartner and Jones 1986). The Philips' combine provides an outstanding early example. By the 1920s this firm had developed an extensive sales network in Europe but continued to confine production to the Netherlands. Then, to combat the trade barriers erected as part of the Depression's rising tide of protectionism, the firm's basic strategy switched to production in the major market areas. This reorientation was so successful that, by 1939, 60 per cent of the labour force worked outside the Netherlands (Pinder 1976).

Both processes blossomed in the post-war period, the rate of penetration being fastest in the oligopolistic basic and final-consumption industries, such as oil and petrochemicals, electronics, cars and household durables. However, incoming investment was distributed far from evenly geographically. Generalizing broadly, the indications are that foreign-controlled firms accounted for between 20 and 30 per cent of industrial production in all the leading European economies by the mid-1970s, and this proportion appears to have changed little in subsequent years (Table 6.4). In contrast, the degree of penetration achieved in many smaller countries, especially in the periphery, was 10 per cent or less. Allowance must of course be made for aberrant cases: in the periphery, the Republic of Ireland pursued a very effective policy to attract foreign investment. Yet the overall pattern indicates that the rise of the multinationals intensified, rather than ameliorated, core–periphery contrasts at the Western European scale. While foreign investment was an important source of growth in the Western European industrial sector, therefore, its overall distribution was arguably not ideal.

Since the mid-1970s, Western Europe's share of multinational investment in the world as a whole has fallen from 40 per cent to less than a third as new markets have become attractive and new opportunities have developed. In addition, manufacturing has lost its near monopoly on multinational investment as large sums have been invested in the service sector (Table 6.5). Even so, Europe must still be considered an important target area for investment by multinational manufacturing companies. The United States has remained a leading source of incoming capital, but Japanese firms are rapidly expanding their European investment and smaller capital flows are coming from many other sources (Commission of the European Communities 1992c; Dicken 1990; de Smidt 1992). In the case of EC countries, one important factor sustaining these flows has been the development of the Community's Single Market, which has been officially operational since January 1993. Perceiving that an EC location will offer access to all EC consumers, and fearing that trade restrictions might ultimately be imposed to impede imports to the Community, many incoming investors have viewed the Single Market as both an opportunity to be seized and a threat to be

Table 6.4 Share of foreign-owned multinationals in economies of major EC countries, mid-1980s (%)

	West Germany	France	UK
Sales	19	27	20
Value added	N/A	24	19
Employment	8	20	14
Exports	24	32	30

Source: Commission of the European Communities (1992c).

Table 6.5 Manufacturing's share of total foreign investment in selected EC countries, 1989

	Total foreign investment (million ECU)	% in manufacturing
Denmark	984	16
France*	6 083	45
Netherlands	5 621	55
Portugal*	555	33
Spain	9 564	40
UK*	12 852	43

* 1988.

Source: Commission of the European Communities (1992c).

minimized. As Fig. 6.2 reveals, foreign investment in the Community quadrupled in the later 1980s as the Single Market movement gained momentum, and growth was impressive in almost all Community countries.

In addition, recent years have witnessed the rise of significant pressures which have encouraged multinational investment for different reasons. In the uncertain economic conditions that have existed since the oil crises, many multinational companies have closed factories in order to remain profitable, but it has also been common for them to embark on new investment in order to ensure that their surviving plants are as competitive as possible. As might be imagined, this process has often been linked with further automation so that, increasingly, capital has been substituted for labour. An outstanding example of multinationals' use of investment to protect themselves from the pressures of the market is provided by Western Europe's oil refining industry, which is examined later in this chapter.

Secondly, in some industries the strength of global competition and the cost of developing new products has forced Western European companies that were not originally multinationals to collaborate in new multinational ventures. The outstanding example of this is to be found in the aircraft industry, in which collaboration between British and French firms originated in response to the cost of developing the world's first supersonic airliner, Concorde. This initiative was followed by the creation of the Airbus consortium, a grouping of British Aerospace (UK), MBB (Germany), Aérospatiale (France) and CASA (Spain) (Watts 1990). The chief motivation for the creation of the consortium was the difficulty any individual Western European aircraft manufacturer would have in developing and marketing an entirely new airliner in the face of overwhelming competition from the USA in general and the Boeing company in particular. The resulting Airbus is now assembled in Toulouse by the consortium's holding company, Airbus Industrie. But the major individual components – wings, fuselages, etc. – are manufactured by the various national companies and are transported to Toulouse when complete.

Although the Airbus venture has demonstrated that a collaborative model of multinational production is possible, several words of caution are appropriate. First, it may be difficult to establish effective logistics for collaborative ventures. For example, if highly complex products are to be manufactured, language barriers must be fully overcome. Second, leading companies in major industries are typically highly independent, and may be backed by governments which see strategic advantages in maintaining their independence. This may be one reason why, for example, Europe's major national car manufacturers have shown little inclination for international cooperation. The pressures in the vehicle industry are as yet insufficient to make collaboration more attractive than independence. It must also be recognized that the terms of collaboration must be satisfactorily negotiated between the partners. Each company must be satisfied that the deal agreed is in its interests, and this must be ensured throughout the life of the consortium. The need for constant monitoring of the terms of collaboration was, in fact, highlighted in 1992, when Germany announced it was to withdraw from a second aircraft consortium engaged in the development of a European fighter aircraft for NATO. In this instance the consortium involved Germany, the UK, Spain and Italy, and Germany's particular problem was the new economic environment created by the country's recent reunification. Faced with the unexpected costs of reunification, and rising development costs of the fighter aircraft, the German authorities opted to economize by withdrawing support for Germany's share of the consortium, thus placing the project as a whole in danger. Given difficulties such as these, it is perhaps not surprising that multinational industrial collaboration in Western Europe is as yet in its infancy.

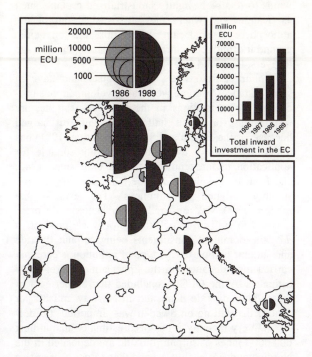

Fig. 6.2
Inward investment in EC countries by multinational companies,
1986 and 1989.

Multinationals – self-interested parasites?

Although multinational companies have been major forces contributing to Western Europe's economic development, they have also provoked controversy which centres on penalties which, it is argued, are imposed on the countries and localities in which international corporations choose to operate. Criticism of even relatively restricted foreign investment flows began to be voiced before the present recession struck, and concern has intensified as economic conditions have deteriorated (Bassett 1984; Blackbourn 1982; Dicken 1992a; Dokopolou 1986a, b; Hayter 1982; McDermott 1979). In essence, it is argued that multinational capital creates 'branch-plant' economies as a range of disadvantages is imposed on the locations and regions selected for investment. Profits, for example, are likely to be remitted to a company's home country, rather than being reinvested to generate further growth in the host economy. Similarly, although they use local labour, firms may depress multiplier effects by choosing to purchase materials and components elsewhere – sometimes from independent producers but often from other branches of the corporation. Before the recession it was argued that multinationals could disrupt the labour market by inflating wage rates and attracting workers from local firms. And it has also been observed that investments frequently exploit the product life-cycle concept: research, development and other high-level activities are commonly retained as headquarters' functions while decentralization to branches takes the form of established products or components. Thus host areas may become technologically subservient to other regions or countries. This subservience will cause employment to be biased towards low-level functions, so that labour demand is unlikely to match the broad cross-section of educational attainment. But the implications of the product life-cycle go beyond this because of the probability that – even in strong economic conditions – the future of branch plant products will be scrutinized as innovation, technology and changing demand require headquarters' decision-makers to re-examine their final product mix. The spatial structures of multinational activities dictate that the crucial decision-making based on this re-examination is largely external to the host region and, in most instances, to the host country. Finally, it has been argued that, by dominating a locality or region, the branch plants of major corporations depress the formation of small firms that are necessary to the long-term health of local economies. This may occur, for example, because branch plants may absorb potential entrepreneurs, or because they do not require numerous local linkages for their survival.

Objections to multinational investment must not be accepted uncritically (Dicken 1986, 1990, 1992a). From a policy viewpoint, the argument that potential benefits can be overlooked has been well summarized by the EC in a recent economic report (Commission of the European Communities 1992c). For example, while it is acknowledged that foreign investment is a contentious issue and entails potential dangers, the survey demonstrates that Japanese investors are not conforming to the assumption that R&D will be retained strictly in the home economy. Instead it is shown that, by the early 1990s, only a fifth of 227 Japanese firms surveyed had not established research or design functions within Europe. More broadly, the Commission's report couches the potentially positive aspects of foreign investment in terms of their possible contribution to continuing economic integration. Multinational enterprises can, it is argued, play an important role in achieving the gains which integration offers. They are able, for example, to deploy resources to take advantage of regional divisions of labour. They may also provide investment in regions which are adversely affected by industrial restructuring, and their organizational structures allow them to increase productive efficiency through greater economies of scale. Partly through this mechanism, but also through their open-ness to new technologies, they are seen as having the ability to provide a competitive stimulus to many other firms. According to this thesis, the pace set by multi-nationals should in many instances force EC compa-nies to sharpen their competitiveness and so improve their ability to defend their markets against the growth of foreign rivals. This is one argument in favour of investment by Nissan, Toyota and Honda in UK car plants; while they may undercut the sales of some indigenous EC car producers, they also set efficiency standards to which other firms should eventually be forced to rise. However, as the case study of the car industry later in this chapter demonstrates, this stimu-lating effect can be produced without direct investment by multinationals in a particular market. Extensive restructuring by EC car manufacturers predates deci-sions by Japanese car firms to build European factories and is a highly constructive response to rising imports.

Industrialization, pollution and hazards

We shall return to the debate on industrial structures later in this chapter when considering trends in major industries and the proliferation of policies to encour-age small and medium-sized enterprises (SMEs). Meanwhile, it is important to recognize that some forms of industrialization entailed environmental impacts that have become major issues since the mid-1970s (Commission of the European Communities 1976, 1987b; OECD 1979b, 1984). Particularly impor-tant were increased energy demand, coupled with the greater scale of industrialization and the emergence of new types of waste product, especially from the chemi-cals industry. Simultaneously, the public's perception of environmental degradation increased as education, pressure groups and – perhaps most important of all – television brought about the widespread awareness of the environment's vulnerability.

Despite the introduction of more stringent legisla-tion and improved technology, air pollution remains a principal offender and is frequently associated with basic industries such as oil refining, chemical manu-facturing and the production of steel and other metals. One dimension of this problem is the substantial impact on local populations, as Pinder (1981) has shown in the case of Rotterdam. But increasing con-cern is now being voiced at the supposed effects of international air pollution, especially by sulphur diox-ide and, to a lesser extent, nitrogen dioxide and hydro-gen chloride (Commission on Energy and the Environment 1981; Environmental Resources Ltd 1983; OECD 1977; OECD 1992). These originate in Europe's principal industrial zones but are carried by

winds to relatively lightly industrialized regions, such as Scandinavia, where they are washed out of the atmosphere as acid rain. When this falls on acidic ground it cannot be neutralized, and ecological damage to rivers, lakes and forests then ensues.

Sulphur emissions are commonplace and have in some instances become worse as development has pro-ceeded. Power stations in the UK, for example, are responsible for about half the sulphur discharged because, as the Health and Safety Executive (1981) has underlined, control technology 'was not suitable for application to large power stations built since 1945, and from these sulphur dioxide is all discharged from suitably designed tall chimneys'. The result is that modern 2000 MW oil-fired power stations are often licensed to emit up to 650 tonnes of waste products *a day*, the dominant constituents being sulphur dioxide and sulphur particles. This type of pollution is certainly carried long distances in the upper atmosphere. Early estimates by the OECD concluded that 88 per cent of the sulphur dioxide deposited in Norway and 80 per cent of that falling on Sweden was 'imported' (OECD 1977). Very recently it has been estimated that almost 600 000 tonnes of sulphur dioxide are deposited in the Nordic countries each year, and that 85 per cent of this pollution comes from other countries (OECD 1992). Major European sources are the former USSR, the UK, the reunified Germany and Poland, but a quarter is of non-European origin (Fig. 6.3).

Despite such evidence, and despite the existence of the Geneva Convention on Long-range Transboundary Air Pollution, in practice the industrialized countries have had little incentive to invest to protect remote environments outside their boundaries. Consequently, questioning the causal link between emission and envi-ronmental impact has been a standard substitute for remedial action (Commission on Energy and the Environment 1981). However, international pressure for reform has been maintained and has been re-inforced by the realization that acid rain is implicated in the degradation of many valued forest environments within the industrial countries themselves. By the mid-1980s, countries were increasingly accepting the need for tighter emission standards (OECD 1983, 1984), perhaps the most impressive convert to the cause of emission control being West Germany. Here a pro-gramme to install emission 'scrubbers' is well advanced at, for example, the lignite-burning power stations near Aachen, where the sulphur content of the fuel is particularly high. Similarly, in the later 1980s the German chemical industry was spending more than 200 million ECU a year on new air pollution control equipment, plus 500 million ECU a year on the opera-tion of air pollution control facilities.

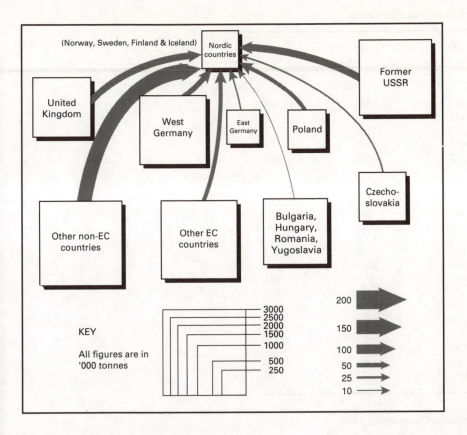

(Norway, Sweden, Finland & Iceland)

Nordic countries

Former USSR

United Kingdom

West Germany

East Germany

Poland

Czecho-slovakia

Other non-EC countries

Other EC countries

Bulgaria, Hungary, Romania, Yugoslavia

KEY

All figures are in '000 tonnes

3000
2500
2000
1500
1000
500
250

200
150
100
50
25
10

Fig. 6.3
Transboundary pollution: sulphur dioxide deposition in the Nordic countries, 1990.

From the vantage point of the 1990s, it is evident that significant progress towards the control of sulphur dioxide pollution was made in the previous decade. Between 1980 and 1988, West German emissions of this gas fell by almost a third, while in Denmark and the Netherlands they were almost halved and in France they fell by nearly 60 per cent (Table 6.6). However, this success was not entirely due to the effects of new control measures at the sources of pollution. For example, industrial restructuring, and particularly the continuing decline of heavy industry, eliminated or scaled down a substantial number of polluters, while France's impressive progress was closely linked to the expansion of her nuclear power programme. The French case, in fact, highlights a continuing dilemma concerning nuclear electricity. On the one hand, the commissioning of new nuclear power stations increases the risk of potentially serious accidents yet, on the other, it also enables the rundown of conventional power stations which are usually the major sources of sulphur dioxide pollution.

Table 6.6 also demonstrates that, while the continuing decline of sulphur dioxide emissions is encouraging, the end of air pollution is by no means in sight. As the Nordic countries are well aware, sulphur dioxide itself remains a significant source of pollution despite its decline, while very little progress has been made towards the control of other gases such as nitrogen dioxide and carbon dioxide.

Water pollution, chiefly in the form of chemical waste, also possesses a long-distance dimension, the effects of which are again difficult to reduce because improvement often necessitates international agreement and active cooperation (Walter 1975; Taylor, Diprose and Duffy 1986). In the West European context, progress towards these goals has been made via the Oslo Convention for the Prevention of Marine Pollution, the London Convention on the Dumping of Wastes at Sea, the Barcelona Convention on the Protection of the Mediterranean Sea and the International Commission of the Rhine against Pollution (see Ch. 11). The full range of international agreements on water pollution has been reviewed by contributors to Fabbri (1992), yet despite their existence it is generally acknowledged that progress towards the elimination of industrial pollution of the marine and riverine environments has varied greatly throughout Western Europe. Industrial wastes still

Table 6.6   Air pollution trends in selected EC countries

	1980	1985	1987
Sulphur dioxide emissions ('000 tonnes)			
Denmark	452	339	248
Germany (West)	3187	2345	2223*
France	3512	1734	1517
Netherlands	462	275	249
UK	4836	3682	3867
Nitrogen dioxide emissions ('000 tonnes)			
Denmark	245	262	266
Germany (West)	2935	2924	2969*
France	1861	1700	1652
Netherlands	166	303	N/A
UK	2264	2118	2303
Carbon dioxide emissions (million tonnes of carbon)			
Denmark	17	17	17
Germany (West)	212	204	201
France	130	103	100
Netherlands	43	41	44
UK	159	153	160

* 1988.

Source: Commission of the European Communities (1990).

enter the ecosystem either by dumping at sea (which may be licensed or illegal) or through direct discharges into rivers and outfall pipes to the sea. Such discharges involve far more pollution sources than does dumping, and chiefly for this reason the scale of the problem is by no means clear. However, it is severe in southern Europe where it is exacerbated by the enclosed nature of the Mediterranean and limited tidal movements. Action such as the Community-supported programme to improve the Bay of Naples has scarcely scratched the surface of the task. Around the North Sea, the available data suggest that direct discharge is most important for the UK and Scandinavia, economies that have not relied heavily on dumping at sea. But the conclusion that the continental seaboard from France through to former West Germany is relatively benign in this respect must be treated with caution because of the absence of reliable data for major estuaries.

This is underlined by recent experience in Rotterdam, where consultants established in the late 1980s that 100 000 tonnes of oil-contaminated residues and 275 000 tonnes of chemical waste disap-

peared each year. Much of this waste was collected by 'floating dustmen' and then dumped illegally, often in local waters. New national legislation and port regulations are now bringing this problem under control, but these regulations will do nothing to ameliorate a second environmental issue that is the legacy of generations of heavy industrialization in Rotterdam's hinterland. Silt dredged from the port has in the past been used to raise land for building purposes, but tests have now shown that this silt is often badly contaminated with heavy metals, to the extent that they are unfit for use in residential building programmes. Most of these deposits have been brought to the port from Germany and Belgium by the Rhine and Maas rivers and, since dredging must continue, safer methods of silt disposal have had to be devised. These include the construction of a storage basin at the seaward end of the port. This will be sealed to prevent the seepage of pollutants into the surrounding sea. While it is arguable that problems such as these are particularly acute in Rotterdam because of the port's scale, their emergence inevitably raises questions about other ports and estuarine areas.

Table 6.7 Lead levels in EC rivers (microgrammes/litre)

		1980	1985
Belgium	Meuse at Agimont	4.0	9.1
	Meuse at Lanaye	20.0	6.7
	Scheldt	25.0	6.2
West Germany	Rhine	7.0	11.0
	Weser	2.0	2.8
	Danube	1.4	2.6
Spain	Guadalquivir	12.7	28.0
	Ebro	5.0	0.0
France	Loire	0.0	—
	Seine	8.0	40.0
	Garonne	10.0	0.0
Netherlands	Maas at Keizersveer	12.0	3.6
	Maas at Eijsden	23.0	6.2
	Scheur	11.0	1.9
	IJssel	9.0	5.0
	Rhine	15.0	4.2
UK	Thames	10.0	9.0
	Severn	40.0	4.0
	Clyde	18.0	8.0
	Mersey	15.0	11.0

Source: Commission of the European Communities (1990).

The problem of water pollution also illustrates the difficulties of achieving successful international cooperation in environmental protection, an issue considered in detailed case studies by Guruswamy, Papps and Storey (1983), Haigh (1986) and Taylor, Diprose and Duffy (1986). As with sulphur dioxide air pollution, however, there is now evidence that certain types of water pollution are falling significantly. This appears to be the case particularly with metal pollutants, although the most recent data unfortunately refer only to the mid-1980s. In most rivers surveyed by the EC, pollution by lead, cadmium, chrome and copper fell markedly after 1980. Table 6.7 illustrates this with respect to lead, the levels of which showed clear reductions at more than two-thirds of the sampling sites. Taking the long-term view, this trend will ease marine contamination and the silt pollution problems of ports such as Rotterdam.

What is not yet clear, however, is the extent to which lower water pollution reflects investment by former polluters to comply with the 'polluter pays' principle (Commission of the European Communities 1976, 1987b; Ellington and Burke 1981; Williams 1990). Substantial anti-pollution investments are being made by the private sector: in the late 1980s, for example, the West German chemicals industry was spending almost 1.5 billion ECU a year on the construction and operation of water purification plants (Commission of the European Communities 1992c). But, to an unknown degree, lower pollution levels are also the consequence of decline in the polluting industries.

What must also be recognized is that many of Europe's worst sources of international water and air pollution are not covered by existing international agreements since they lie in the former Communist countries. These countries pose major challenges with respect to environmental degradation, yet only the former East Germany – now part of the EC – is currently in a position to expect large-scale investment for this purpose. In general, pollution reductions in the former Eastern bloc, many of which will be beneficial to the West, may well depend heavily on the decline of archaic and uncompetitive industries that is now in progress.

The European Community and pollution control

The leading organization working towards international cooperation in pollution control has been the EC. March 1987 to March 1988 was designated the

Table 6.8 EIB loans for environmental protection, 1991

	million ECU	(%)
Water conservation and management	1070.8	64
Waste management	53.7	3
Air pollution control	384.6	23
Other measures	177.1	10
Total	1686.2	100

Source: European Investment Bank (1992).

Community's Year of the Environment, but this was only one high-profile episode in a programme of activity initiated in the early 1970s. Two Action Programmes adopted in 1973 and 1977 aimed to stimulate remedial efforts to control existing industrial pollution. A third Action Programme launched in 1983 underlined the benefits to be gained by preventive action to limit the environmental impacts of new projects. The major measure introduced in this context was a Community directive (1985) requiring that major infrastructural and industrial projects would undergo a form of environmental impact assessment. A fourth Programme, which covered the period up to 1992, re-emphasized the importance of prevention and ambitiously urged that environmental interests should be taken into account by all economic and social policies (Commission of the European Communities 1987b). In practice, the Commission's powers to encourage protection of the environment lie in the issue of directives, regulations and decisions, and the proliferation of these gives a good indication of its activity in this field. Before 1973, when the Community's environment policy was initiated, only two or three directives, regulations or decisions relating to the environment were produced each year. Between 1973 and 1981 the average rose to nine. And since 1981, when the Commission's Environment Directorate was established, the average has exceeded twenty.

Another positive form of action that has been taken is the mobilization of European institutions to encourage environmental protection investment. The European Regional Development Fund now routinely considers the environmental implications of projects put to it for funding, and the EIB makes extensive loans related to environmental protection. Between 1987 and 1991, 15 per cent of all EIB loans had environmental objectives of one form or another, the Bank's largest single target being the reduction of water pollution (Table 6.8). This emphasis on the aquatic environment is reflected in the major schemes in which the EIB is engaged with other agencies. These include an Environmental Programme

for the Mediterranean and projects for the Environmental Rehabilitation of the Baltic and the Protection of the Elbe.

Although the Commission is deeply involved in encouraging environmental protection, and although its supranational status makes it a unique vehicle for the attack on international pollution, its record to date has been criticized (Klatte 1986; Kromarek 1986). Even if it is accepted that uniform emission standards throughout the Community are not desirable (Taylor, Diprose and Duffy 1986), permissible pollution levels still vary unsatisfactorily between member countries. In some respects the Commission's powers are poorly defined, and the publication of directives is not equivalent to their effective implementation. It is widely believed that the European Commission is much more deeply committed to protection measures than is the Council of Ministers. And the financing of environment policy is poorly organized. At present the principal prospect that more effective Community action can be achieved lies in the provisions of the Single European Act (Lodge 1986; Williams 1990).

Hazards

As well as generating pollution problems, some forms of industrialization that have achieved rapid growth have brought in their wake considerable potential hazards (Health and Safety Executive 1979). For example, because of the volatility of the products handled, oil refineries and petrochemical installations inevitably entail the danger of explosion. This is so despite their emphasis on fail-safe systems and automation, as was strikingly shown when an explosion and fire destroyed the Nypro (UK) chemical complex at Flixborough, Lincolnshire. Even though the works occupied a greenfield site, and even though the incident took place outside normal working hours, twenty-eight out of the thirty-two workers present were killed and 2000 distant buildings were damaged (Department of the Environment 1975). But hazard risks are not confined to blast and fire resulting from explosions. Another potential danger is the release of dangerous substances as a result of an explosion or of a less spectacular plant failure. As with blast hazard, the immediate problem is likely to be restricted to a limited area. This occurred in 1976, for example, when dioxin – a poison 10 000 times more toxic than cyanide – was released from the Hoffman-La Roche plant at Seveso in northern Italy (Ellington and Burke 1981). Yet there is clearly scope for wind-borne hazardous emissions to be spread more widely. Although this long-range dimension is not a major problem in the chemicals industry (because escapes are usually diluted naturally by air movements) the risks have long been apparent to the anti-

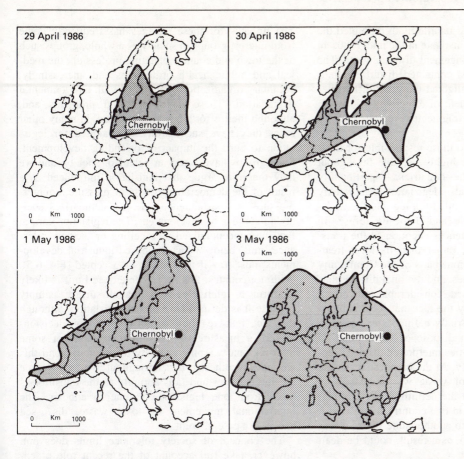

Fig. 6.4
The Chernobyl nuclear cloud, 29 April to 3 May 1986.

nuclear power lobby. That this concern is not misplaced was graphically demonstrated on 26 April 1986 when an explosion and fire at the Chernobyl power station in the Ukraine released into the atmosphere up to half a one-megawatt reactor's contents (Lopriano 1986). Within three days the radioactive cloud from this disaster had reached Sweden, Denmark and West Germany, and in eight days the whole of the continent – and 300 million people – were affected (Fig. 6.4). Because of the long-term nature of radiation damage, the possible consequences of this accident – higher cancer rates and mortality – will not be apparent for years. The reactor itself has been contained in a concrete shell, and work is continuing to deal with contaminated soil in the area. Meanwhile it is not known when, or if, the 100 000 people evacuated from a 30 km radius of the site will be allowed to return. Defenders of nuclear power argue that higher safety standards in Western Europe make a comparable accident extremely improbable (Nuclear Energy Agency 1986). Critics point out that many other nuclear plants in Eastern Europe are poorly maintained, that this Europe-wide hazard was created by a reactor that was not particularly large and that a similar accident in Western Europe

might be of larger magnitude and would almost certainly affect areas of higher population density. As was noted in Chapter 5, the spatial relationships between nuclear power stations, major population centres and national frontiers are such that even a restricted release of radiation could well have a major impact outside the country of origin.

In addition to illustrating the problem of local impact, the Seveso incident demonstrates that accidents involving dangerous chemicals can have other serious international implications if control in the aftermath is insufficiently rigorous (OECD 1985a). In this instance a contract for the destruction of residual dioxin, combined with 2 tonnes of inert material and sealed in forty-one drums, was placed with a West German firm, Hannesman. The consignment was known to have left Italy in 1982, but it then disappeared amid rumours that it had not been destroyed. In May 1983, after the director of a French waste disposal firm – Spelidec – had spent two months in detention, the drums were discovered in a disused abattoir near St Quentin in northern France. Only then was Hoffman-La Roche forced to undertake the safe incineration of the dioxin-contaminated material.

The Seveso incident and its aftermath prompted the European Commission to instigate in the mid-1980s an investigation into the management of toxic waste. The final report (Haines 1988) contained disturbing findings, many of which highlighted the great difficulties surrounding the development of effective toxic waste control in a politically fragmented region such as Western Europe.

The training available to those responsible for dealing with toxic waste was highly variable throughout the Community. In some countries – particularly Greece, Italy and Portugal – legislation was weakly imposed, allowing uncontrolled tipping and the indiscriminate mixing of toxic and domestic refuse. In addition, various circumstances encouraged the transport of dangerous waste, rather than its local treatment. For example, in various parts of the Community a lack of treatment facilities, and regional differentials in treatment costs, caused considerable movement within countries. Similarly the fact that local authorities are responsible for strictly defined areas encouraged some contractors to dispose of toxic waste in districts where the authorities interpreted disposal regulations relatively leniently. At the broadest scale, the cross-border movement of waste was increased by international variations in the definition of toxic materials. Thus a contractor in one country might find it economic to move waste to another state where it was not classified as toxic and, as a result, could be dealt with cheaply.

This investigation therefore exposed a highly complex set of problems causing, among other things, the growth of pollution havens at both the international and intranational scales. The Commission is still attempting to untangle this web in order to establish a high standard of toxic waste control throughout the EC.

Hazards, pollution and industrial location theory

The concept of pollution havens is but one step from the introduction of hazards and pollution into industrial location theory. Since the Second World War the European economies have witnessed the introduction of many items of legislation intended to ameliorate the impact of industrial projects on the surrounding environment. In most countries, for example, planning legislation has controlled the siting of industrial activity; it is now commonplace for regulations intended to moderate a project's visual and aural intrusion to be applied; and, although pollution continues to be a problem, in many parts of Europe official attitudes to emissions into the atmosphere and surface water are far more strict than in the immediate post-war period.

The impetus for these changes has frequently come from interest groups in society as a whole, groups such as the town and country planning profession, the medical and public health professions and, increasingly, voluntary organizations concerned with environmental degradation. These have identified problems and, although their efforts have often failed, in many other cases they have successfully campaigned for legislation to curb the impact of offending development. Such groups have acted on behalf of society, and we may therefore argue that legislation introduced as a result of the democratic process of pressure, counterpressure and parliamentary debate summarizes society's tolerance to development. The legislation can, in fact, be conceptualized as a ceiling or 'society tolerance limit' below which proposed industrial developments must stay if they are to be accepted (Fig. 6.5). Within a country, this society tolerance limit is likely to form a relatively even spatial surface, simply because it is defined by national legislation, although the fact that regulations allow planning decision-makers a measure of discretion ensures that some inter-regional warping is likely to occur. Beyond this we may hypothesize that throughout Western Europe the surface rises and falls abruptly as the transition is made from one legislative regime to another. The international movement of toxic waste, discussed above, is a case in point.

The concept of society tolerance limits does not, however, take full account of the recent role of environmental pressure groups in the development decision-making process. Indeed, a very pronounced feature of the period since the late 1960s has been that many groups have become active primarily *because* tolerance limits set by the legislation in force have been perceived to be inadequate. On specific issues, therefore, these groups have attempted to lower the tolerance surface around endangered localities, in some instances to ameliorate the impact of existing industries, and in others to ensure that proposed developments are either excluded completely or are only allowed to proceed in a greatly modified form. But the goal of lowering the society tolerance surface is not achieved simply by opposing the undesirable implications of industrialization: it can only be attained by the successful exertion of influence on those with decision-making power.

Strategies for achieving this influence have taken many forms, but two broad approaches (which are not mutually exclusive) can be identified. In the first, advantage is taken of the fact that, as a statutory requirement, most of Western Europe's planning systems now incorporate some form of public participation. Within this framework it is open to protesters to argue the case for a

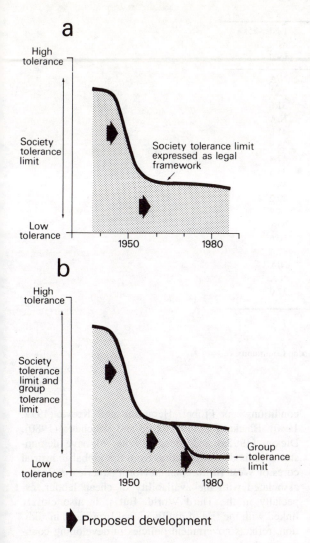

a

High tolerance

Society tolerance limit

Society tolerance limit expressed as legal framework

Low tolerance

1950   1980

b

High tolerance

Society tolerance limit and group tolerance limit

Low tolerance

Group tolerance limit

1950   1980

◆ Proposed development

Fig. 6.5
The concept of tolerance limits to development.

non-existent or weak, or because the statutory system is perceived to restrict the opposition by employing quasi-judicial investigative methods to form a barrier between objectors and decision-makers. This alternative is direct confrontation with industry and/or political decision-makers, confrontation intended to be so intense that the former will withdraw to less hostile investment environments and the latter will perceive clear political advantages in opposing development.

When schemes are blocked through either route the effect is to bring into operation a new and lower group tolerance surface (Fig. 6.5). This may represent attitudes within the local community or the interests of vociferous groups external to the locality in question. But the consequent lowering of the tolerance surface cannot be considered permanent. Permanence can only come through legislative modifications (which will spread the new lower surface through the national space) or through the introduction of new local regulations or plans. Unless this happens it is quite possible that the group surface will evaporate as time elapses, circumstances change and issues die down.

From growth into crisis

Although environmentalism remains an important influence on industrialization, since the mid-1970s its progress has been matched by a resurgence of concern for Western Europe's industrial future. In this connection it is important to recognize that, despite recessions following the two oil price shocks, industrial output trends have not been entirely adverse. Production recoveries have been made in most countries, but this contrasts with industrial employment, which continues to decline. Between 1981 and 1990, most countries lost 10 per cent or more of their manufacturing workforce, Spain maintained its workforce and only Denmark recorded a significant rise in industrial employees. This general decline has been paralleled by the rise of large-scale unemployment, an important feature of which is the prevalence of youth unemployment (Table 6.9). Although high unemployment is not simply a consequence of industrial restructuring, the emergence of both phenomena has underlined for individuals and governments the reality of the problems to be faced.

Superficially, the re-emergence of the pre-war unemployment spectre might suggest that the problems of the 1980s can be explained simply in terms of recession triggered by the energy price shocks of the 1970s. Important though it is, however, the recession is a relatively late arrival superimposed on the consequences of other major forces that have been gathering strength since at least the early 1960s. Just as much as the onset

restrictive application of legislation, and in many instances they do so. But as Lucas (1979) has emphasized in his study of the French anti-nuclear movement, what is presented as a participation exercise may offer objectors little or no direct contact with decision-makers. Similarly in the UK, there is growing dissatisfaction with a public inquiry system which may make participation in major issues a highly expensive exercise, with no guarantee that a conclusion favourable to the opposition will be upheld by the relevant government minister.

The second approach is quite different and may be adopted either because participation procedures are

Table 6.9 Unemployment, 1975–90 (%)

	1975	1985	1990	Under-25s as % of total
Austria	2.0	4.8	3.6*	N/A
Belgium	6.8	13.5	11.6	35.1
Denmark	5.1	9.3	7.2	31.0
Finland	2.2	6.3	3.4†	N/A
France	4.1	9.7	10.2	41.8
Germany (West)	4.7	9.3	7.1	28.8
Greece	—	5.1	7.8	47.4
Ireland (Rep)	12.2	17.4	18.2	38.5
Italy	5.9	10.6	9.6	59.0
Luxembourg	—	1.7	2.9	49.2
Netherlands	5.0	12.7	10.5	34.8
Norway	2.3	2.5	4.9†	N/A
Portugal	4.5	8.5	8.8	50.6
Spain	4.0	22.0	21.8	47.4
Switzerland	0.3	1.0	0.6*	N/A
Sweden	1.6	2.8	1.4	N/A
UK	4.1	11.9	11.4	37.0

* 1988.

† 1989.

Sources: International Labour Office (1986); Commission of the European Communities
(1992e).

of deep recession, these forces demonstrate that it is essential to view Western Europe in its global context if today's industrial malaise is to be understood.

One of these long-term factors has been the improvement of maritime transportation and the associated development of ports (Pinder and Hoyle 1981). Here the fundamental point is that the technologies which created the crude oil supertanker have been adapted to revolutionize the transport and transhipment economics of many other cargoes. In many instances, of course, these developments have benefited Western Europe: to quote just three examples, bulk-ore carriers, specialized chemicals carriers and container ships have all helped to sustain industrial momentum. Yet improved transportation can prove to be a double-edged sword: the problem is that, in breaking down global trade barriers, new generations of specialized ships have made Western Europe an attractive market for producers around the world. Without the ocean-going car transporter, Western Europe would export fewer vehicles but would also be less exposed to the rigours of global competition.

Beyond this, however, competition has intensified as a result of the new international division of labour. This is the spread of manufacturing, at the global scale, to countries offering advantageous production conditions. As Fröbel, Heinrichs and Kreye (1977, 1980), Blackaby (1979), Torre and Bacchetta (1980); Dicken (1992a) and Steed (1978a, b) have demonstrated, this decentralization from the global industrial cores of Western Europe and North America is closely associated with the availability of cheap labour, especially in the Third World. But it is also closely linked with the rise of multinational firms and, in addition, reflects government policies in developing countries, which encourage manufacturing to engage in import substitution and to develop exports. Interesting though competition from developing countries may be, however, the reality is that the main pressures on Western Europe's manufacturing sector come from other developed areas (Dicken 1990). This is well illustrated by the rise of Japan, to which we shall return towards the end of this chapter. Before then, however, consideration must be given to what is now a high-profile aspect of economic development in Western Europe – the small-firm sector.

Small firms and economic recovery

One striking reaction to economic crisis in Western Europe has been widespread emphasis on the promotion of small and medium-sized enterprises (SMEs). As

Mason and Harrison (1990) have shown, concern to develop the small-firm sector spread virtually throughout Western Europe in the 1980s and it remains a feature of the present decade. This trend reflects the desire to accelerate employment creation and create balanced industrial economies in which problems of decline are offset by the employment and products of small firms which mature into successful enterprises.

In placing this emphasis on SMEs, Western European countries have – either wittingly or unwittingly – followed in the footsteps of the EIB. This organization was created in 1958 by the original members of the EC specifically to promote economic development and, especially, to assist in the reduction of regional imbalances. As early as the mid-1960s, the EIB identified many of the problems described earlier in connection with multinational investment, the outcome being the introduction in 1969 of a Global Loan industrial finance scheme targeted solely at SMEs (Pinder 1986). Since then, more than half the EIB's loans to industry have been to SMEs, and many firms which now qualify for assistance from programmes established by their own governments are also eligible for EIB Global Loan support. Moreover, although this support was originally available only in problem regions, its availability has recently been widened as general economic circumstances have deteriorated. In 1991 almost half the Bank's investment in SMEs was for purposes not directly connected with the alleviation of regional problems.

Despite the substantial support available from national governments and the EIB, however, the contribution which SMEs are making to economic recovery appears to be disappointing. Mason and Harrison (1990) were able to conclude that there had been a substantial increase in both the absolute and relative size of the small-firm sector in most of Western Europe since the mid-1970s. This shift was explicable partly in terms of a wide range of factors, including rationalization by large firms, which released employees with entrepreneurial ability and redundancy money to invest. A change in social attitudes to entrepreneurship was also important, as was government support in the form of incentive schemes and liberalization of the regulations constraining business. Yet it was not clear whether the resurgence of small firms was a permanent or a cyclical phenomenon, and the evidence suggested that SMEs made only a modest contribution to local and regional economic development processes. Only a few firms had been able to achieve dramatic growth, and the very large majority of businesses were not concentrated in profitable or growing sectors. Most reinforced existing economic structures rather than producing progressive diversification, and the jobs created often paid relatively low wages, were far from secure, offered few fringe benefits and, in many instances, were part-time. Similarly Pinder (1986) concluded that EIB Global Loan investment protected, but did not expand, employment and was in no sense concentrated on sectors of key significance for future industrial development.

The fact that small firms cannot be considered a cure-all for the ills of the industrial sector re-emphasizes the important role which major firms and leading industries still play in the Western European economy. For this reason it is appropriate to return to these industries and, in particular, to examine their responses to the increasing pressures they have experienced since the 1970s. Three industries – oil refining, petrochemical production and car manufacturing – have been selected to illustrate themes of current significance. The first spotlights the consequences of recession, defined as a slump in anticipated demand. The second illustrates the transition from recession-generated crisis to one based on the pressures of global competition. The third provides a case study of adjustment forced from the outset by intense competition between major world manufacturing power blocs, and also underlines the challenges now posed by the Single European Market. All three demonstrate that – far from being confined to traditional industries – major difficulties confront many of the most successful post-war activities on which Western Europe's recent prosperity has been heavily dependent.

Oil refining

Western Europe's post-war economic take-off stimulated – and was in turn stimulated by – the spectacular expansion of oil refining. Refinery capacity in 1976 – the peak year – was almost 1050 million tonnes, compared with 230 million tonnes in 1960. Most of the growth was concentrated in the European core (Fig. 6.6a), although a significant number of projects were implanted at coastal locations in the Mediterranean periphery, the planning theory being that in a variety of ways these would act as catalysts for local economic development. Soon after the 1973 oil price shock it became apparent that the industry's policy of growth must be abandoned (Bacchetta 1978), and after 1979–80 – when the second price shock hit the European economies – the crisis in refining became intense (Molle and Wever, 1984a, b). In Western Europe as a whole, refinery utilization rates fell from 69 per cent in 1979 to 58 per cent in 1981. By then, the capacity surplus was equivalent to more than thirty average refineries, and most of the industry was making substantial operating losses on its basic refining activities. One estimate (*Petroleum Economist*, August 1981) was that these losses averaged $18 a tonne, but in some areas they were considerably worse. For example, in West Germany, where the indigenous capacity surplus

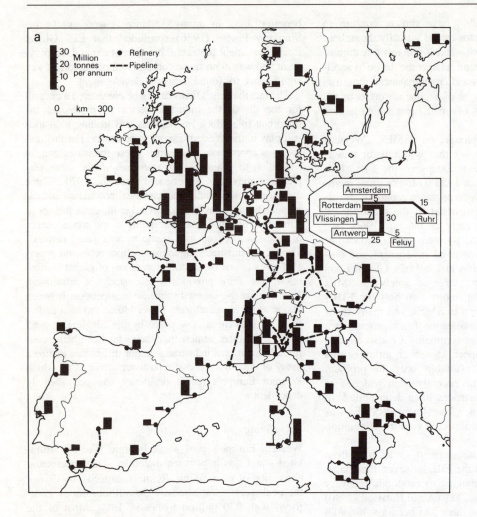

a

30
20
10
0

Million
tonnes
per annum

● Refinery
--- Pipeline

0   km   300

Amsterdam
Rotterdam   5
Vlissingen   7   30   15   Ruhr
Antwerp   5
                25   Feluy

Fig. 6.6
Oil refining: (a) refining capacities at peak year (1976); (b) capacity and throughput, 1980 and 1981.

was exacerbated by product dumping from other European countries, losses in 1981–82 were probably at least $50 a tonne. Similarly, losses in the French refining industry were in the region of $3 billion in 1981.

Understandably, investment in new refineries or major expansion schemes virtually ceased in this new economic climate and, as yet, there is no sign of a recovery in construction. But, in addition to reducing construction, fundamental changes in the extent and nature of the refiners' West European operations have been imperative to return the industry to a sound financial footing (Pinder and Husain 1987a).

The most obvious restructuring strategy pursued in the 1980s has been the rapid reduction of surplus capacity. By 1991 the industry's capacity was almost 30 per cent lower than in 1981, and contraction had affected most parts of Western Europe (Fig. 6.6b). This was achieved in two ways. Firstly, no less than forty-six refineries were closed in this period, including some relatively large under-

takings. The largest shutdown to date has been the Ravenna refinery (annual capacity 12.3 million tonnes) owned by the Italian company SAROM. But, in general, companies have opted to close small refineries; in 1979 the average capacity of all refineries was 6.5 million tonnes, while the average size of closed refineries was only 4.2 million tonnes. This partly reflected the fact that small refineries were often vulnerable because they failed to achieve sufficient economies of scale. But many were also prime targets for closure because the technologies they employed were insufficiently advanced (Pinder and Husain 1987a, b). We shall return to the role of technology in the industry's survival strategy below.

Secondly, in addition to closing refineries, companies reduced unwanted capacity by adopting a far less obvious strategy known as downrating. This entails maintaining refineries in production, but substantially reducing their capacity by taking out of commission surplus refining units. In sharp contrast to closures,

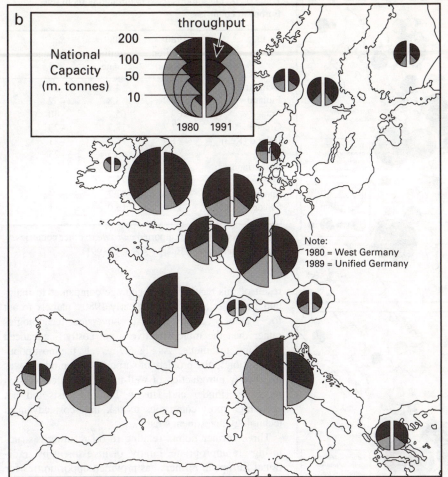

Fig. 6.6b

this strategy was chiefly applied to relatively large refineries which could be contracted without severe reductions of scale economies in the refining process. The average size of refineries downrated was 10.6 million tonnes before contraction and 8.1 million tonnes afterwards. Altogether, the downrating strategy was responsible for 47 per cent of Western Europe's total capacity reduction in this period, the remaining 53 per cent being accounted for by closures. The effects of this disinvestment on refinery utilization rates is apparent in Fig. 6.4b. In most countries, and especially those with major branches of the industry, throughput as a proportion of total capacity has risen significantly. In 1991 the utilization rate for the industry as a whole was 84 per cent, well above the 80 per cent target adopted by most refining companies.

Impressive though it was, disinvestment was not the only strategy open to the refiners in their attempts to improve the industry's economics. In the early 1980s one of the industry's major handicaps was that overproduction of relatively heavy and low-value products (such as high-sulphur fuel oil) was particularly serious, primarily because of the limitations of traditional distillation technology. But the application of more versatile cracking processes to the heavier fractions allows lighter products to be obtained, and the fact that refineries already equipped with these technologies remained profitable triggered a major investment wave. Dozens of projects to expand the capacity of conversion facilities had been completed by the early 1990s, and most involved the introduction of either thermal or catalytic cracking technologies (Fig. 6.7). As a result, fuel oil's share of total refinery output fell from 34 per cent in 1979 to 20 per cent in 1989.

Although the pace has slackened, investment is still continuing. Vielvoye (1985) has argued that investment

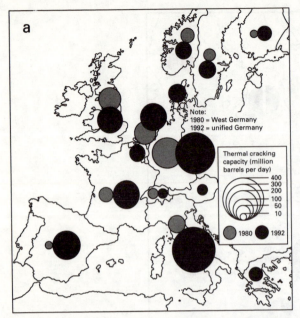

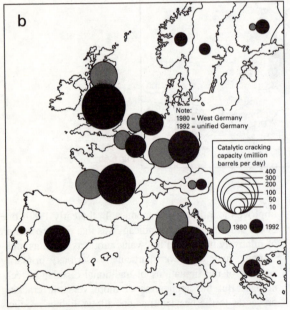

Fig. 6.7
The growth of oil refinery conversion capacity, 1980–92:
(a) thermal capacity; (b) catalytic cracking capacity.

Table 6.10 Oil imports passing through ports in selected European countries

	1980 (million tonnes)	1989 (million tonnes)	% change
Denmark	5.7	4.3	–24
Finland	12.9	8.8	–32
France	109.5	65.7	–40
Greece	17.7	13.7	–23
Ireland (Rep)	1.9	1.5	–21
Italy	89.1	68.5	–23
Netherlands	49.7	47.7	–4
Portugal	8.3	9.7	+17
Spain	47.4	47.7	+4
Sweden	17.9	16.0	–11

Note: Data exclude oil imported *en route* to other countries.
Source: International Energy Agency (1991).

result of this has been to encourage companies to maintain a trend established in the early 1980s, namely to opt for relatively cheap thermal conversion technologies rather than the more effective – yet costly – advanced 'cracking' techniques. As Fig. 6.7 reveals, the most popular solution to the problem of surplus heavy products was the speedy introduction of well-tried thermal conversion processes (Pinder and Husain 1987a). Crisis did not encourage most companies to risk high-cost advanced technology investment strategies.

Three further points relating to oil refining restructuring are appropriate. Firstly, disinvestment by traditional European refiners has provided opportunities for inward investment from outside Europe. The prime example is the foothold gained by the Kuwait Petroleum Corporation by purchasing Gulf Oil's marketing network and its refineries in Rotterdam and Denmark. Secondly, however, most closures and cutbacks have had more serious implications for the localities affected. Closures clearly entail the loss of employment, even though the industry's capital intensity means that typical redundancies (300–400) are less severe than is often the case when deindustrialization occurs. In addition, the business tax base of local authorities may be severely hit, while reduced oil imports have obvious implications for port revenues. The effects on ports have been very uneven, but in several countries the flow of oil through the port system has been reduced by at least a fifth (Table 6.10). Where decline has been severe, most ports have been very unsuccessful in generating alternative traffic. Refinery closures also pose a physical planning problem more normally associated with old industrial areas in major conurbations. Given the lack of demand from

will need to be maintained until the end of the century to counteract falling demand for relatively heavy refinery products. What must be recognized, however, is that current investment is taking place after the oil price collapse of 1985–86. Consequently the profits to be made from the conversion of heavy products are now less attractive. One industry, how can redundant sites of perhaps 2 km² be recycled for the good of the economy and the community? Satisfactory solutions to this problem have scarcely begun to be formulated.

Thirdly, although the crisis in Western Europe's refining industry can largely be attributed to the recession, the indications are that the global spread of refining capacity will become an additional complicating factor by increasing significantly the volume of high-grade products traded internationally. This would set the scene for a further price squeeze that the European refiners now fear. The threat of imports is greatest from Middle Eastern and North African countries. Even though declining oil revenues forced Saudi Arabia to cancel a number of projects in the mid-1980s, the Middle East and Africa have expanded their capacity impressively in the last decade (Table 6.8). In 1985, Western Europe imported 30 million tonnes of products, 20 per cent more than anticipated by the European Commission. Imports were expected to reach 40 million tonnes by 1990, but in reality this figure was easily outstripped. More than 80 million tonnes of products were imported in 1991, almost an eighth of total consumption. This clearly demonstrates that, although Western European refiners acted decisively to contain the crisis in the industry, and achieved considerable success, new challenges are emerging from the process of globalization.

Petrochemicals

In many ways Western Europe is the cradle of the petrochemicals industry. It is the home of four of the world's five largest chemical concerns – Bayer, BASF, Hoechst and ICI – while the two largest indigenous oil companies (Royal Dutch Shell and BP) both have extensive petrochemical interests. In addition, the sector's post-war growth naturally stimulated the entry of many other firms, indigenous and foreign. Yet plentiful capacity, and price cutting as producers strove to maintain their market shares as the recession deepened, manœuvred the industry towards financial crisis. Against the background of rapidly rising crude oil prices, in the West German market the price of polystyrene rose by only 75 per cent between 1973 and 1982, while that of polypropylene actually declined. Because of their bias towards high-volume, standard products, the effects were felt particularly badly by oil companies: BP lost nearly $1000 million on chemicals between 1980 and 1982. But European chemical companies in general saw relatively healthy returns diminish rapidly in the early 1980s. And the European companies' overall performance would have deteriorated even more rapidly if losses in high-volume

products had not been offset by better returns on foreign activities and high-technology chemicals.

The strategies adopted by petrochemical producers in response to crisis have been similar, but not identical, to those employed by oil refiners (OECD 1985b). Capacity cutbacks, involving closures and downrating, have been frequent. Early examples included decisions by BP and ICI to close seven plastics installations in the UK, and by BASF in West Germany and DSM in the Netherlands to curtail polyethylene capacity by 20 per cent. In 1986–87 the ICI and Enichem partnership closed plant capable of producing 300 000 tonnes of PVC a year. The aim of this contraction was to halve PVC over-capacity in Europe. Investment has also been central to the recovery strategy. Although the number of construction projects significant at the world scale slumped in the crisis period, many revamping schemes have been undertaken in existing installations and others are in progress or planned. These schemes are primarily cost-cutting exercises, designed to achieve higher plant efficiency. Specific goals are the more economical use of raw materials, flexibility allowing plants to switch to cheaper raw materials and improvements in the design and control of production processes. Progress in this latter area has been particularly closely related to the development of advanced computer software to facilitate design, process control and operator training (Pathe 1986).

Unlike the refining industry, however, the petrochemicals sector has actively pursued a third survival strategy – ownership restructuring. This has taken several forms. In Italy, a government-orchestrated rescue plan reorganized the industry under ENI and Montedison. Mergers of interests have been undertaken, such as the PVC production partnership between ICI and Enichem noted above. This joint venture created the European Vinyls Corporation. Similarly, BP and Atochem (a subsidiary of Elf) have merged their polypropylene production interests in France. But the most impressive aspect of ownership reorganization has occurred as firms have manœuvred to specialize in their strengths and dispose of weak elements in their product ranges. In the UK, an outstanding example of this strategy was provided by a major exchange of assets between BP and ICI. The effect was that BP withdrew from the PVC market while ICI ceased production of low-density polyethylene (LDPE). Outside the UK, ICI has sold much of its LDPE business to Atochem but has bought Atochem's interest in ethylene oxide derivatives. In the Netherlands, meanwhile, Shell has purchased the Hoechst polystyrene plant; and in West Germany, BP has acquired marketing control of Bayer's polyethylene output. In this way BP has become the second-largest marketer of polyethylene in

Western Europe. From these examples it is evident that the petrochemicals industry has been forced into a state of flux that would have been inconceivable before the oil price shocks.

As in the refining industry, restructuring was initially triggered by declining demand. But in the 1980s the petrochemicals industry has faced increasing global competition much more severe than that as yet experienced by oil refining. This challenge chiefly reflects the expansion of Middle Eastern petrochemical capacity and, above all, the completion of twelve substantial petrochemical plants in Saudi Arabia (Chapman 1992). Although this country aims at all world markets, sales in Europe have proved particularly attractive. Within a year of large-scale shipments starting in 1985, Saudi Arabia was the largest marketer of every chemical it sold in Western Europe. This dramatic rise was a consequence of low production costs and an absence in Europe of effective tariff barriers. Developing countries such as Saudi Arabia enjoyed modest duty-free quotas for sales in Europe and tariffs applying to sales above these quotas were low. In contrast, the USA allowed no duty-free quotas and applied tariffs which controlled the import of many products, even from neighbouring Mexico and Canada. In 1986 the EC countries reacted by imposing tariffs on many imported petrochemicals, effectively raising their cost by 13 per cent. Yet the signs are that this has not halted the influx. In 1992 the basic petrochemicals industry was still considered to be highly competitive and experiencing growing competition. This came primarily from the Middle East, but also from new producers in South-East Asia (Commission of the European Communities 1992c).

While this is another indication of low overseas production costs, it is also a consequence of the reorganization of capital in the world petrochemical industry in the 1980s. Thus, Saudi Arabian expansion has been partly achieved through joint ventures with existing major companies. For example, the Kemya plant is a partnership between the Saudi Sabic company and the Exxon Corporation. Similarly, Mobil and Yanpet have joined forces to operate a plant on the Red Sea coast. Joint ventures have facilitated penetration of the European market in two ways. One is that the major companies' marketing experience has been available for the recruitment of industrial consumers. Thus Mobil established a marketing company in Brussels and built a network of customers before its venture with Yanpet came on stream. The other is that companies have in some instances cut back European production in favour of imports from their overseas ventures. A case in point is that of Exxon's subsidiary Essochem Europe, which has shut down an ethylene cracker at Cologne and sold another in Sweden; as an alternative, ethylene-based products now come from the Sabic/Exxon Kemya plant. Integration with the world production system has therefore made a significant contribution to the pressures faced by the Western European branch of the industry.

The West European car industry in its global context

The development of the car industry has been interpreted by governments and economic planners in terms of growth-pole theory (Perroux 1955; Pinder 1983), and there is ample justification for this view. The end process, final assembly, is essentially a 'propulsive' activity which not only creates a large, well-paid labour force, but also animates a host of component firms and producers of basic inputs such as steels, alloys, rubber and machine tools. Dicken (1992b) has recently explored in detail the structure and far-reaching significance of the industry. In some traditional foci of the industry (the West Midlands, the Paris region and industrial northern Italy are good examples), many of the linked activities have clustered in geographic space around the assembly firms. But from the macro-economic planner's point of view a more significant characteristic is that the propulsive assembly firms and their linked activities have formed powerful groupings in 'economic space' – in other words, they have become aspatial poles of development penetrating national economies and assisting their progress to higher development levels. In the EC, which has generated the overwhelming majority of Western Europe's car production, the consequences of penetration by this single growth pole were clear by the early 1980s. Car assembly employed almost 2 million workers; in linked activities the dependent labour force was almost twice as great; and the industry absorbed 20 per cent of the Community's steel and machine-tool production (Commission of the European Communities 1981b).

By the early 1980s, car exports still earned the Community as much as they did for Japan, while net export earnings from the industry covered 20 per cent of the oil import bill. Even so, at the global scale, shifts in the balance of power between major sectors of the industry posed substantial challenges for the traditional European industry. At that time Spain was not a member of the EC but was becoming a significant exporter to EC markets. An aggressive Spanish expansion policy, involving large-scale investment by EC and US companies, rapidly raised production to 900 000 cars a year by the early 1980s and more than 1.5 million in the late 1980s. By that stage Spain was, of course, a member of the EC, but the fact that more

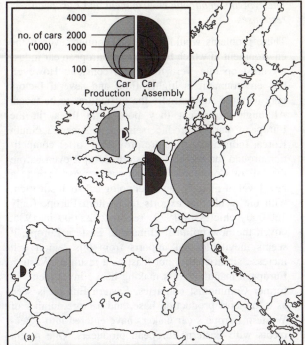

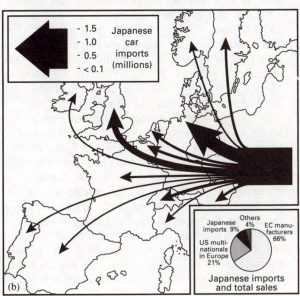

Fig. 6.8
The European car industry and Japanese competition
(a) Western European production; (b) Japanese imports.

mere 45 000 cars in 1970, imports from this source rose to 743 000 in 1980 and exceeded 1.2 million a year in the early 1990s (Fig. 6.8). Moreover, this penetration was achieved despite two obstacles. The first was strong and effective protectionist measures taken by France and, above all, Italy to exclude Japanese imports. Without these measures, Japanese sales in Europe would by now be several hundred thousand higher each year. One reason why Nissan, Toyota and Honda have built factories in the UK is that the vehicles they produce are not classed as imports, and can therefore be sold throughout the Single Market without such restrictions. Secondly, large-scale imports were achieved despite a major deterioration of the exchange rates for European currencies against the yen. In 1991 the European Currency Unit (ECU) was valued at only 163 yen compared with 370 in 1975.

As is well known, much of Japan's success can be attributed to productivity advantages. Although the efficiency of European car plants varied enormously in the 1980s, productivity in a Japanese assembly plant was typically three times better than the best European practice and up to six times better than in the UK (Commission of the European Communities 1981b). Above all, this reflected the ability of Japanese producers, developing in the recent past, to limit the size of their workforces by substituting automation technology for labour. But cost control has also been linked to the question of wage rates. Although these rose rapidly in 1976–78, they subsequently stabilized and are now less than half the prevailing West German rate, which is the highest in Europe.

One additional factor boosting Japanese sales has been the strategy of making quality an attraction to the customer. Many models eclipse their forerunners in terms of comfort and road-holding, and the challenge to European producers has been intensified by installing as standard equipment traditional 'extras' such as radios, electric windows and sun roofs.

Impressive though they are, Japan's advantages and strategies do not by themselves explain in full the contrasting conditions experienced by these two major branches of the industry. They fail, in fact, to highlight the full framework of constraints within which Western Europe's industry must operate. To a degree these constraints arise from labour relations. The impact on productivity of the UK's highly complex union system is legendary and, at least until recently, in Western Europe in general the unionization of labour has often acted as a brake on the reduction of manning levels.

Only recently have the Western European producers made rapid productivity gains. Similarly, employment protection legislation – highly desirable though it is for the individual – has raised the cost of rationalization

than half her production was exported to other EC markets placed pressure on many long-established centres of vehicle production.

However, the chief threat to the European industry was perceived to be the Japanese car industry. From a

Table 6.11 Car producers and the EC market

	Market share	Profits (billion ECU)
Fiat	14.8	2.4
Ford	11.6	1.1
General Motors	10.9	1.6
Peugeot SA	12.6	1.5
Renault	10.3	1.3
Volkswagen	15.1	0.5
Japanese plants in Europe	2.0	N/A
Japanese imports	8.9	N/A
Other European producers and imports	13.8	N/A
Total	100	

Source: Commission of the European Communities (1992c).

programmes. These constraints have been the offspring of societal attitudes, but other highly significant limitations can be traced back to the political geography of Western Europe. The most important effect of this factor has been to retard reorganization at the European scale. Trans-boundary mergers of national car industries have been shunned and reorganization by US manufacturers has by no means eradicated their early preference for simultaneous operations in several European host economies. Political fragmentation has therefore perpetuated organizational fragmentation, with consequent losses in economies of scale. This does not imply that the industry has fossilized, yet mergers have chiefly taken place within rather than between countries, and individual firms still command relatively restricted market shares. Six leading firms account for three-quarters of Western Europe's output (Table 6.11), whereas only two supply the same proportion in both the USA and Japan.

A second, associated, limitation caused by fragmentation has been that R&D is relatively inefficient. This problem, it is true, has been of limited significance for US firms in Europe since their operations are part of a world system within which innovations can be readily transmitted. Indeed, R&D experience in Europe has been central to US manufacturers' attempts to switch to new, more fuel-efficient, product ranges at home. European manufacturers, however, have inevitably tended to duplicate efforts in this field beyond the point at which the pursuit of alternative approaches to specific problems is desirable

The evolving economic environment and European responses

These problems would be serious enough, even if the environment in which the Western European car industry must operate were to remain constant. However, this environment is highly fluid and several factors may increase substantially the difficulties to be faced. Although at present they pose little threat in the European market, producers in Latin America, South Korea, India and Africa are providing stiffer competition around the world. In Eastern Europe, even before the fall of Communism, the industry was reaching a new level of maturity and technological sophistication, with the result that exports to Western Europe (only 160 000 vehicles in 1979) reached 223 000 in 1985. Given the new political climate in Eastern Europe, it seems inevitable that imports from this region will increase significantly, not least as a result of Western European manufacturers making heavy investments in former Communist countries. Eastern Europe is seen as a low-cost production base, and a large number of Western Europe's car makers have entered into agreements with Eastern European producers. One recent estimate (Commission of the European Communities 1992c) is that exports from this source will rise to 330 000 cars a year by 1995. Examples of new West–East links in the industry are Fiat's joint venture with FSM in Poland, which should produce 300 000 minicars a year by 1995; investment by General Motors in Hungary; and redevelopment of Russia's Yelabuga car plant, again in conjunction with Fiat.

In addition, the US branch of the industry, hit by severe crisis in the early 1980s, has implemented a series of projects intended to improve fundamentally production efficiency, as well as the mismatch between product ranges and market demand. In embarking on these projects, the primary aim of the US industry was to defend itself against Japanese rather than European producers. Yet the implications for Western Europe are none the less substantial: highly efficient competition increases the difficulties of sustaining exports to the USA. Also, US manufacturers may compete more effectively with European products in markets around the world.

Against this background, the European Commission has encouraged the industry to pursue rapid restructuring within the overall context of free competition. By the early 1980s the Commission's view was that it was impossible and, in the interests of productivity undesirable, to avoid the loss of several hundred thousand jobs. However, it was argued, if the industry adopted a constructive, cooperative stance, employment decline could be achieved by orchestrated rationalization rather than uncoordinated *ad hoc* closures. In addition, technology lay at the heart of the Commission's view that the industry could survive and continue to fulfil its growth-pole function in the West European industrial economy. In this connection the basic argument was that the need to optimize production costs to combat competition, plus the technological improvements called for by energy conservation and the car's environmental impact, provided the industry with the opportunity to become a powerful force in the growth of high-technology industries. Seizing this opportunity, it was argued, would generate new forms of high-grade employment, encourage the spread of automation to other industries and protect Western Europe from external control in this vital technological field.

Since the early 1980s, therefore, the industry has been exhorted to revolutionize its technological basis and intensify active cooperation in long-term research and development programmes. Governments have been encouraged to be more liberal in funding research. And there has also been pressure to harmonize national legislation relating to vehicles. Progress in this sphere should increase the prospects of establishing a more uniform European market which will encourage economies of scale. Beyond this, correctly formulated legislation should ensure that European manufacturers are not automatically excluded from major global markets on safety or environmental grounds. But although legislation has the power to influence the product, how efficiently that product is produced remains the ultimate responsibility of the industry itself.

It is now evident that the pace of constructive change has accelerated markedly in response to these economic and political pressures. Various responses to the growing challenge can be identified, including mergers such as those by Peugeot and Citroën (Peugeot SA) in France and Volkswagen–Audi in Germany; joint ventures with Japanese producers (Rover and Alfa-Romeo); and development into true multinational companies. Renault, for example, has assembly plants in the Republic of Ireland and Spain, produces engines in Portugal, is now closely integrated with Volvo and has purchased its first shareholding in the American industry to use as a platform for manufacturing in the USA. In addition, Western European firms have taken the first steps towards collaborative activity intended to gain economies of scale in the production of components such as engines and gearboxes. There is, for example, a network of production agreements between Peugeot, Renault, PSA, Rover, Fiat, Volkswagen and Volvo.

Despite the growth of collaboration, however, the greatest progress has been achieved by restructuring undertaken within firms. This has naturally entailed a swing to automated production, coupled with labour shedding in old plants and low manning levels on new production lines. Thus Renault's plant at Palencia, northern Spain, has an annual output of 100 000 vehicles but only 3300 employees. The company's Seville plant has a workforce of 1600 yet manufactures 500 000 gearboxes a year. But, in addition to embracing new production technologies, European producers have responded to the global challenge by emulating the Japanese strategy of quality improvement. In this context, one goal has been greater mechanical reliability – long since a strong point of Japanese products. The second goal has been the introduction of new models matching imports in terms of performance, equipment and comfort.

Various indicators suggest that these European restructuring efforts are now bearing fruit. First, the industry is now far more automated than in the early 1980s. Employment has been reduced by more than 400 000 jobs, a fifth of the total, with dramatic effects on productivity. Since the early 1980s, productivity has increased by more than 40 per cent in the Western European industry as a whole. Second, European production, after stagnating in the 1970s, has staged a marked recovery (Table 6.12 and Fig. 6.8). Output in the late 1980s was a third higher than at the start of the decade, a comparable performance to that of Japan (+30 per cent) and far better than that of North America (+10 per cent). Third, continued expansion of the international trade in cars has not simply meant a growing stream of imports. Despite the publicity given to Japan, EC exports have regularly exceeded imports, usually by around 500 000 a year. Fourth, in contrast to the earlier situation, production capacity shortages are now to be found in many parts of Western Europe. This has encouraged general interest in Eastern Europe as a production area, and it also lies behind the possible reopening of closed car plants. General Motors, for example, may reactivate its idle factory in Antwerp, while Peugeot is considering this option at Valenciennes. An important consequence of these trends has been a clear improvement in the financial position of many European manufacturers. Most were achieving modest profits by the late 1980s (Table 6.11), compared with heavy losses incurred in earlier years.

Partly as a result of these factors, the rapid loss of European markets to overseas producers that was widely anticipated has been averted. Although the absolute volume of Japanese imports has grown by 500 000 a year since 1980, this is small in comparison with the total number of new cars purchased (more

Table 6.12 West European car production*

	1980 ('000)	1989 ('000)	% change
Austria	7	7	0
Belgium	214	316	+52
France	3488	3532	+1
Germany (West)	3521	4547	+29
Italy	1445	1973	+36
Netherlands	81	135	+67
Spain	1029	1662	+62
Sweden	235	384	+63
UK	924	1229	+33
Total	10944	13795	+26

* Assembly excluded; see Fig. 6.8.

Source: Commission of the European Communities (1981d, 1992e).

than 12 million a year). As a result, Japanese imports to the EC account for only 9 per cent of the market, with imports from other countries taking only a further 4 per cent. Conversely, 87 per cent of the cars sold are produced in the EC, and three-quarters of these are manufactured by European companies (as opposed to US producers, such as Ford and GM, operating in Europe).

In retrospect, Japanese imports have proved to be an important stimulus to the European branch of the industry, and the latter's responses have done much to secure its present position. This is not to say, of course, that competition in Western Europe can be ignored. Indeed there is every prospect that, in the long term, Japan will once again begin to eat into Western European markets. This is because the flow of Japanese imports is not simply held back by the increased attractiveness of the European car industry's products. In addition, in order to avoid very strict controls, Japan has entered into voluntary restraint agreements with five EC countries: the UK, Italy, Spain, Portugal and France. These national agreements cannot be retained now that the Single Market is operating and, with great difficulty, the European Commission has had to negotiate alternative arrangements. These allow for the level of Japanese imports to the EC to be frozen at the current level of 1.2 million a year until 1999. Given a likely decrease in demand, this means that the Japanese market share is likely to rise to 16 per cent by the end of the century. Moreover, after 1999, restrictions on Japanese imports will be removed completely (Dicken 1992b). In the meantime, therefore, it is imperative that efficiency gains in the Western European industry continue as rapidly as possible.

Perhaps the best guarantee that this will occur is provided by the fact that, although the focus is usually on Japan, there is also intense competition between European producers to sell in other national markets. In volume terms this competition is now much greater than that from Japan, and it has undoubtedly contributed a great deal to the emergence of a fitter Western European car industry. As yet, however, the market for European producers is far from perfect. For example, Italy exports many more cars to France and Germany than she does to the UK; France has a much larger share of the Spanish market than does Germany; and Belgium's exports are overwhelmingly towards France and Germany. These and other patterns reflect a wide range of economic, political and cultural influences, but they are now being challenged by the existence of the Single Market. It may well be that it is this which stimulates even greater competition between the European producers and, in the process, equips them to deal with the threat of unrestricted Japanese imports in the next century.

Retrospect and prospect

An increasingly popular framework on which to set the remarkable dynamism of the Western European industrial system is provided by Nicolai Kondratieff's theory of long waves in economic life (Freeman 1984). This hypothesizes that innovations form the basis for extended phases of economic expansion which carry economies to new scales of development and levels of sophistication, but then lose impetus and give way to down-wave periods. In these, direction is lost and uncertainty prevails. Western Europe's impressive post-war industrial growth, the foundations of which lay in technological discoveries made in the pre-war and wartime periods, can clearly be related to this concept. Similarly, the economic deterioration since the late 1970s bears all the hallmarks of a down wave.

Yet this economic interpretation avoids substantial geographical issues that have emerged in this discussion. Although the expansionist era was in macroeconomic terms an unprecedented success, it was associated with unbalanced growth at both the national and international scales. The rise of new forms of industrial organization – especially multinational firms – raised the questions of external dependence and control. Improved legislation by no means eradicated local or long-range industrial pollution. And, almost unanalysed until it had reached an advanced stage, the industrial sector was exposed to greatly intensified competition as the effects of the long wave spread from global core economies to the industrializing periphery.

All these problems have been carried forward into the down wave, a period in which resources to solve them are becoming increasingly scarce, partly because of the growing desire of governments to curb public expenditure. While resources are limited, however, new policy priorities aiming to initiate a new Kondratieff upswing can be identified. For example, as the discussion of SMEs has indicated, strategies designed to encourage the foundation and expansion of small firms were originally formulated by the EC, but in the 1980s they have burgeoned at the national level (Haskins, Gibb and Hubert 1986; Mason and Harrison 1990). As computer technologies advance, the spotlight is focusing on the need to raise awareness in all industries of the potential benefits of computer-assisted manufacturing (Commission of the European Communities 1985a; Woods 1987).

In addition, within the EC, the creation of the Single Market is intended to accelerate the industrial sector into a new growth era. At present, the eventual outcome of this initiative is far from clear. On the one hand the pace of economic integration – measured by mergers and takeovers – is accelerating appreciably and is not simply confined to the activities of high-profile companies or industries. In 1989 more than 700 cross-border acquisitions were made in Western Europe, involving more than a dozen industrial groupings (Table 6.13). This echoes Watts's (1990) finding that a leading player in the takeover field is the far-from-glamorous food industry.

On the other hand, however, there is considerable concern about the potential spatial impact of the Single Market, as several contributors to Bachtler and Clement (1992) make clear. In this context one major fear is that the Single European Market (SEM) with its emphasis on free-market principles, will inexorably work to the advantage of the European core and further disadvantage the periphery (see Ch. 12). Related to this is the possibility that progress towards a single currency will seriously handicap industries in peripheral regions because their governments will no longer be able to improve their competitive positions through currency devaluations. This trend towards the loss of autonomy was highlighted very clearly in 1992, when Italy and the UK left the exchange rate mechanism (ERM) because of the restrictions and pressures which membership placed on their over-valued currencies.

In addition, the 1992 ERM crisis also demonstrated the pressures and uncertainties that have been spread through Western Europe as a result of the fall of Communism in the East, even though this offers the West new market and investment opportunities. In particular, the rapidly escalating costs of German reunification have led the Bundesbank, fearful of inflation, to

Table 6.13 Cross-border industrial mergers and acquisitions, 1989

	Number of deals	Cross-border acquisition value (million ECU)
Food, food retailing, drinks and tobacco	121	7623.1
Automotive and aircraft	93	5061.9
Electronics, computers, etc.	116	3869.0
Paper, printing and advertising	101	3126.4
Chemicals and plastics	61	2948.9
Pharmaceuticals	57	941.4
Engineering	81	917.8
Fashion and textiles	39	831.1
Mining and steel	39	368.5
Packaging	33	300.7
Domestic white goods	11	76.5

Source: Commission of the European Communities (1992c).

impose high national interest rates. Other countries, especially those with weak currencies, have been obliged to follow suit, two important outcomes being reduced consumer spending and a brake on investment. Both trends have hit the manufacturing sector, inhibiting new projects and, therefore, the sector's long-term development prospects. In addition, the withdrawal of Italy and the UK from the ERM for an unspecified period has called into question the whole future of this monetary system.

As the twenty-first century approaches, therefore, it is by no means evident that a new era of trouble-free industrial expansion is about to begin. Although some industries, such as oil refining and car manufacturing, have shown great resilience and ability to adapt, the political and economic environments in which the industrial sector must operate have created major new uncertainties. Successfully adapting to these will be a severe challenge for Western Europe's industries in the coming years.

Further reading

Aldcroft D H 1978 *The European Economy.* 1914–70 Croom Helm, London

Bachtler J, Clement K (eds) 1992 Special issue: 1992 and regional development. *Regional Studies* **26**: Part 4

Commission of the European Communities 1985 Advanced manufacturing equipment in the Community. *Bulletin of the European Communities, Supplement* 6/85

de Smit M 1992 International investments and the European challenge. *Environment and Planning* A **24**: 83–95

Dicken P 1990 European industry and global competition. In Pinder D A (ed) *Western Europe: challenge and change*. Belhaven, London, pp 37–55

Dicken P 1992 Europe 1992 and strategic change in the international automobile industry. *Environment and Planning A* **24**: 11–31

Environmental Resources Ltd 1983 *Acid Rain: a review of the phenomenon in the EEC and Europe*. Graham and Trotman, London

Freeman C (ed) 1984 *Long Waves in the World Economy*. Frances Pinter, London

Haigh N 1986 *European Community Environmental Policy, Volume 1, Water and Waste in Four Countries*. Graham and Trotman, London

Mason C M, Harrison R T 1990 Small firms: phoenix from

the ashes? In Pinder D A (ed) *Western Europe: challenge and change*. Belhaven, London, pp 72–90

Taylor D, Diprose G, Duffy M 1986 EC environmental policy and the control of water pollution: the implementation of Directive 76/464 in perspective. *Journal of Common Market Studies* **24**: 225–46

Wabe J S 1986 The regional impact of deindustrialization in the European Community. *Regional Studies* **20**: 23–36

Watts H D 1990 Manufacturing trends, corporate restructuring and spatial change. In Pinder D A (ed) *Western Europe: challenge and change*. Belhaven, London, pp 56–71

Williams R 1990 Supranational environmental policy and pollution control. In Pinder D A (ed) *Western Europe: challenge and change*. Belhaven, London, pp 195–208

7

The service sector

The sectoral transition

It is abundantly clear that each of the economies of Western Europe is now established in the post-industrial era, despite the enduring imprint of manufacturing activities old and new and the fact that the greater share of the land surface is still devoted to agriculture. Indeed, it is sometimes remarked that the so-called industrialized economies should be renamed service economies. Varying forms of service activity have always existed but their range and importance in Western Europe have increased markedly since mid-century as levels of development and personal well-being have been enhanced, even when economic crisis and rising unemployment since 1973 are taken into account. As central place theory informs us, low-order 'traditional' services, such as retailing, distribution, schooling, basic health care and local administration are spread widely through the settlement hierarchy, while top-level, specialized and innovative activities cluster in a small number of highly developed, information-rich metropolitan centres, notably national capitals and 'world cities' (Hall 1984). It is at these focal points that international finance and banking, marketing, advertising and the media are particularly prominent (Howells 1988; Ochel and Wegner 1987). Departing from theory, the promotion of tourism in many coastal, mountainous and culturally rich environments has boosted the significance of service facilities for visitors in many Mediterranean and Alpine areas (see Ch. 10).

With respect to both employment and financial importance, the production of food and manufactured goods has been overtaken by the generation and delivery of services to support the production of goods and to enhance the quality of life. Three-fifths of the gross value added in the eighteen economies derives from services, which employ a comparable share of the total workforce. Over the last two decades, the absolute number of jobs in both the primary and secondary sectors has contracted sharply in Western Europe as a result of rationalization in farming, forestry, mining and traditional manufacturing and because of the challenging interaction of an array of social, political, financial and organizational processes that are summarized conventionally by the harsh term 'deindustrialization' (Goddard and Champion 1983). By contrast, the number of employees in service activities in the eighteen countries rose from 58 630 000 to 88 308 000 between 1971 and 1990, by which time service-sector employment was very much greater than the primary (10 035 000) and secondary (48 180 000) sectors combined. This transition from the production of goods to the delivery of services has implications for every man, woman and child in Western Europe. It permeates all aspects of human well-being, affecting economic dynamism, employment opportunities, educational requirements, class and gender relations, regional disparities and a host of other characteristics relating to the quality of life. Once-vibrant activities have become 'sunset industries' and service-related 'sunrise' employment has brought new opportunities to some localities but has denied them in others. The economic geography of Western Europe is being painfully refashioned and so too is its social geography, as old skills are no longer valued and new ones are required in an age of computing and information technology (Marstrand 1984). Indeed, the services revolution has infiltrated the total geography of our eighteen states so thoroughly and forcefully that it is simply taken for granted.

'Service employment' and its synonym the 'tertiary sector' represent a poorly defined, catch-all category which lumps together most activities that do not directly involve producing farm crops, extracting minerals or manufacturing commodities (Daniels 1991; Illeris 1989). Services therefore range from international banking to tourism, from government and administration to entertainment, and from health care

Table 7.1   Origin of gross domestic product, by sector, 1990 (%)

	Goods	Market services	Non-market services		Goods	Market services	Non-market services
Austria	44	41	15	Italy	46	40	14
Belgium	36	49	15	Netherlands	39	47	15
Denmark	34	43	24	Norway	46	40	14
Finland	46	37	17	Portugal	48	41	12
France	41	45	14	Spain	47	43	10
Germany (West)	45	41	14	Sweden	39	36	26
Greece	48	42	10	UK	42	42	16

and education to distribution and retailing. Some service jobs are emphatically in the private sector, others depend on the state; some are filled predominantly by men (e.g. transport), others involve mainly female staff (e.g. domestic and hotel services); some are immensely well paid, others place their workers well below the poverty line (Seabrook 1985). City financiers, civil servants, opera singers, bus drivers, kitchen staff, hotel porters, dustmen and lavatory cleaners are all service employees! Thus the services sector stretches from the most comfortable and well-rewarded members of society to the most exploited 'guestworkers' suffering the crudest discrimination.

Trying to classify this great variety of activities has become something of a craft industry among social scientists in recent years. A fundamental distinction between 'market' (or private) services and 'non-market' (state-organized) services is clear enough and effectively separates central and local government and state education and health services from the enormous array of privately funded retailing, leisure and financial operations (Table 7.1). Likewise, another distinction between 'producer' (or intermediate) and 'consumer' (or final) services provides a helpful although complex classification. In the case of producer services (such as wholesaling, business or planning consultancy, public relations and market research), the output is used exclusively by other industries, whereas consumer services (including retailing, passenger transport, building societies and leisure functions) deal directly with the public (Daniels 1982, 1985). Some researchers have distinguished 'quaternary' activities which embrace a high-order component among the broad array of services. Typical quaternary activities would include the legal professions, accountancy, management consultancy, banking, insurance, central and local government, higher education and research (Goddard 1975).

The number of service workers throughout Western Europe rose by 51 per cent between 1971 and 1990 but markedly different rates of growth were in evidence between nations, regions and individual localities.

Variations in census definition between countries may account for a fraction of these differences but the real explanation lies, first, with the level of economic development in each country prior to the onset of the energy crisis and, second, with the national mix of service activities, each of which displayed its own distinct growth trend. For example, long-urbanized, services-rich economies, such as those of West Germany, the Benelux countries and the UK, displayed slower rates of growth than those of the services-poor developing economies of the Mediterranean where office employment and tourism-related jobs were increasing rapidly from a notably low base line.

According to United Nations' statistics, half of all service jobs in Western Europe involve 'community, social, personal and other services' which are strongly represented in the Nordic 'welfare states' of Denmark and Sweden (Fig. 7.1). Employment in 'trade, restaurants and hotels' covers almost one-third of the total, being very significant in tiny Luxembourg and in popular tourist destinations (Spain, Switzerland, Italy). The remaining one-fifth of service jobs are divided equally between 'transport, storage and communications' (well represented in Greece, Iceland, Norway, Finland and Austria) and 'finance, insurance, real estate and business services' (which are particularly significant in France, the Netherlands, Denmark, the UK and in Switzerland, which has the rare distinction of having more bank branches than dentists).

These four groups of service have developed in very different ways since 1975. Taking Western Europe as a whole, the fastest rates of growth involved the financial and business sector which expanded substantially in almost every country. However, an important distinction must be drawn between the high-earning-power employees in big business operating in national capitals and 'world cities' (notably London and Paris) and the more modest implications of additional main street banks and building societies in provincial towns (Daniels 1983; Green and Howells 1987). The finan-

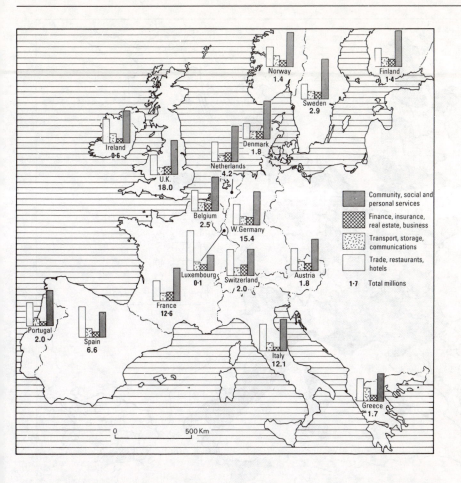

Legend:
- Community, social and personal services
- Finance, insurance, real estate, business
- Transport, storage, communications
- Trade, restaurants, hotels
- 1·7 Total millions

Map labels:
Norway 1.4, Finland 1·4, Sweden 2.9, Ireland 0·6, Denmark 1.8, Netherlands 4.2, U.K. 18.0, Belgium 2.5, W.Germany 15.4, Luxembourg 0·1, Switzerland 2.0, Austria 1.8, France 12·6, Portugal 2.0, Spain 6.6, Italy 12.1, Greece 1.7

0   500 Km

Fig. 7.1
Tertiary employment by category (%), *c* 1988

cial group was followed quite closely by rapid job creation in community, social, personal and other services, especially in the Nordic states. On average, jobs in trade, restaurants and hotels increased only modestly across Western Europe, but stronger growth occurred in Mediterranean (Italy, Portugal, Spain) and Alpine (Austria) nations that were exploiting their tourist facilities vigorously. Finally, total employment in transport, storage and communications has risen only slightly since 1975, with increases in some countries being offset by losses in others.

At a finer scale of analysis, a distinction must be drawn between countries that have traditionally concentrated high-order finances, services and administration in their capital cities only (e.g. France, Spain and the UK) and those that have supported them in a more deconcentrated fashion (Reitel 1976). The best example is undoubtedly the array of ministries and administrations in each state (*Land*) of the Federal Republic of Germany. Recent measures to devolve some political responsibilities to the regions of France and especially to Spain may tone down the tertiary supremacy of their capital cities. However,

there is an opposing tendency which retains top-level activities in capital cities to ensure that they maintain or even enhance their international role. With 70 per cent of more of their gross value added being derived from service activities, sixteen regions (Fig. 7.2) form the emphatically services-rich parts of the EC. Roughly half of these areas house major administrative and/or financial functions in the form of national capitals (e.g. Copenhagen, Madrid, Rome) or 'world cities' (e.g. London, Paris, Brussels, Randstad Holland) (Burtenshaw, Bateman and Ashworth 1991). In addition, the tiny city-state of Hamburg, with top-level commercial shipping and publishing activities, also falls into this class. Such core areas are the leading nodes of West European capitalism and form the most affluent components of their national territories and – in some cases – of the whole of the EC. The remaining services-rich regions are strongly geared to tourism (including Provence, Corsica, the Canary Islands and the Balearic Islands). Unlike the first group of regions, these Mediterranean holiday playgrounds are particularly vulnerable to economic fluctuations elsewhere in Western Europe (Williams and Shaw 1991).

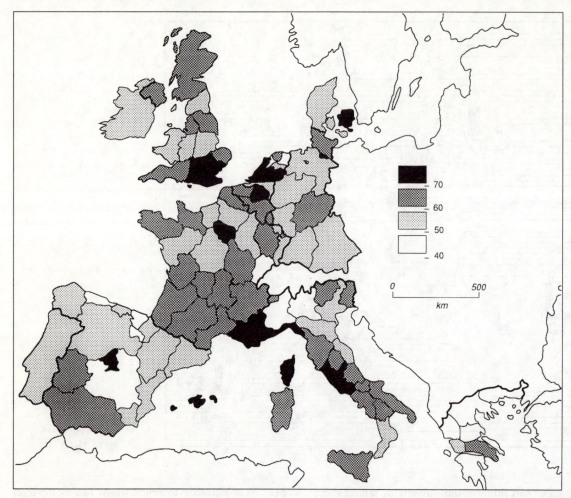

Fig. 7.2
Proportion of gross value added derived from services (1988).

Extra accommodation is constructed when demand is buoyant, but when bookings slump cut-throat 'bargains' have to be offered to avoid hotels standing empty or planes operating below capacity. Out-of-season unemployment is endemic, and domestic and catering staff are notoriously poorly paid, since the financial margin is drawn exceedingly tight. A job in the service sector is by no means always rewarded with a pot of gold, nor for many people is there even a glimpse of the rainbow.

Geographical expressions

The geography of Western Europe has been reshaped at both the regional and local level as a direct result of the services revolution. The economic potential of many core regions has been reinforced but, in addi-

tion, new opportunities have been realized in provincial areas, thanks to the development of tourism, the relocation of office work and the creation of completely new service functions. Detailed land-use plans have been transformed not only in long-established central business districts but also in greenfield sites whose accessibility and economic significance have been enhanced by the massive increase in private motoring that has affected sizeable sections of all West European societies over the past quarter century. These themes are interwoven in the discussion on many pages of this book (e.g. Ch. 8 on Urban Development and Ch. 10 on Recreation and Conservation); a small but significant sample has been selected for particular attention at this point.

Offices

The office is the vital ingredient and the most powerful symbol of the services revolution, with additional office space being provided in a host of ways (Cowan 1969; Daniels 1975, 1979). Changing the function and internal arrangement of existing warehouses, factories, shops or residences forms the simplest solution which may, indeed, be the only one possible in an urban conservation area (Damesick 1979; de Smidt and Wever 1984). But in many business districts, not only individual buildings but whole blocks have been demolished and replaced with modern office premises. Complete redevelopment offers the chance of providing exactly the floorspace and facilities that a business client may require. It also offers the developer the opportunity of maximizing the financial return from his investment. In this way the detailed fabric of central districts of London, Brussels and almost every large city in Western Europe has been subject to piecemeal transformation and the installation of high-rise office blocks in place of low-rise buildings (Bateman 1985). (By virtue of a ban on construction of skyscrapers, the historic heart of Paris is an exception to this rule) The risks involved in such operations are considerable, with the slump in demand for floorspace during the mid-1970s leaving extensive areas and even whole office buildings unoccupied for months and, in some notorious instances, for years on end.

A more radical response to the demand for modern office accommodation has been the construction of completely new business districts some distance from the traditional central business district (CBD). The Part-Dieu district in eastern Lyons and Hamburg's City-Nord are interesting provincial developments which were designed to provide good facilities for major firms. However, the most striking example in the whole of Western Europe is the creation of 1 550 000 m^2 of offices in a suite of controversial towers at La Défense in the suburbs of Paris, some 9 km north-west of Notre-Dame. Over the past thirty years, 70 000 jobs have been installed in the offices, shops and other service activities of La Défense. Its success has further distorted socio-economic imbalances in greater Paris, where the deindustrialization of eastern working-class districts contrasts with the growth of tertiary employment in more emphatically middle-class neighbourhoods on the west (Clout 1988). A major challenge for Parisian planners and politicians is to stimulate new employment opportunities in offices, leisure activities, research institutes and high-tech industries on the eastern side of the city (Tuppen and Bateman 1987). The construction of a massive office complex at Canary Wharf in London's Docklands has a similar strategic aim of installing jobs and new opportunities in an area of serious economic decline (Lenon 1987; Brownill 1990).

The utility of relocating offices away from congested and expensive national capitals in order to stimulate additional employment in the provinces has been appreciated by planners in several West European countries and special laws have been passed and agencies set up to this effect. The most comprehensive schemes involved the work of the Location of Offices Bureau between 1963 and 1978 which encouraged firms to decentralize their activities and their workforce to cheaper premises in provincial cities or in outer parts of south-east England (Marshall 1985). Routine, paper-shuffling operations in government, banking, insurance and many other activities were relocated in quantity, taking full advantage of postal, telephone and telecommunications services to maintain links with clients. Roughly comparable policies shifted a number of routine functions from Paris to the French provinces, but success was greater with government departments than with commercial firms. Likewise, some government agencies have been shifted from Stockholm and from Randstad Holland (Öberg and Oscarsson 1987; de Smidt 1985).

By contrast, top-level office functions (in both the public and private sectors) and many of a particularly innovative nature have remained firmly established in Western Europe's capitals. Conventional wisdom maintains that such cities provide the only environments where top-level managers and directors can have face-to-face contacts with their counterparts in finance houses, banking and government (Donnay 1985). A second strand of reasoning argues that the status of a 'world city' – and the nation it symbolizes – would be reduced if high-order functions moved away. The rather ambivalent attitude of the French government to decentralizing office activities may be explained in this way.

The introduction of new technology in the form of telephones and visual telecommunications has revolutionized practices on London's stock market and has done away with some time-hallowed face-to-face contacts, but the attractiveness of the City of London for financial dealing remains strong, although there are powerful challenges from other 'world cities' Deregulation of trading accelerated demands for new, spacious offices that could house large computing facilities within the 'square mile', its immediate environs and as far eastwards as Docklands. (In the present recession such offices are in oversupply in London.) Modern telephone and facsimile systems have transformed many aspects of office activity and have certainly reduced some locational constraints; however, it is unlikely that advanced telecommunications will remove the friction of distance and provide complete spatial freedom, thereby producing a measure of economic convergence between metropolitan cores and provincial peripheries (Gillespie and Goddard 1986). For unlike the ubiquitous telephone service, really advanced systems of electronic communication will be market-led and will

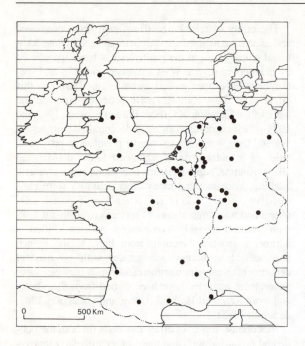

Fig. 7.3
Main science parks in Benelux, France, West Germany and the UK.

probably serve only areas of established economic advantage and the routeways between them. It is most unlikely that sufficient investment will be forthcoming to equip all peripheral areas or 'sunset' industrial zones.

Innovative developments

During the past couple of decades, many important new trends have added to the complex impact of the services revolution on the geography of Western Europe. Three examples relating to research, leisure and retailing provide a view of these innovative developments.

The first involves the creation of science or technology 'parks' where R&D are undertaken, important services are generated and perhaps also a range of high-tech industries are to be found (Benko 1991; Komninos 1992). Inspired by the success of Silicon Valley in California and Boston's Route 128 in New England, developments of this kind have been perceived as a panacea to many economic and social ills and have been attributed with remarkable powers of job creation and income generation (Hall and Markusen 1985). An information-rich environment, usually adjacent to a dynamic university or other establishment undertaking vibrant and useful research, is accepted as a prerequisite for a successful science

park. An 'attractive environment', expressed in landscape, climate, mountains or cultural activities, forms a secondary but still important contributory factor. There is, in fact, considerable diversity of approach, with various initiatives emerging from universities, city councils and regional planning organizations. In the UK, the Cambridge science park opened in 1973 and has been emulated by a dozen other British universities or development agencies, but with varied results. Although not a science park in the narrow sense of the term, the real British success story involves the M4 corridor to the west of London, with good access to government research institutions, universities, an international airport and the national motorway network (Fig. 7.3).

The science park idea germinated early in Belgium, with four universities having created them by 1974; by contrast, developments came much later in the Netherlands. The quality of research in universities and other institutions, coupled with the attractions of the Alpine environment, gave rise to a 'Silicon Bavaria' phenomenon, with 40 per cent of West Germany's computer software suppliers located around the city of Munich. German technology centres have been established to stimulate the growth of 'future' activities in a wide range of locations including areas of industrial decline (e.g. the steel towns of Dortmund and Saarbrucken), already prosperous cities (e.g. Bonn, Heidelberg and Stuttgart), and West Berlin which has received substantial financial support from the federal authorities and where there are close links with the technical university. French science parks at Grenoble (3600 jobs) and Nancy (800 jobs) and the constellation of high-tech activities in the Bièvre Valley to the southwest of Paris, each benefit from established university expertise. However, the large development at Valbonne-Sophia Antipolis is an entirely new creation that originated in the early 1970s following government decisions to focus a range of 'future' activities on the Côte d'Azur. The state invested heavily to establish the

Table 7.2 Top ten sites visited in the UK, 1991*

Madame Tussaud's, London	2 547 000
Tower of London	2 298 000
Alton Towers, Staffordshire	2 070 000
Natural History Museum, London	1 534 000
Chessington World of Adventures	1 515 000
Blackpool Tower	1 426 000
Royal Academy, London	1 309 000
Science Museum, London	1 303 000
London Zoo	1 250 000
Kew Gardens, London	1 196 000

* Number of visitors where there is an admission charge.

Fig. 7.4
Major theme parks (after Cazès 1988).

the Netherlands is Western Europe's largest theme park which attracts about 2 400 000 visitors each year, followed closely by Phantasialand, near Cologne (2 300 000 visitors) (Cazès 1988) (Fig. 7.4).

In March 1987 a contract was signed for EuroDisneyland to be constructed on 1940 ha of farmland within the perimeter of Marne-la-Vallée new city to the east of Paris. Fourteen sites in France and Spain had been considered before this decision was reached. The seasonality of Mediterranean tourism outweighed the far better climatic prospects of Iberia. Despite a much cooler and damper climate, Marne-la-Vallée has the advantage of being relatively close to large concentrations of population in other states of north-west Europe, is served by a major international airport (Roissy-Charles de Gaulle) and was made all the more appealing by generous financial offers from the French state (Fig. 7.5). Work progressed apace enabling

infrastructure of Sophia Antipolis, where it relocated the prestigious Ecole des Mines, other research institutions and the Air France international reservations centre which required innovative telecommunications systems. Now over 4000 people are employed on the technology park and more than half are specialists who have moved from other regions to work in the attractive environment of the south of France. Local people work mainly in low-qualified positions and private investors have been somewhat reluctant to follow the example of the public sector. Innovation transfers mainly involve links with international agencies; indeed, the impact of Sophia Antipolis on the regional economy has been rather small. The hard truth is that it takes much more than a university, some land and some sunshine to produce a Silicon Valley!

A second new development reflects the fact that the great majority of West Europeans have more free time and greater amounts of money to devote to non-essential items than ever before. Providing facilities for leisure time use has indeed become a major 'industry' with important spin-offs for catering and the hotel trade and with very real appeal to visitors from overseas, many of whom are no longer satisfied with the round of castles, cathedrals and coasts but seek innovative attractions which entertain or educate and sometimes manage to do both. For example, the top ten sites frequented by visitors in the UK during 1991 included a number of modern attractions and theme parks as well as castles, zoos and gardens (Table 7.2). At present, De Efteling in

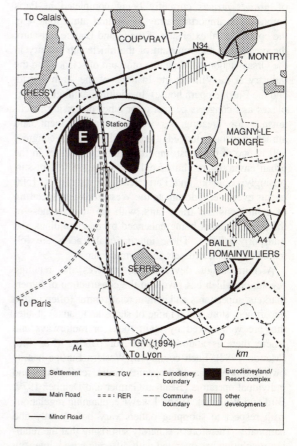

Fig. 7.5
EuroDisneyland.

EuroDisneyland to open in April 1992. Access is by express metro from central Paris and by motorway. In addition EuroDisneyland has a station on the TGV (high-speed train) network with services from many parts of France and neighbouring countries. If the promoters' estimates prove correct, EuroDisneyland will attract 10–12 million visitors per annum, perhaps rising to 17 million, of whom half will come from beyond France. Six themed hotels (e.g. 'New York', 'Far West'), providing 5000 bedrooms, have been constructed, together with camping and self-catering accommodation, shops, offices and congress facilities. Perhaps 12 000 jobs will be created. However, it has to be noted that some theme parks have encountered difficulties in recent years and have failed to reach their visitor targets and estimated financial turnover. Theme parks also provide an imaginative way of restoring derelict land with the 'World of the Smurfs' (northern Lorraine) and 'Wonder-World' (Corby, East Midlands) both being developed on the sites of former steelworks.

In a rather different style, it is notable that the suite of *grands projets* currently being completed in Paris embraces an important array of cultural attractions (e.g. the controversial extension of the Grand Louvre museum, the creation of the museum of the nineteenth century in the Orsay station building, the museum of science and the industry at La Villette, the Institute of the Arab World, and the Bastille opera house). In addition, exhibition complexes and congress centres offer other promising mechanisms for creating service-sector employment and attracting income; indeed, it may well be that the congress hall is one of the most important tertiary facilities in the urban environment (Labasse 1984; Law 1986). Not surprisingly, London and Paris (each hosting over 1500 events every year) dominate the West European spectrum of exhibitions and trade fairs, with the attractiveness of the French capital being enhanced by new buildings and transport facilities. These international leaders are followed by Milan, Brussels and Birmingham.

A third major development involves the retailing revolution which has favoured the construction of hypermarkets, superstores and commercial centres (often with a department store and a range of shops and banks) at sites that are well served by main roads or motorways and where there is abundant space for parking (Fig. 7.6) (Metton 1982). Each country in Western Europe has its own traditions regarding opening hours, retail chains and individual operators (Beaujeu-Garnier and Delobez 1979; Bird and Witherick 1986). Likewise, planning regulations with respect to shopping outlets vary both locally and nationally. Therefore it is hardly surprising that the retailing revolution has operated with differing speed, intensity and form from country to country. Even now, the trend is

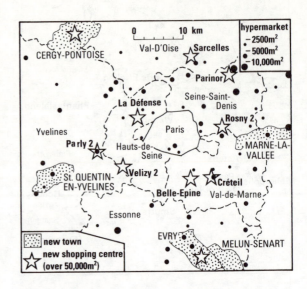

Fig. 7.6
Major shopping centres and hypermarkets in the Paris region, 1981 (after Metton 1982).

less developed in the UK than in France and several other nations; however, the largest shopping complex in Western Europe has been constructed on what was formerly a power station's waterlogged ash dump alongside the River Tyne at Gateshead. The site forms part of the Tyne and Wear enterprise zone and has a potential clientele of 1 300 000 within 30 minutes' drive. The completed Metro Centre boasts over 200 000 m² of floorspace which is occupied by shops, department stores, restaurants and support facilities. Despite national variations, there is no doubt that with high rates of car ownership, more families having freezers, and increasing numbers of women in full- or part-time employment outside the home, shopping for many households has been transformed from a daily ritual to a weekly, fortnightly or monthly bonanza.

The competition can be can be deadly for conventional main-street shopkeepers who see a sizeable proportion of their clientele being attracted to hyper-markets, retail warehouses and do-it-yourself superstores (Jones 1984). In France the situation became so controversial that city-centre traders engineered special legislation in 1973 to dampen down the proliferation of large per-urban shopping facilities (Dawson 1979,1980). However, the clock cannot be turned back and for very many car-owning West Europeans, convenience shopping means an occasional trip out of town. City centres can only hope to recapture trade by ensuring a range of attractive outlets

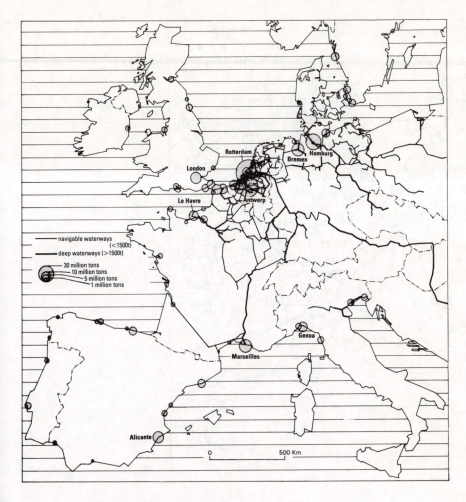

Fig. 7.7 Transport systems: (a) ports and navigable waterways; (b) railways (after DATAR, Paris).

Table 7.3 Major ports and airports in Western Europe, 1986

Port	Cargo (million tonnes)	Airport	Passengers (million)	Freight ('tonnes)
Rotterdam	224.4	London*	50.1	726
Marseilles	86.6	Paris*	33.0	713
Antwerp	80.3	Frankfurt	19.8	784
Le Havre	53.5	Rome	12.9	179
Hamburg	50.6	Amsterdam	11.7	451
Genoa	42.2	Madrid	10.8	170
London	41.7	Stockholm	10.6	60
Tees-Hartlepool	33.5	Copenhagen	10.0	131
Milford Haven	30.6	Palma	9.9	25
Dunkirk	30.2	Athens	9.6	89

* More than one airport.

(pedestrian streets or precincts, good department stores or indoor shopping complexes, and boutiques) and by providing adequate cheap parking space. But even a glittering, purpose-built shopping centre in the heart of a city is not without its problems. It can easily become a place where young unemployed people congregate and may acquire the reputation of being slightly dangerous. Its car parks can become places of petty theft and crime, so more security guards are needed. And, of course, its very existence 'steals' activity and

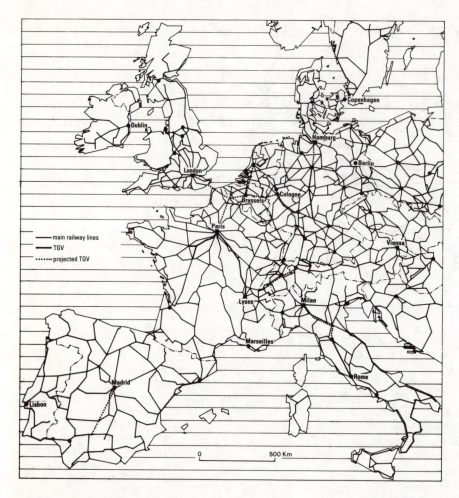

main railway lines
TGV
projected TGV

Fig. 7.7b

trade from old-fashioned shopping streets which all too rapidly display boarded-up windows, charity shops and declining pubs and cafés. In the evening the outer doors of the new shopping complex are locked, since it is 'defensible' private property and not a public thoroughfare. Increasingly the life and death of city centres and shopping streets is being decided in the boardrooms of urban development corporations and major retailing companies.

Transport and trade

Road and rail

With 157 000 km of functioning railway line, 30 000 km of inland waterway, 35 000 km of motorway, 332 000 km of main road, a host of oil and gas pipelines, numerous air and ferry services, and a variety of major ports and airports, the densely peopled states of Western Europe display the most complex

network of connectivity of any comparably sized section of the earth's surface (Table 7.3; Fig. 7.7a). Indeed, that distinction is far from new; it was valid in earlier centuries although the range of components was strikingly different. The pattern has evolved substantially since mid-century and important changes have occurred with respect to both volume and mode of transportation (Caralp 1989). Undoubtedly the most significant has been the growth of private motoring which has endowed individuals and families with completely new forms of spatial choice and travel convenience but has also generated serious problems for all kinds of public transport.

In 1990 there were 142 million private cars on the roads of Western Europe (four times the 1960 total) and over 18 million commercial vehicles. As well as the privilege of personal mobility, mass car ownership involves serious problems of congestion, pollution and consumption of land for new highways and parking facilities. No fewer than 46 800 people were killed and

1 690 000 injured on the roads of the Twelve in 1990, but mercifully these figures represent sizeable reductions since the mid-1970s. Construction of new motorways, highways and bypasses continues apace although with somewhat less vigour than before the energy crisis. None the less, motorways are still rare in some regions although high densities serve West Germany, the Benelux countries and parts of northern Italy. But all motorways were planned by individual nations which placed the requirements of their core areas at the top of their agenda and paid less attention to the needs of peripheral areas adjacent to neighbouring countries. The EC has long argued the need to plug the gaps and has advanced loans through the EIB to assist construction of Alpine motorways (Brenner Pass and Val d'Aosta) and the links between Paris and Brussels, Lorraine and the Saar, and between France and Italy along the Mediterranean. Several new challenges result from German unification. Existing autobahns and main roads in the eastern *Länder* need improving to meet current requirements and east/west routes need to be enhanced, especially through what previously had been border localities. Improved road and rail facilities through east-central Europe will be essential if the two parts of the continent are to be brought together effectively in the immediate future.

Over the past thirty years, Western Europe's railways have lost passengers to air services and to the use of private cars, and increasing volumes of freight have been transferred to movement by lorries. Railway authorities have reacted in differing ways to these critical changes. Most have successively rationalized their networks and the services operated on them, closing little-used stations and passenger services and retaining some lines simply for the movement of freight. Investment has been concentrated on improving track and rolling stock on a limited number of major axes. The Beeching Report (1963) on Britain's railways was a well-known analysis of this kind. A more recent (1987) example is a review of traffic flows and financial turnover on Swedish railways. This made grim reading, with market forces identifying the predictable logic of retaining only a few trunk lines and closing all passenger services in the northern two-thirds of the country. By contrast, other authorities – such as those in Italy – have accepted the necessity for heavy and sustained subsidy on the basis of the social role that the railway plays in local as well as inter-city travel.

A third reaction has been for railway authorities to rise to the challenge posed by road and air transport by introducing innovative and greatly accelerated services along a small number of carefully selected routes. The rationale for this kind of policy has been boosted by growing concern over the amount of land being taken for motorways and airport extensions and by the telling fact that trains are far more efficient energy consumers than either cars or planes for inter-city journeys on the West European scale. Thus rail can compete particularly effectively with road and air over journey distances of 200–1000 km.

French and West German railway authorities have each introduced its own type of high-speed train using existing tracks. In addition, a couple of new sections of line have been built in western Germany in order to shorten the north/south route between Hamburg and Munich and to link Mannheim and Stuttgart. Their construction has been particularly desirable since West German railways simply inherited part of the much larger pre-war German network which focused emphatically on Berlin and was not particularly well provided with north/south links. By far the most impressive progress has been achieved in France with the opening in 1981 of a TGV (*train à grande vitesse*) service on the heavily used route between Paris and Lyons (Fig. 7.7b). The rail distance between the two cities was trimmed from 510 to 425 km by slicing a special track with neither tunnels nor level-crossings through the Paris Basin and the hills of western Burgundy, whose gradients can be negotiated successfully by the powerful TGV locomotives. Travelling time from city centre to city centre has been cut from 3 hr 50 min to 2 hr, which compares very favourably with total travel time by plane. Equally important are the improved services which fast TGV trains provide beyond Lyons with, for example, Geneva being only 3 hr 20 min and Marseilles 4 hr 20 min away from Paris. In planning their TGVs, the French deliberately democratized the service, allowing second-class passengers to travel on the fastest trains for the first time in a generation. Many passengers have been wooed back to using the train and some postal services between Paris and the south of France are now carried by TGV rather than by plane. The rule of thumb that says that air travel will attract a majority of travellers between two cities if the competing centre-to-centre rail journey takes 4 hr or more is having to be rethought.

TGV Sud-Est is a clear success story, but it is just the first part of a much larger plan. Some 285 km of new track were built for the TGV Atlantique which provides even faster services than those operating between Paris and Lyons, bringing Nantes and Rennes just 2 hr travelling time from the capital and Bordeaux only 3 hr away. Already the TGV has adopted an emphatically international dimension with the Paris–Berne service being inaugurated in spring 1987. Just a few months earlier, the transport ministries of France, Belgium, West Germany, the Netherlands and the UK had accepted a Franco-German draft report on the desirability of a high-speed inter-city network. Two years later a Community of European Railways report, covering three West European states, envisaged an expanded network of 7000 km of TGV services in France (including 2300 km of new lines), 4500 km of high-speed inter-city services in Germany and 2200 km of new track for the Italian Alta Velocita system. In addition, Spain ordered TGV-type trains from France for its own high-speed service on a new track between Madrid and Seville. (This duly came into operation in time for Expo '92 and provided a $2^1/_2$ hr service between the two cities in place of the previous 6 hr journey time.) A number of 'corridors' were recognized along which the operation of high-speed trains would be desirable, including Lille–Paris–Lyons–Marseilles, an 'Atlantic' route (Lille–Paris–Bordeaux–Madrid–Lisbon), and an X-shaped double-corridor starting in Amsterdam or Hamburg, converging on the Rhine–Main region, then continuing to Milan–Rome–Naples or Munich–Vienna. Seen from this mainland perspective, the UK market is unquestionably peripheral; it will only be able to lock on to the continental network when the Channel Tunnel and necessary high-speed links in Britain are completed.

The Channel Tunnel

A number of major improvements to Western Europe's motorways and rail services have been achieved in recent years but some important missing links remain. A bridge has been proposed to span the Straits of Messina (between the Italian mainland and Sicily) and work is under way on a fixed link across the Great Belt between the Danish islands. This involves a tunnel and a bridge to handle railway traffic (due for completion in 1993) and a motorway bridge that will be finished in 1996 (Vickerman 1991). The question of a link across the Sound between Denmark and southern Sweden remains unresolved. By far the most notable omission has been some form of fixed link between northern France and south-eastern England. A Channel tunnel was first proposed in 1751 and the twenty-sixth scheme (this time for a publicly financed tunnel) was abandoned by an incoming British Labour government for reasons of cost. But since that time the volume of cross-Channel traffic has increased substantially, with many more cars and coaches using ferries and hovercraft for holiday travel and for freight movements which have been facilitated by containerization and technical improvements (but also risks!) in roll-on roll-off ferries (Vickerman 1986, 1987). In January 1986, a Socialist president and a Conservative prime minister announced the agreement of their governments to a proposal from a Franco-British consortium for a privately financed 37 km long rail tunnel to run between Calais and Dover and to be completed in 1994. Cars and lorries are to be loaded on to special trains that

will shuttle through the tunnel. Long-distance passenger trains will operate directly between the two capital cities.

The principal benefits of the fixed link will be a reduction in the time – if not the money – costs of travel and the encouragement for more journeys for passengers and freight. The main financial costs are to do with construction and operation, but there are profound social and environmental implications as well. After so many dashed hopes in past decades, the present scheme has not attracted investment as readily as had been hoped and has given rise to profound opposition in some quarters in Britain. Environmental groups in Kent oppose the further desecration of 'the garden of England' by terminals, marshalling yards, new roads and predictably overcrowded motorways. Motorists did not see much point in a project that would simply give them the choice of loading on to a ferry or a train. Local politicians fear that construction workers would not be recruited locally and that after 1993 many ferry services would be forced to close, with consequent loss of jobs, especially in Folkestone and Ramsgate. MPs from northern England fear that the tunnel would simply increase the nation's north/south divide. Academic analyses suggested that eastern Kent may benefit during the construction phase from an accumulation of goods, capital, people and services, but that the long-term local effect might involve large job losses and poor prospects for further industrial or commercial development (Gibb 1986). By contrast, and as northern MPs fear, the economy of inner Kent and south-eastern England as a whole may benefit from its enhanced accessibility. French attitudes to the tunnel have been far more optimistic, since northern France is not part of 'the home counties' but is an industrial and former mining region with a disturbing history of employment decline. It is, by definition, closer than southeast England to major continental markets and is in receipt of substantial regional development aid from Paris and from the EC. There is abundant land for terminals, improved rail and motorway links and the local population has made important strides in enhancing the skills of its labour force for employment in high-tech and other new economic activities. It would seem, in fact, that northern France has everything to gain (Chisholm 1986; Bruyelle 1987, 1988; Holliday and Vickerman 1990). The tunnel will be open to train traffic in 1994. In October 1991 the UK government announced that the passenger terminal for the rail link from the Channel would be at Stratford on the northern fringes of Docklands and would contribute to wider plans for the regeneration of the eastern side of London. Just as the Channel Tunnel has proved highly controversial, so the idea of establishing a fixed link between Denmark and Sweden has given rise to great differences of opinion. Would a road and rail bridge across the Sound from Copenhagen to Malmö be appropriate, or

Table 7.4 Imports and exports for the EC, EFTA and Western Europe, by commodity class, 1990 (%)*

	Foodstuffs	Mineral fuels	Raw materials	Machinery	Others
Imports					
EC 12	10.1	8.1	6.5	32.5	42.8
EFTA	5.6	6.0	5.0	37.2	46.2
W. Europe	9.5	7.8	7.8	33.2	43.2
Exports					
EC 12	10.0	3.3	3.4	37.7	45.6
EFTA	3.0	7.3	6.2	32.0	51.5
W. Europe	9.0	3.9	3.8	36.9	46.4

* Commodities according to the Standard International Trade Classification: food, beverages and tobacco; mineral fuels, lubricants and related materials; crude materials, oils and fats; machinery and equipment; others.

would that seriously damage the environment and hence a rail tunnel between Helsingor (Denmark) and Helsingborg (Sweden) be preferable? The debate continues.

A common transport policy?

Western Europe's transport network has experienced numerous technical improvements in recent years but it still comprises a set of quite distinctive national systems. Within the Twelve transport involves 7 per cent of jobs and an equal share of GNP, 30 per cent of energy consumption, and 40 per cent of public investment. Transport policy was given a chapter of its own in the Rome Treaties and the EC has been trying to organize a common market for land transport ever since 1957 and for air and sea transport since 1973. Many agreements have been reached on individual items but the overall result has been modest. This state of affairs has been condemned by the European Parliament, and in 1983 the European Commission identified the following policy targets: to improve integration of national transport policies; to create a better climate of competition between and within different forms of transport; to enhance efficiency and productivity by eliminating bottlenecks and bureaucratic constraints; and to support major infrastructure projects of Community-wide importance (Whitelegg 1988). Particular attention has been devoted to projects for reducing traffic congestion (especially where that relates to crossing mountain ranges and water surfaces) and to schemes that will open up peripheral areas. In an attempt to avoid marine accidents, safety standards were laid down for vessels at sea and entering port. In addition, numerous schemes to promote road safety were implemented and finances allocated to many parts of the Community to improve many kinds of transport infrastructure, including roads and railways

in Greece, the Tyne–Wear metro railway in the UK, and ferries operating in the Irish Sea. The need to integrate the Spanish economy more closely with those of her neighbours in the EC has emphasized the need to improve trans-Pyrenean rail links (including a fast service between Barcelona and Perpignan) and to accelerate the expansion of the Spanish motorway system.

Only in the last fifteen years has the EC become active in the realm of air transport, which directly employs 300 000 in the Twelve and a further 200 000 ancillary staff. The EC is playing an important role to encourage the liberalization of routes and the cutting of passenger fares by involving a range of competing companies rather than the traditional 'duopoly' arrangement whereby two national airlines each command roughly half of the traffic between any two countries. Full 'deregulation' on the American pattern is not possible since the twelve states regard their national airlines as status symbols of their respective governments rather than simply as commercial carriers of passengers and freight. For this reason, the transport ministers were remarkably reluctant to facilitate competition and price cutting on a wide scale; however, they set 1992 as the deadline for action. Bilateral 'open skies' agreements between the UK and the Netherlands and a relaxation of pricing rules for flying between the UK and the Republic of Ireland gave rise to some remarkably cheap flights. Special deals have also been arranged for a range of services between the UK and six other West European countries.

Trade

From the world perspective, the eighteen small countries of Western Europe are rich, markedly industrial and emphatically urban. These characteristics derive in part from domestic resource endowment but have also been

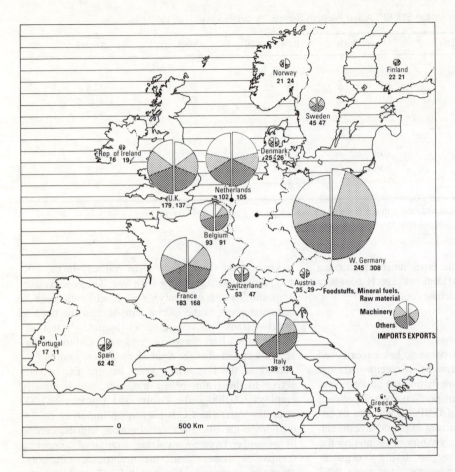

Fig. 7.8
Exports and imports by commodity classes, 1989 (hundred million ECU).

shaped and intensified across the centuries on the basis of trade. Some nations established vast colonial empires to supply raw materials and absorb manufactured products; others involved themselves in transporting other nations' commodities; few were content with being engaged in purely local transactions. Since mid-century the balance of world commercial power has shifted, with decolonization and industrial development modifying the nature of the relationship between Western Europe and the states of the developing world, and new industrial and commercial superpowers emerging in eastern Asia to challenge the role of the West.

The eighteen states of Western Europe still live by

trade and together account for almost two-fifths of the total value of the world's commercial transactions, commanding a notably greater share than either the USA (14 per cent) or Japan (8 per cent). As Chapter 2 discussed, the EC (one-third of world trade) and EFTA (about 5 per cent) evolved in quite distinct ways and are very different institutionally, although both have the central objective of stimulating and facilitating trade between their respective members. The idea of establishing a 'common market' or customs union was of course at the heart of the Treaty of Rome (1957). Three years later the Stockholm Convention gave rise to EFTA, which aimed to create free trade in industrial

Table 7.5 Imports and exports for the EC and EFTA, by value 1990 (%)

	EC 12	USA	Japan	Rest of world
Imports to EC from	58.2	7.8	4.3	29.7
Imports to EFTA from	59.5	6.5	5.4	28.6
Exports from EC 12 to	60.0	7.5	2.0	30.5
Exports from EFTA to	56.6	7.5	2.4	33.5

Table 7.6 Percentage of intra-Community by trade, by value, 1989 (%)*

	Bel/Lux.	Denm'k	Germany (West)	Greece	Spain	France	Ireland (Rep.)	Italy	Neths.	Portugal	UK
Bel/Lux.	—	6.3	14.5	6.2	5.6	16.3	3.3	8.7	22.1	5.8	8.7
Denmark	0.9	—	3.5	2.0	1.2	1.3	1.3	1.7	1.8	1.4	3.6
Germany (West)	32.5	45.6	—	31.9	28.3	32.8	11.4	37.4	40.1	21.2	31.9
Greece	0.2	0.4	1.2	—	0.3	0.6	0.1	2.0	0.3	0.2	0.6
Spain	2.0	1.7	3.8	3.0	—	6.6	1.5	4.3	2.0	21.4	4.3
France	20.7	8.6	22.4	11.1	24.7	—	5.6	25.9	12.8	17.2	17.3
Ireland (Rep.)	1.0	0.9	1.6	0.9	0.9	1.6	—	1.1	1.6	0.5	6.9
Italy	5.9	6.7	16.6	23.9	17.2	18.1	3.5	—	5.7	13.4	10.9
Netherlands	25.4	14.4	22.4	11.0	6.6	10.0	7.2	9.7	—	8.0	14.2
Portugal	0.6	1.9	1.4	0.5	4.3	1.5	0.4	0.6	0.9	—	1.6
UK	10.8	13.4	12.5	9.5	10.9	11.2	65.8	8.5	12.5	10.9	—

* Calculated on basis of import statistics

goods for all members (with special arrangements for farm products and fish); to assist in forming a single trade market throughout Western Europe; to promote the growth of world trade; and to encourage economic expansion and better living standards. Austria, Denmark, Norway, Portugal, Sweden, Switzerland and the UK were its founding members, with Finland becoming an associate member in 1961 and Iceland joining in 1968. Successive enlargements of the EC trimmed back the dimensions of EFTA, whose six current members have no common policy on anything except trade, unlike the complex links that bind members of the EC one to another. None the less, the EFTA nations were associate members of the EC and free trade in industrial goods between seventeen countries became possible in 1977 and extended to the eighteenth nation when Spain entered the EC at the start of 1986. Tariff walls were removed within the Community but a host of non-tariff complications remained, such as differing technical and other standards between member states and the existence of complicated frontier controls which involved haulage firms in great financial loss and time-wasting as their lorries were subjected to lengthy formalities. The removal of such barriers in the EC has been central to the objective of establishing a Single Economic Market by the end of 1992 (see Ch. 13).

The structure of imports and exports by commodity classes shown in Table 7.4 summarizes the main features of commercial activity in the EC, in EFTA and across the full range of West European nations. Foodstuffs and raw materials together make up just over one-sixth of the total value of trade (imports plus exports) for the eighteen countries, with a slightly smaller proportion being accounted for by mineral fuels (basically oil) and raw materials. Machinery,

equipment and an enormous range of manufactured products (listed as 'other' commodities) make up no less than 70 per cent, being relatively more significant in the total trade of EFTA than in that of the EC.

The commercial profile of each country is unique, being conditioned by the relationship between its specific resource endowment, its level of economic development and the resultant mix of activities, but several key points emerge from the complex information shown in Fig. 7.8. Food imports are less significant in the trading profile of each of the EFTA members, whereas foodstuffs account for a quarter of the value of exports from Denmark, Greece and the Republic of Ireland. Mineral fuels (especially oil) represent a particularly large component in the import bill for Finland and the Mediterranean countries (with between a quarter and a third of the cost of all imports to Greece, Italy, Portugal and Spain deriving from these commodities); by contrast, the export of mineral fuels is an impressive – if temporary – feature in the trade of three nations surrounding the North Sea oil province. The importance of timber products emphasizes the role of raw materials in Finland's trade profile, while machinery and related equipment figure very prominently in the range of exports from West Germany and Sweden.

During the past quarter century the volume and value of commercial transactions by and between West European nations have risen substantially. Thus the share of EC imports from member states increased from 40 per cent in 1962 to 58 per cent in 1990, and the proportion of exports rose from 43 to 60 per cent over the same period (Table 7.5). As Tugendhat (1987) has stressed, for each of the older member states the rest of the Community is not just its principal export market and source of imports, but is overwhelmingly so. Thus the Benelux nations (and also the Republic of Ireland) are particularly strongly committed to intra-EC trade with over two-thirds of both imports and exports (measured by value) involving other EC countries. By contrast, at least four-fifths of the commercial contact of the four southern member states involve 'the rest of the world' (particularly Mediterranean countries) and Denmark retains substantial links with its Scandinavian neighbours immediately to the north.

Not surprisingly, the EFTA nations collectively derive the greatest share of their imports (59.5 per cent of value) from the EC, to which they send 56.6 per cent of their exports. Given their location in the heart of central Europe, it is not surprising that particularly high proportions of the trade of Austria and Switzerland involve the EC states that are their nearest neighbours. Likewise, facts of geography and history combine to maintain significant exchanges among the Nordic members of EFTA. Focusing exclusively on the Twelve, the matrix of intra-Community trade (Table 7.6) identifies the enormous commercial weight of West Germany which commands a quarter of the value of all imports in the EC and three-tenths of all exports and is the leading trading partner with no fewer than six member nations (Denmark, France, Italy, Netherlands, Portugal, UK). Tradition and sheer proximity explain why 47 per cent of Irish trade within the EC is with the UK and why 35 per cent of Dutch trade within the Community is with West Germany and a further 19 per cent with Belgium. Further south, a centuries-old commercial alliance helps account for one-fifth of Portuguese commercial activity within the Community being with the UK. As with most elements of the contemporary human geography of Western Europe, one finds that complex aspects of continuity with the past underlie the drama of recent change.

Uncertainties

In some respects the distinction between the service sector and modern high-tech manufacturing has become blurred; however, as this chapter has shown, the broad shift from manufacturing to service-based employment has had profound consequences throughout Western Europe. Serious social and geographical polarization has resulted, with members of the 'services elite', such as 'yuppies' working in the City of London, enjoying large financial rewards (but considerable uncertainty at a time of recession). The under-

class of lowly service workers, on whose labour the comfort and routine well-being of the high fliers depend, experience much less reward (Townsend 1987). Likewise, some regions have enjoyed far greater propensity than others to flourish as a result of the services revolution. By its very dynamism, diversity and seemingly inexhaustible capacity to accommodate, the services sector may appear to have provided an escape route in many localities, although it must be noted that many service activities are staffed by women and young people and may comprise part-time rather than full-time jobs. The mismatch between the 'old' work lost by men in their prime and the 'new' work of the tertiary sector can be both considerable and traumatic. Day-to-day relationships in very many families are having to be rethought.

There is no doubt that plenty of potential remains for applying human ingenuity to the creation of new brands of service employment in the future, but one is forced to question whether the 'services bubble' really has an infinite capacity for growth. In recent years we have been accustomed to talk of sunset industries, but already there are sunset services! As many West Europeans have been required to understand, jobs in public services (e.g. education, health care and local and national administration) are very vulnerable to changes in government financial policy. Even more worrying is the harsh truth that any future decline in international financial dealings and in West European economic vitality in general would produce a widespread reduction in spending power that would send massive shock waves throughout the full array of market services. The ultimate uncertainty remains the ever-fluctuating relationship between the number of new jobs created by modern technology and the volume of old ones destroyed as a result of it.

Further reading

Bateman M 1985 *Office Development: a geographical analysis*. Croom Helm, London

Beaujeu-Garnier J, Delobez A 1979 *Geography of Marketing*. Longman, London

Daniels P W 1985 *Service Industries: a geographical appraisal*. Methuen, London

Dawson J A (ed) 1980 Retail Geography. Croom Helm, London

Hall P G, Markusen A 1985 *Silicon Landscapes*. Allen and Unwin, London

8

Urban development

Old cities: new problems

Past and present

The countries of Western Europe are richly endowed with urban settlements whose diversity derives from important differences in planning practice as well as variations in culture and provides a powerful source of fascination for residents and visitors alike (Lichtenberger 1970; Burtenshaw, Bateman and Ashworth 1991). Most towns and cities originated as classical or medieval foundations which were enlarged subsequently for a complicated set of political and economic reasons, most notably during successive waves of industrialization that transformed parts of north-west Europe during the nineteenth and early twentieth centuries but had little impact in southern Europe until after 1950. The physical expansion of towns and cities has formed the main manifestation of urbanization in recent decades, although an important array of completely new towns has also been built in the UK and more recently in France.

Ravages of war and the implementation of slum clearance and modernization schemes in inner-city areas have erased many historic townscapes; however, most cities still contain important collections of old buildings. Now the dual task of conserving and rehabilitating them is receiving sympathetic attention in many countries, for a combination of aesthetic, social and political reasons. As Paul White (1984) insists, 'diversity' is the best word to characterize the housing situation of Western Europe, which ranges from the shanties of Lisbon to the neat middle-class suburbs of Zurich, and from the massive post-war housing estates (*grands ensembles*) around Paris to the tall, carefully conserved seventeenth-century houses of old Amsterdam.

National definitions of 'urban' settlements are remarkably idiosyncratic but it is clear that most residents in Western Europe live in towns and cities. The latest inter-national collation of statistics (c 1990) identified no fewer than 144 cities of more than 250 000 inhabitants apiece (Fig. 8.1). Two world-cities in highly centralized countries dominate the list (Paris 8 510 000; London 6 378 000) and are followed by newly unified Berlin (3 335 000) and three capitals in southern Europe (Athens 3 027 000; Madrid 2 991 000; Rome 1 821 000). A further score of cities contain between 1 million and 2 million residents apiece (Table 8.1).

Myths and realities

Historically, major towns and especially capital cities have tended to display a remarkable propensity to flourish as centres of innovation and generators of employment and wealth, so much so that self-sustaining growth appeared to be almost one of their intrinsic characteristics. There were, of course, exceptions to the rule, such as the city of Vienna which stagnated following the dismemberment of the Austro-Hungarian Empire after the First World War. None the less, the majority of regional development programmes in recent decades have accepted unquestioningly the assumption that official policies could be devised and implemented to steer new employment and housing to ailing provinces or to greenfield sites without causing irreparable damage to the total habitat, economy or society of metropolitan areas.

Unfortunately, such confidence has proved to be increasingly misplaced since the optimistic days of the 1960s. Large cities throughout Western Europe have tended to lose at least some of their traditional strengths, and new human problems have arisen which undermine much of the buoyancy that such places habitually displayed. The root causes of these difficulties are complex in the extreme, reflecting the declining health of national and regional economies since the early 1970s, the changing social structure and human aspirations of city dwellers, and the growing

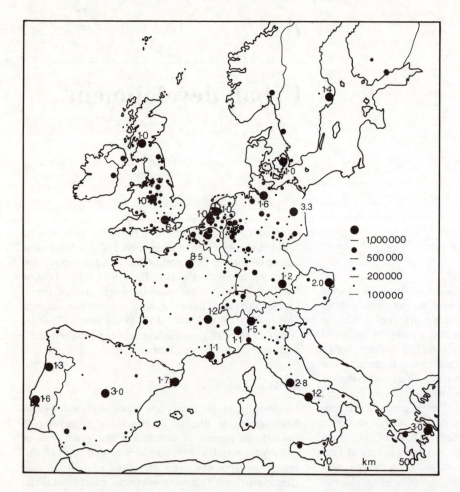

Fig. 8.1
Major urban settlements in
Western Europe, *c.* 1990.

realization that the built environment of the city is the product of political decisions which may be challenged and even overturned. Strikes and riots have, of course, occurred in urban areas in the past but their proliferation in recent years points in a particularly clear way to the problems of urban life and helps paint a scenario for the final years of the present century that is very different from that conceived by past generations of city administrators and regional planners. In short, the urban geography of Western Europe in the 1990s is more problematical and emphatically more controversial than that of the 1960s (Cheshire and Hay 1989; Thornley 1992).

Deconcentration

The wide view

In recent years, the continuing 'growth' of cities has tended to take on more dispersed forms, with new

housing being installed in peripheral estates (sometimes beyond the city's administrative limits) and in more distant 'rururban' localities which offer pleasant residential environments and from which daily commuting is possible by private or public transport (Lichtenberger 1976). In the heart of the city, slum-clearance schemes and urban-renewal programmes have swept away old housing areas and have contributed to a decline of the local residential population. Such deconcentration or 'counter-urbanization' has been in evidence in the USA for at least a couple of decades and it is arguable that it simply represents another phase in the complex set of processes labelled as 'urbanization' (Berry 1976, 1978; Fielding 1982; Champion 1989).

In any case, deconcentration is by no means a new phenomenon in Western Europe, with the 'City' of London having recorded its peak residential population as early as the mid-nineteenth century and the introduction of cheap season tickets on Belgium's railways in 1869 allowing workers to commute to major cities

Table 8.1   Top twenty cities in Western Europe, *c.* 1990 (millions)

Paris	8.510	Milan	1.549
London	6.378	Stockholm	1.462
Berlin	3.335	Oporto	1.315
Athens	3.027	Naples	1.208
Madrid	2.991	Munich	1.206
Rome	2.829	Lyons	1.170
Vienna	2.044	Marseilles	1.080
Barcelona	1.668	Turin	1.060
Lisbon	1.612	Rotterdam	1.038
Hamburg	1.595	Amsterdam	1.035

from 'suburban' towns and villages dispersed in many parts of that country.

Comprehensive research in Sweden has shown that population growth during the 1960s was dispersing away from major cities in favour of medium-sized towns, lower down the urban hierarchy, and there were plenty of additional examples of individual cities elsewhere in Western Europe 'turning themselves inside out' during that decade. Such evidence raised very real fears that what had been assumed to be North American problems of downtown decay and exurban sprawl were destined to occur on this side of the Atlantic as well.

In a major piece of comparative research, Hall and Hay (1980) showed that as early as 1950 some 86 per cent of Western Europe's 270.5 million inhabitants were living in metropolitan areas, while twenty years later the proportion of a substantially greater total (316.2 million) had risen to 88 per cent. Seventeen 'megapolitan' growth zones emerged clearly from the statistics for the 1950s and 1960s, including important axial urbanization along the Rhine, in the Rhône–Saône valleys, and in Provence. However, none of these zones was located in the outer peripheral regions of Western Europe; only one was focused on a giant city (namely Paris); and the early impact of deconcentration meant that the urban cores of London and the Randstad cities did not form part of the megapolitan zones adjacent to them. Likewise, the fastest rates of population growth in the countries of north-west Europe were being recorded in medium-sized cities, rather than in very large cities or in the 'old' industrial conurbations (Cheshire, Carbonaro and Hay 1986).

By contrast, the most dramatic rates of population growth in southern Europe did involve major cities (e.g. Barcelona, Bilbao, Madrid, Valencia, Milan, Rome and Turin) which expanded in response to powerful rural–urban migration that produced enormous

losses in the countrysides of the Meseta and the Mezzogiorno (Wynn and Smith 1978). By the mid-1970s it was becoming evident that urbanization was operating in different ways in various parts of Western Europe, with a number of city regions in the UK and north-western parts of the European mainland undergoing absolute or relative decline in the cores, while growth continued in their hinterlands; conversely, in southern Europe people and jobs were still becoming more concentrated in urban core areas. Since the mid-1970s the trend towards inner-city decline has become even more pronounced and has recently started to have an impact in Mediterranean cities, notably in Italy (Dematteis 1982; Pacione 1987).

In tighter focus

A complementary analysis of population trends in over 150 functional urban regions in Western Europe confirmed the North American experience that centralization, and subsequent decentralization of both population and employment, is the process which characterizes the evolution of urban systems through time (Drewett 1979). Rates of population growth in European urban cores declined progressively between 1950 and 1975, although the precise process of change varied from country to country and from city to city. At a national scale, the major cities of Denmark, France, the Netherlands, Sweden and Switzerland switched from initial modest growth to decline by the latter years. In Belgium and the UK the trend was negative through the whole quarter century, with the rate of decline accelerating with the passage of time; by contrast, city populations continued to grow in Austria and Italy but rates of increase decelerated over the years.

Of course, national trends such as these convey only part of the story, since each functional urban region evolved in its own particular way, but it is possible to state first, that by the 1970s, relative centralization of population was occurring around only a score of major cities in Italy, southern France and other parts of southern Europe; and second, each major urban core in the UK and virtually every one in the Benelux, West Germany, the Alpine states and northern France was displaying population decline. The traditional, taken-for-granted scenario of growth in the inner city had become sharply at odds with the facts of life across much of Western Europe (Uhrich 1987).

Publication of results of the 1981 UK census provided clear confirmation of that statement (Robert and Randolph 1983). Between 1971 and 1981 the population of England rose by a mere 0.4 per cent; Scotland had a loss of 2.1 per cent, while Wales increased its total by 2.2 per cent. Every large city in the land lost

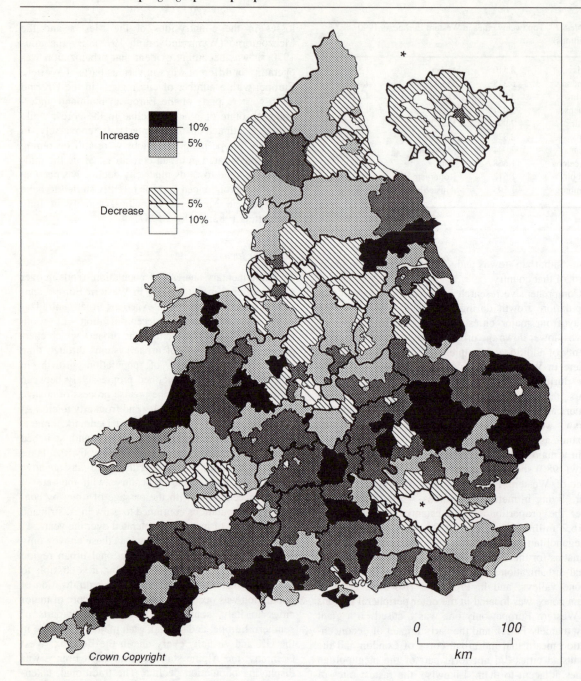

Fig. 8.2
Population change in England and Wales, 1981–91.

population, with the largest absolute losses being from Greater London (−756 300) and Glasgow (−219 150). The 1991 census revealed similar trends of urban population decline during the 1980s. Greater London lost a further 318 300 people (−4.8 per cent), Strathclyde −185 897 (−7.7 per cent), the West Midlands conurba- tion −146 800 (−5.5 per cent), Greater Manchester −139 900 (−5.4 per cent) and Merseyside −136 300 (−9.0 per cent). Within London the boroughs of Lambeth (−10.5 per cent), Brent (−10.0 per cent) and Hackney (−9.0 per cent) lost the greatest proportions of population. During both the 1970s and 1980s greatest relative growth occurred on the outer fringes of metropolitan areas and especially in the less urbanized areas of eastern and southern England, the south-west and central Wales (Fig. 8.2).

Expansion and renewal

Making good

Just as the distribution of Western Europe's urban population has been transformed over the past three or four decades, so the spatial structure and physical appearance of her major city regions have undergone profound change (Hajdu 1979). In the second half of the 1940s and throughout the 1950s, great efforts were made to reconstruct the war-torn centres of Antwerp, Hamburg, London, Rotterdam and literally hundreds of other towns and cities. Mass housing was constructed for the homeless according to the guidelines of hastily prepared plans. Money and materials were in short supply and the greater share of these building programmes remained firmly in the hands of public authorities. In some cities, such as Caen and Le Havre, the urban patterns were changed fundamentally as ownership plots were consolidated and large new blocks with wide intervening streets were laid out. On the European mainland the tradition of multi-storey living was continued and apartment houses were the usual result both in inner-city districts and on the city margins. By contrast, new terraces of family houses proliferated across many local-authority estates on the margins of large towns in the UK. In a minority of war-damaged cities, a diligent effort was made to respect old property boundaries and building forms, and this was certainly the case in the old cores of many German cities (Hajdu 1978). For example, the completely destroyed Hoherstrasse in Cologne was rebuilt on its old pattern as a narrow, crooked street with a strong historic character, and important sections of old Frankfurt were reconstructed to convey their traditional appearance; however, subsequent commercial pressure has changed adjacent skylines beyond recognition.

By the second half of the 1950s, wartime losses had largely been made good and the need to plan future urban growth came to the fore. The large volume of post-war marriages and the resultant baby boom meant that many couples with young children had no immediate alternative to living with parents or other relatives but were in urgent need of homes of their own. Important rural–urban migration on the European mainland intensified housing demand in cities while some countries, most notably France, needed to catch up on very low rates of home construction during the 1930s. The multiplication of new housing schemes raised the dual threat of chaotic conversion of land to urban uses and of inadequate provision of transport, schools, local shops and other necessary services on the new estates. Such was certainly the reality in the early years of the *grands ensembles* that mushroomed around Paris at this time, and similar difficulties were encountered in many other countries as well.

Master plans

Physical planning appeared to offer the solution for coping with these problems and a crop of master plans, laden with axes of expansion, growth nodes, transport corridors, new towns and even green belts, duly appeared. Every master plan varied in detail, according to the site of its central city and the administrative, political and cultural contexts within which it was fashioned (Hall 1967, 1977). For example, around Stockholm the construction of large estates of suburban apartments, such as Farsta and Vällingby, was coordinated with the building of an underground railway system that had been started in 1945 and exceeded 100 km in 1980. Each group of estates was provided with a commercial precinct, within which movement by pedestrians and vehicles was kept separate, and large parking facilities were installed to cope with the rising tide of private car ownership. Thus, the new suburb of Skärholmen, to the south-west of Stockholm, was completed in 1968 and was equipped with shopping facilities to service a clientele of 300 000, as well as being provided with the largest car park in Scandinavia. The last so-called 'underground suburbs' within Stockholm municipality were constructed during the 1970s, but similar models have been employed in surrounding local authorities which are served by extensions of the underground, suburban surface railways and the network of motorways installed in the late 1960s and 1970s to facilitate journey-to-work movements throughout Stockholm's metropolitan region. Such estates of apartment blocks fringed many Continental cities as master plans were implemented, but in Britain they tended to be built as a result of slum-clearance schemes in inner-city locations, with pairs or terraces of single-family houses being the usual formula on the margins of cities (Duclaud-Williams 1978; Pacione 1979).

Master plans had to be modified in response to subsequent changes in demography, economic health and political conditions with, for example, the 1965 *Schéma directeur* for the Paris region undergoing revision in 1969, 1975 and the early 1980s (Moseley 1980; Flockton 1982). It was revised once again in 1990 to make provision for the facilities that would be needed if the city-region were to enhance its role in the international economy of the Single European Market (see Ch. 12). These master plans were implemented with varying degrees of success in response to the fact that the concept of physical planning tended to be accepted more readily in the countries of north-west Europe than in the Mediterranean south, where piecemeal, chaotic urban sprawl forged ahead in defiance of official plans.

Urban sprawl

By way of example, the city of Rome grew very rapidly in the 1960s and early 1970s, with recent migrants finding accommodation wherever they could: in crowded tenements, caves and the hastily built shacks and houses in the *borgate* which mushroomed haphazardly around the old city regardless of the 1962 development plan. Some 800 000 people live in the *borgate* and they are organizing themselves to demand proper municipal services: paved streets, public transport, schools, health facilities and, in some localities, even the most basic of necessities like electricity and drains. Not surprisingly, the costs of providing such facilities after suburbanization has occurred are astronomical. Now the administration is trying desperately to compel developers and private individuals to respect official plans so that this kind of chaos may be avoided in the future.

Since the mid-1950s, Lisbon and Oporto have been beset by comparable problems which have resulted from the recruitment of large numbers of workers from the countryside for Portugal's rapidly expanding industries (Williams 1981). Illegal housing settlements (*bairros clandestinos*) sprang up and were further swelled by the arrival of repatriates from Angola and Mozambique. The *bairros* varied in quality from quite respectable houses to miserable shanties but all shared in the distinction of being poorly serviced and existing in contravention of land-use plans. The authorities have cleared the worst peripheral slums, although greater emphasis of late has been placed on providing services. Similarly, the *barracas* around Barcelona, Madrid and other Spanish cities have sprawled alongside railways and across wasteland, hillslopes and patches of public land (Naylon 1981). Many of their shanties lacked electricity, water and mains drainage and during the 1970s numerous residents' action groups came into being to influence the authorities and ensure that such shortcomings be set right. Towards the eastern end of the Mediterranean the recent sprawl of greater Athens (3 027 000) exhibits comparable features. The city continues to grow but one in every five of its dwellings has been constructed without planning permission. Arguably Athens has few slums but about half of all dwellings depend on cesspools until the construction of a new sewerage system is finished.

Downtown transformation

Priorities and ambitions

Each master plan was ambitious and expansionist, emphasizing the allocation of new housing to peripheral locations but paying much less attention to changing economic and social conditions in the city core and the inner suburbs (Michael 1979). Each programme took it for granted that city centres would continue to flourish, thereby drawing people and jobs to them and generating ever-growing quantities of wealth that would be available for wise allocation elsewhere. To enable central areas to perform these roles even more vigorously in the buoyant climate of the 1960s, a number of modernization programmes were announced for clearing parts of the central areas of many West European cities to enable the installation of office accommodation on a massive scale, with shopping facilities, cultural equipment and new apartments occasionally being included. At this time, rising rates of car ownership afforded greater flexibility in journey-to-work behaviour than ever before in Western Europe, but car ownership remained less widespread than in North America where decentralization and suburbanization of employment were forging ahead very rapidly. In Western Europe, many people continued to rely on public transport for travelling to work and railway lines and bus routes remained strongly focused on central areas. Unlike the demise of many downtown areas in North American cities in the 1960s, city centres in Western Europe continued to flourish as seats of government, administration and commerce, still requiring a large and growing labour force to service them.

Indeed, some cities took on a range of additional functions following the course of national and international events. Brussels, Luxembourg and Strasbourg acquired new roles following the creation of the EEC and other supranational organizations. At the same time, Paris and several other capital cities were being vigorously promoted by their national governments as 'ideal' locations for Europe-wide headquarters of multinational corporations which were increasing their power dramatically. In addition, administrative and commercial offices dealing with routine matters were being established in many provincial cities as part of official policies for decentralizing paperwork as Western Europe's economies became emphatically 'post-industrial' (Tuppen 1977).

The office booms that characterized many of Europe's cities in the 1960s were designed to satisfy the ambitions of city politicians and the requirements of businessmen at one and the same time. Planners and administrators were keen to 'modernize' their cities and sanctioned the use of public funds for provision of new roads and other infrastructural features. None the less, each of these schemes also depended heavily on massive injections of capital from finance houses, pension funds and big business in general. Thus the Stockholm city plan of 1962 advocated the wholesale demolition of the city's central business district (CBD), with a quarter of the cleared area being proposed for new thoroughfares and walkways, a quarter for multi-level parking, and the remaining half for new commercial buildings. The plan was not implemented in its original form but the CBD was duly demolished and rebuilt with high-rise office blocks, shopping precincts, an impressive 'culture house' and a vast lower-level concourse which is filled with animation during the day but acquires a much more sinister appearance at night.

Waves of destruction

The collective impact of programmes such as these was to unleash a new wave of destruction in many city centres in the name of 'slum clearance' or 'urban renewal' which swept across many of the old dwellings that had provided cheap accommodation and many of the workshops which had offered local sources of employment. Whole communities could be destroyed in a matter of months or even weeks! Some residents were rehoused in public-authority houses or apartments on suburban housing estates, others simply sought cheap accommodation wherever it was to be found, while those with sufficient capital behind them purchased property in more desirable parts of the city, undertaking new journey-to-work movements by public transport or by car.

Many inner-city firms were already beset by functional inefficiencies stemming from trying to operate in old, cramped but usually cheap workshops which were only accessible along narrow city streets. Often such firms were small and had little capital behind them; many simply went out of business when urban-renewal schemes got under way since they were not able to cover the relocation costs that would be involved in moving to alternative and almost inevitably more expensive accommodation out in the suburbs. The office blocks and shops that rose where humble workshops had once stood could not provide replacement work for those who had lost their industrial jobs, just as the new expensive apartments that rose from the ruins of modest tenements could not provide replacement housing for those of limited means. In some neighbourhoods, communities were dispersed; in other they were substantially diluted as middle-class

newcomers moved into new expensive apartments or invaded and restored what remained of the old housing stock in adjacent areas, thereby placing it beyond the reach of the working-class population.

Popular protests and gentrification

Such transformation of inner areas in many cities during the buoyant years of the 1960s soon gave rise to strong opposition. Conservationists lamented the planned erosion of historic townscapes and opposed schemes that would involve further demolition. They advocated careful restoration of existing properties rather than total destruction and construction of new buildings. Such arguments could, of course, easily open the way for 'gentrification' unless actions were taken by municipalities to subsidize rents in rehabilitated housing and thereby ensure that established residents could continue to afford to live there. At the same time, citizens' action groups criticized the destruction of traditional working-class communities and the loss of long-established forms of local employment (Pickvance 1976). The effective control of many city centres by the forces of international capital aroused deep-rooted local hostilities, as did the social transformation of some inner-city areas which paralleled the construction of luxury apartments and the gentrification of old housing stock that escaped demolition (Castells 1978a, 1983).

This kind of opposition became forceful throughout Western Europe during the critical years of the late 1960s and has continued to be expressed in various forms ever since. Economic recession, energy crisis, high rates of inflation and devastating unemployment dampened the demand for new office space for perhaps a decade after the mid-1970s and certainly reduced the supply of public capital for inner-city modernization schemes. Redevelopment projects were slowed down or even halted and many historic sections of the built environment were thereby granted a reprieve. By the mid-1980s the demand for modern office space had picked up dramatically in London and some other major cities but was certainly not experienced in all urban centres. As a result of major redevelopments in and around the City and the massive construction boom in Docklands, London is now oversupplied with office premises. During the period between the affluent 1960s and the present day, profound human problems have appeared in many of Western Europe's inner-city areas, and planners have been made painfully aware of the complex economic, social and political implications of any programme that they propose.

The process of gentrification, whereby newly

incoming middle-class households restore deteriorated urban property, especially in former working-class neighbourhoods, has continued apace in many West European cities throughout the past quarter century (Smith and Williams 1986). Instead of settling for an outer suburban existence, these well-paid professionals and service-sector workers were rediscovering the qualities of living in truly urban neighbourhoods. Ordinary, often poorly maintained, houses were purchased cheaply by the pioneer gentrifiers who themselves undertook the restoration of their property or engaged small builders to do so. Areas that were well served by rapid-transit systems and contained homes that displayed 'period charm' were the prime targets for gentrification during the first wave. Since property values were continuing to increase, the restoration of housing continued to be a good capital investment for individual households who were duly joined by developers and building companies that moved in to renovate old tenements and to convert empty warehouses into attractive apartments. Many old buildings in London's Docklands, commanding splendid views of the river Thames, have been transformed in this way. Similar waterfront redevelopments have taken place in many other port cities in Western Europe (Hoyle, Pinder and Husain 1988).

The urban space of less obviously appealing neighbourhoods also began to be transformed with, for example, London's gentrification affecting not only Islington and Clapham but also Hackney and Balham; while Parisian gentrification extended from the evident attractions of the Marais to more distant quarters in the city's east end. As well as enhancing the quality and visual attractiveness of housing, gentrification greatly boosts its price and brings with it a new middle-class range of up-market services and shops, including delicatessens, wine bars, boutiques and the mandatory proliferation of estate agents, which take the place of ordinary retail outlets. This transformation of values associated with the invasions of the 'yuppies' effectively excludes less well-paid households from purchasing a home in a gentrified neighbourhood. The obvious and highly contentious consequence of gentrification is the displacement of the working class, a process which can be intensified so easily by well-intentioned conservation policies.

Conservation

Responsibilities

Legal powers for protecting individual buildings of architectural merit or historic interest were being acquired by government agencies and local authorities in every state of Western Europe from the latter years of the nineteenth century onwards, but only recently has attention been directed to conserving groups of houses, complete streets and neighbourhoods rather than separate buildings. Individual nations have approached this challenge in various ways and a very great deal normally depends on initiatives taken by local authorities as they set about implementing detailed land-use plans. However, a number of types of approach may be distinguished.

Urban conservation has been a very centralized affair in France and the Netherlands. For example, Dutch legislation of 1961 provided protection not only for individual buildings and whole neighbourhoods but even surrounding stretches of landscape, while one year later the French 'Malraux Law' empowered central and local agencies to define 'safeguarded sectors' where a blend of restoration and rehabilitation would be applied to whole neighbourhoods or selected parts of them (van Voorden 1981; Stungo 1972). Just over sixty safeguarded sectors have been designated which represent only one-sixth of the potential identified in 1962, with only one additional area being defined after 1976. Local residents have reacted to these schemes with varying degrees of enthusiasm; indeed, many resented the likelihood of being rehoused in the outer suburbs and opposed the process of gentrification that would probably occur in the newly safeguarded sector. In spite of opposition and reservations about the social implications of this approach, architecturally impressive conservation has been accomplished in Avignon, Chartres, Lyons, the Marais quarter of Paris and neighbourhoods in many other cities. Further legislation was introduced in 1976 to encourage closer cooperation between state-appointed architects and local councillors in the remodelling of safeguarded sectors. A year later, new urban-improvement programmes were started which channel financial aid to owners and tenants who wish to improve the basic amenities of their housing property (Kain 1982). These simpler rehabilitation schemes have turned away from previous preoccupations with prestigious architecture and centralized, technocratic approaches to conservation; instead, they recognize local needs for housing, employment and services and place strong emphasis on residents' involvement and community initiative. Investment is being directed to meeting human needs rather than conserving bricks and stone.

In Germany and Switzerland, legislation for urban conservation is enacted and applied in varying ways in the component *Länder* and *cantons*. Elsewhere, enactments have been made with regard to individual cities in Austria, Belgium, Denmark and Italy, including

Salzburg, Bologna, Venice and the Christianhavn district of Copenhagen, where work was started in the early 1960s and has continued under Danish 'slum clearance' legislation of 1969 which embraces modernization of houses and wider improvement of neighbourhoods. Finally, in the UK, local councils were empowered to establish conservation areas under legislation of 1967 and they have been stimulated to do so by local amenity groups. Further legislation in the early 1970s gave local authorities new planning-control powers over all buildings in conservation areas. Bath, Chester, York and Edinburgh's Georgian 'new town' have been subject to particularly comprehensive conservation schemes (Ford 1978).

Local initiatives

In some instances, urban conservationists have had to enlist international aid but many individual cities, such as Bruges and Bologna, have been able to achieve a great deal through their own initiative. For example, the old city of Bruges contains some 2000 historic houses which accommodate a residential population only half as great as that at the end of the nineteenth century. The conservation programme involves rehabilitation of old buildings, installation of a new sewer system to prevent further pollution of the canals, managing a number of pedestrian axes, establishing community facilities for nine local neighbourhoods and providing special parking areas on the margins of the old city.

The conservation work being undertaken in the heart of Bologna is more intriguing since it is the product of the city's Communist administration, which has been in power for many years (Appleyard 1979). It has gained a world-wide reputation for sensitive restoration of old buildings and for controlling rents in rehabilitated properties, thereby ensuring that the existing (usually working-class) residents can afford to continue living in restored housing. Those whose homes are being renovated are rehoused for a short while in a small stock of carefully built new apartments which harmonize with the ancient surroundings of the inner city. Participation by local residents in this effective but very expensive scheme has been encouraged by the administration, which argues that 'conservation is revolution' and that the forces of capitalism threaten to destroy the historic buildings and the traditional communities of the city alike.

The Venetian dilemma

Venice has posed even more fundamental conservation problems which have required international aid; indeed, the city's very survival was threatened following a rise in local sea level in the order of 35 cm since the 1870s. Shrinkage of the polar ice cap provided the general background to the specific problem of terrain under Venice sinking by 6 mm per annum as increasing volumes of groundwater were extracted for use by residents and by the 200 or so industrial concerns located around the Venice lagoon. Extraction of methane gas caused further problems, as did the excavation of deep channels to allow tankers to reach nearby oil refineries which belched out air pollution. This exacerbated the damage caused to Venice's ancient stonework by thick layers of pigeon droppings. Industrial pollution and sewage destroyed plant and animal life in the water of the lagoon, causing sand islands to break and more sea water to enter the lagoon. Tidal action intensified and, together with disturbance caused by fast-moving water craft, provoked enormous damage to poorly maintained seawalls and foundations. General dampness in buildings and fluctuating water levels in the canals became part of the Venetian way of life. Even persistent low water created problems since it failed to allow sewage to disperse, while removing lateral support from exposed foundations; high water naturally caused devastating floods.

In November 1966, high water rose 2 m above normal peak levels, wreaking havoc in ordinary homes and causing incalculable damage to fabrics, paintings, frescoes and statues in palaces and churches. International aid was sought to devise technical solutions for stabilizing land and water levels, and appeals were launched for finance to conserve damaged art treasures and crumbling buildings (Pacione 1974). Not only did 400 palaces of architectural and artistic importance need attention, but roughly half of all the buildings in the old city demanded repair. Many ground-floor apartments were found to be quite unsuitable for habitation and half of all the houses showed signs of damp. Even more serious was the fact that the residential population of old Venice was declining fast, actually falling by more than half between 1951 and 1981 to 83 000. The younger generation had chosen to move out, leaving the decaying city to the old and poor, as well as to the tourists in their hotels and the rich in their restored town houses. Since the mid-1970s, the national government has allocated considerable sums to flood protection, social housing, urban conservation and restoring the ecological balance of the lagoon. Administrative delays have meant that not all the money has been spent and many would agree that the lagoon is in a poorer condition than in 1966. Even so, 8 million visitors come to Venice each year!

Achievements

Much has undoubtedly been achieved in the realm of urban conservation in Western Europe and the designation of 1975 as European Architectural Heritage Year did a great deal to make the principle even more widely known. For example, the everyday environment of many cities has been improved by the establishment of traffic-free streets (Hajdu 1981). West Germany, the Netherlands and Denmark pioneered such schemes with, for example, Copenhagen's Stroget being closed to traffic as early as 1962. These ideas were duly emulated by other countries such as France, where the first pedestrianized street was designated in Rouen in 1970; ten years later, over 200 French towns and cities had implemented their own schemes. Italy, which is so rich in old towns with narrow streets flanked by historic buildings, was a latecomer in pedestrianization. Now it is pointing the way ahead and many of its schemes prove that carefully planned and implemented pedestrianization can be popular and enhance both the local economy and the city environment. Yet in spite of much successful conservation, it must be stressed that large sections of Western Europe's historic city centres have suffered irreparable damage as modernization has been implemented with little or no regard for the total urban environment. The piecemeal destruction of Georgian Dublin is a case in point, as a large proportion of its finest urban landscapes has been swept away in order to install speculative office blocks (Kearns 1982). Some of these were rented by government departments during the 1970s but recent financial cutbacks have trimmed that demand and have left an array of unwanted office accommodation. The central city now contains a disturbing collection of temporary car parks but precious few signs of new construction (Horner and Parker 1987). None the less, Dublin was honoured as 'European City of Culture' in 1991.

Similarly in Madrid, which received that title in 1992, some individual features of the townscape have been strictly conserved, but whole neighbourhoods of architectural interest have been razed to permit office blocks and car parks to be installed (Wynn 1980). Unfortunately, legislation to protect historic buildings or neighbourhoods is not enough. Environmental pollution, associated with exhaust fumes, oil-fired central heating and indeed acid rain, is causing irreparable damage to the historic built environments of traffic-clogged cities such as Athens and Rome, as well as the urban centres of north-western Europe that are readily identified with nineteenth-century industrialization.

Modernization

Controversy in Brussels

Conflicts between modernization and conservation have been experienced in every city in Western Europe during the past quarter century, but the city of Brussels provides a particularly striking example of the differences between economic and social principles and between the power of international capital and the interests of local people. The city enhanced its international status dramatically following the creation of the EC and attracted many new administrative and commercial functions during the 1960s. No fewer than 165 major American companies had selected Brussels for their European headquarters by 1968, when the city contained over half of all company head offices in Belgium (Goosens and de Rudder 1976). Office activities have become strongly concentrated in the city centre where half the jobs and no less than one-eighth of all employment in Belgium are to be found (Hengchen and Melis 1980). Sizeable sections of the old city have been razed to the ground to make way for scores of new office blocks, of which the EC's Berlaymont building is the best known.

In fact, the planned destruction of Brussels goes back to 1909 when work was started on a 4 km junction between the northern and southern railway termini, which would incorporate a new central station. Over a thousand tenement blocks were demolished in several working-class districts and roughly 100 ha functioned as a vast building site until the scheme was completed in 1952. Districts surrounding the old termini gradually deteriorated and were only selected for urban renewal in the 1960s. The emergence of Brussels' 'European' role and the fact that the World Fair was held there in 1958 triggered off a massive phase of modernization. Inner ring roads were built, new offices constructed and a start was made on an underground railway in 1965. More road schemes and massive office development were announced, including the 'Manhattan' project which envisaged the clearance of 50 ha in the northern inner city and the eviction of over 10 000 inhabitants prior to the construction of an administrative and commercial complex that was to provide over 60 000 jobs.

Such projects aroused widespread protest from the citizens, who not only opposed the destruction of so much of old Brussels but also feared that sections of the inner city would become depopulated. Already there was plenty of evidence to show that the working-class districts which remained standing in the inner city were developing as reception areas for large numbers of poorly paid immigrant workers, especially

Italians, Moroccans, Spaniards and Turks (de Lannoy 1975). Americans and EC staff emulated the middle-class Bruxellois by moving to homes in more agreeable suburbs.

In the following decade, many new office buildings were constructed in inner Brussels and further schemes were conceived for the Marolles district and a 30 ha site opposite the southern railway terminus. Institutions of the EC continued to consume office space, particularly in response to the enlargement of the Community from six to nine members in 1972. In addition, the Belgian government acquired large new offices in the capital to house its growing administration services. However, the financial crisis of the mid-1970s and the collapse of demand for commercial office space meant that several programmes were either shelved or halted part way. Many newly built offices in the Manhattan district remained empty, with developers suffering great financial loss.

At the same time dozens of action committees came into being to challenge the way the city was being planned and to question the way that office development and local conservation schemes both tended to work against the interests of those working-class residents who remained in the inner city. For example, buildings in conservation areas, such as the Marolles and Sablon neighbourhoods, attracted boutiques, restaurants and avant-garde cultural activities as well as providing some renovated apartments which commanded high rents that could only be paid by middle-class incomers. Action committees roundly condemned the gentrification of these and similar areas and the effective eviction of working-class residents who have been unable to meet drastic rent increases. In recent years attempts have been made to involve the local population in conservation schemes and to provide moderate-rent housing in renovated buildings, thereby attempting to ensure that the forces of urban conservation do not work entirely to the benefit of middle-class newcomers (Bateman 1985). The entry of Spain and Portugal to the EC in 1986 absorbed empty office space in Brussels and required the provision of new blocks but under a much tighter planning regime than in the anything-goes atmosphere of the 1960s and early 1970s.

Housing problems of an evolving society

Supplies and demands

Despite the construction of massive housing estates on the periphery of Western Europe's cities and the fact that rates of natural increase and migration from the countryside and from abroad have all declined, shortage of accommodation remains a desperate problem for many sections of urban society (Bassett and Short 1980). Virtually all post-war housing has been built to house families that would be able to raise loans to purchase private property or would qualify for housing by local authorities or by housing associations.

Many other groups in society also require accommodation, but relatively little attention has been paid to coping with their needs. West Europeans are surviving much longer and many live alone in their old age; there are many more one-parent families; more marriages end in divorce; for a host of reasons, more young adults decide to leave their parental home; and increasing numbers of people are choosing to live alone or in domestic arrangements that do not approximate to conventional two-parent families, for which almost the entire housing stock was built. The majority of newcomers to the cities of Western Europe still need cheap and compact accommodation and that point is of special urgency for the young and the jobless.

With high rates of unemployment and a proliferation of modest incomes many people simply do not command sufficient resources to buy property or rent anything than the cheapest room. Local rules vary but many people in the groups mentioned above may not qualify for a place on a public housing list. Urban-renewal schemes have swept away much cheap accommodation upon which such people relied in the past, while legislation affording measures of rent control or security of tenure may have been designed to help the tenant but also serve to make leasing out of accommodation a much less attractive proposition for landlords.

Housing difficulties have given rise to new brands of street politics in the cities of Western Europe over the past twenty years, as residents seek to compel town councillors, urban administrators and planners to tackle problems in more effective ways. For example, vociferous action groups have sprung up in many Spanish and Italian cities: to try to improve services in recently built working-class estates around Milan and Turin; to draw attention to the appalling housing conditions in the back streets of Naples; and to demonstrate how rents have been purposely inflated in central Rome with the result that landlords keep apartments empty for months on end until sufficiently affluent tenants can be found, even though the city endures a massive accommodation crisis. Indeed, most cities have generated a wide range of neighbourhood movements which, in the case of Madrid, have objectives as diverse as securing the installation of basic facilities in shanty-towns, opposing the demolition of downtown areas for new office blocks, and seeking to improve schools and public open spaces in pleasant middle-class suburbs (Castells 1978a).

Mismatch: the case of Amsterdam

The detailed operation of housing allocation varies from country to country and from city to city; however, a general mismatch between the fixed structures of the built environment and the evolving features of society is both clear and widespread. Such problems have been particularly acute in the city of Amsterdam, where new approaches to urban renewal and inner-city housing are being implemented. During the 1950s and 1960s the rapidly growing population of the Netherlands required additional housing on a massive scale and attention was directed entirely to new estates of family dwellings located on the fringe of cities (Anderiessen 1981). However, single people, childless couples and immigrant workers were also seeking housing and were only able to find it in rented rooms in inner-city areas.

Amsterdam contains a large historic core and extensive nineteenth-century neighbourhoods where such accommodation exists, but supply is outstripped by demand and attempts at controlling rents have not been adequate. Students attending the city's universities and colleges, young workers in tertiary employment, younger and poorer members of the sizeable gay community, immigrants and repatriates from former Dutch colonies compete unequally for a dwindling housing stock which has been eroded by slum-clearance and renovation schemes and by emphatic gentrification in some districts, where housing costs have soared beyond the grasp of poorly paid people. It is true that families with children have been moving out of inner Amsterdam and that the residential population has been declining for over two decades, but sufficient low-cost housing for incomers is simply not available and the completion rate of public housing declined appreciably in the late 1970s. For example, only 280 new units of public housing were built in the inner city in 1979 at a time when 53 000 had registered officially as needing accommodation.

Squatting

Squatting in empty buildings had been known in Amsterdam as early as 1965 and became the solution for growing numbers of Amsterdammers as housing problems became more serious and more inner-city properties were made vacant either by officials, prior to urban-renewal schemes, or by developers who were biding their time in anticipation of more lucrative deals.

The city's squatters came from a wide variety of social, occupational and educational backgrounds and held a range of political views, with experience of severe housing stress often being their only common bond. They were characteristically young (usually between 20 and 35 years of age) and single, with men making up two-thirds of the total. Some had come together simply in order to obtain a roof over their heads; for others, living communally was an expression of particular social or political objectives. By 1979, it was estimated that there were between 20 000 and 70 000 squatters in Amsterdam and squatting was encountered in almost every town in the Netherlands and in major cities throughout Western Europe. For example, London contained perhaps 30 000 squatters out of an estimated national total of 50 000 (Kearns 1979).

In Amsterdam the situation became particularly tense in 1979–80 because the squatters believed that many houses were soon to be cleared by the authorities so that urban renewal could begin. They were capable of marshalling hundreds of others to keep officials out of occupied property and had enlisted a considerable amount of public sympathy, since almost every inhabitant of Amsterdam had experienced housing problems of one kind or another. Swift action by police and city officials in 1980–81 removed squatters from many properties and legislation was passed requiring owners to register empty buildings, thereby affording them protection under the civil code. Squatting was thus made an offence in the Netherlands, and Amsterdam's officials were given new powers to expropriate buildings which remained empty for more than a few months. Legislation regarding criminal trespass was tightened up in several other countries but it was also made possible for squats to be 'authorized' by property owners, thereby allowing squatters to remain in occupation until the building was required for other uses. Nevertheless, cheap inner-city accommodation remains in desperately short supply and squatting continues to be one of the contributory causes for conflict between young people and urban authorities.

Despite the fact that squatting is also illegal in German cities, the number of squats in West Berlin rose dramatically during the first half of 1981 so that one-third of the city's 600 technically 'empty' houses were occupied in this fashion. The decaying nineteenth-century houses of Kreuzberg, close to the Wall and sheltering a large community of Turkish 'guest-workers', had the largest concentration of squatters in the city and it was there that street violence broke out in spring 1981 as police prevented a group of young people taking over yet another empty house (Hass-Klau 1982). Planners' attempts to rehabilitate this truly 'marginal' neighbourhood met with little success until very recently when gentrification took hold.

In the heart of Copenhagen a vast collective squat has simply been tolerated by the authorities, with more than 1000 squatters living in the former barracks of Christiania which stands in sharp contrast with the elegant churches, restored houses and new office blocks of neighbouring Christianshavn. Christiania conveys many conflicting images: groups of squatters and large dogs congregate beyond the stalls which offer varied merchandise and refreshments to tourists, while in the distance there is a collective bakery, workshops for recycling 'junk', and small 'eco-farms' where chickens and goats are raised and less usual forms of pot plant may be glimpsed on sunny window ledges. Murals adorn concrete walls but timber fences have been removed for use as building material or simply for firewood. The so-called 'free-city of Christiania' has existed since 1972 but opposition is growing not only in Denmark but also in Sweden and Norway where social workers complain that a share of the drugs entering their countries comes via Christiania. Their governments have made strong representation to the Danes to close it down.

A flexible approach to housing policy

Squatting admittedly, is an extreme case, but it is undeniable that the question of housing has become a major problem and a major issue in almost every city in Western Europe. Harsh experience has shown that local needs deserve careful consideration before urban-renewal schemes or modernization programmes are initiated by city authorities and developers. Amsterdam's older neighbourhoods share numerous characteristics with other European cities (e.g. poor housing, high building densities, many poor people and inadequate cultural facilities) but some sites also suffer from the distinctive problem of having many houses whose timber-pile foundations have rotted beyond hope of replacement. In such difficult circumstances, complete demolition of buildings and subsequent reconstruction is the only possible solution. Before 1971, two fairly distinct policies were implemented in the city's older neighbourhoods: with rehabilitation and reconstruction being practised in nineteenth-century residential districts. (The city's population had doubled from 255 000 to 510 000 between 1870 and 1900.) Since the early 1970s urban renewal has become a much more flexible operation in appreciation of social conditions and the quality of buildings on each site. 'Social welfare' and, more recently, 'ecological urban renewal at affordable cost' have become key concepts.

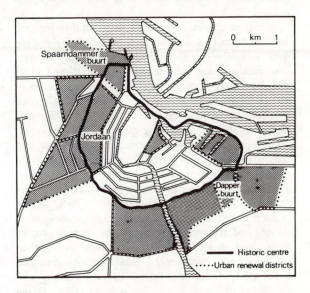

Fig. 8.3
Urban renewal districts in Amsterdam.

The City of Amsterdam authorities have designated eleven renewal districts which contain 126 000 dwellings, some two-fifths of the housing stock. The municipality is committed to maintaining and improving as much existing housing as possible. Where this is not feasible old buildings are replaced by new ones which respect historic construction lines. Special attention is paid to the interests of local inhabitants and the desire to maintain the residential character of each area. Work within the renewal districts demonstrates the flexibility of Amsterdam's policies which are subsidized from central government (Whysall and Benyon 1981). For example, the Dapperbuurt district (38 ha) was constructed in the final quarter of the nineteenth century to house commercial and industrial workers (Fig. 8.3). Roughly 10 000 people now live there in 5000 dwellings, the average occupancy rate having fallen substantially from 2.9 in 1960. The poor condition of many foundations means that the great majority of buildings are having to be replaced by new facilities for housing and shopping. Community action groups have succeeded in ensuring that blocks of small apartments are constructed, as well as family dwellings, and that work is phased in such a way that local residents may be rehoused in new moderate-cost apartments in their old neighbourhood. Rehabilitation of existing housing is the dominant approach in the Spaarndammerbuurt district, although some sites will be cleared to provide open space and for the construction of cultural facilities as well as new apartments.

A mixed approach is also being employed in the Jordaan district sited between the western canals of the inner city. The neighbourhood is composed of houses constructed in the seventeenth, eighteenth and nineteenth centuries; many were in poor repair a decade or so ago but no less than a quarter have been designated as being of historic interest. An early plan for this mixed middle- and working-class neighbourhood of 20 000 people proposed considerable demolition and the installation of new traffic schemes and public open space. There was strong opposition to this project, which threatened to erode the Jordaan's character and reduce its residential accommodation by half. Recent plans respect the 800 listed buildings, practise conservation and reconstruction in roughly equal proportions, and make provision for small as well as medium-sized apartments. However, it must be acknowledged that in-migration by young adults who live in small households (often without children) and enjoy incomes well above the average for inner Amsterdam, stimulated marked gentrification of some sections of the Jordaan during the 1970s. Up to 1986 15 000 dwelling units had been renovated and 12 000 new ones built to replace old housing in Amsterdam's urban renewal districts. Central government covered roughly three-quarters of the costs of such operations.

Malaise of the inner city

Despair

Modernization, conservation and gentrification have brought profound and highly controversial changes to many central districts in Western Europe's cities, but such innovations have been far from ubiquitous and many inner-city localities have entered into a profound malaise (Herbert and Smith 1989). Planning policies and myriad decisions taken by residents and employers have conspired to change such localities for the worse. The precise combination of problems varies from district to district but, in general terms, substantial sections of Western Europe's inner cities are characterized by substandard housing and deterioration of the built environment; many of their residents suffer multiple deprivation. Residents who have chosen to move out normally belong to professional, managerial and skilled blue-collar groups, while those who remain tend to be less skilled and less affluent members of society. They have been joined by labour migrants from abroad and, in the case of Mediterranean Europe, by people who have left the countryside in search of urban work.

Unlike the pleasant suburbs, inner-city districts contain relatively few medium-sized, two-parent households but an abundance of old-age pensioners, single people, one-parent families and large households composed of recent immigrants. Many inner-city households depend on low or insecure sources of income, of which social security payments represent a notable share. Economic recession and unemployment in recent years have made the situation even worse. Poor housing, low incomes, sharply declining job opportunities and diminishing local authority resources combine to exacerbate the deprivation experienced by many inner-city dwellers, imparting a collective sense of hopelessness and alienation (Harrison 1983; Heinritz and Lichtenberger 1986).

Violence

Such conditions do not, of course, attract private investment for creating new jobs, renovating the built environment or providing community services; by contrast, they do provide powerful causes for the outbursts of popular unrest which have occurred in many West European cities in recent years. Violent demonstrations over poor housing, inadequate social facilities, high rates of youth unemployment, fears of police harassment and profound disillusionment with Western society occurred almost as a regular aspect of life on the streets of Amsterdam, West Berlin and Zurich in 1980 and early 1981.

In April 1981, violence broke out in Brixton (south London), but this was merely a prelude to widespread rioting which took place three months later in Brixton, Toxteth (Liverpool), Manchester and a score of other British cities. Each disturbance derived from its own specific set of tensions and frustrations, but the most serious occurred in inner-city areas which had been losing population and prosperity for decades (Lawless 1981). The people they housed were poorly paid and comprised large numbers of immigrants and their British-born children. Rates of unemployment were high, especially among young people who had grown accustomed to the notion of jobs for all, only to find that work was denied them when they came on to the labour market. Levels of dissatisfaction were highest among young immigrants and young people born in Britain of immigrant parentage.

It literally took the riots to reveal the gravity of these problems among Britain's inner-city dwellers (Hamnett 1983). Similar harsh lessons have doubtless been learned by politicians elsewhere in Western Europe and North America. Far from being endowed with eternal strength and self-sustaining vigour, the decaying inner districts of many cities have emerged as some of Western Europe's most intractable problem areas. Their peaceful survival will require a mixture of economic support, social skill and political act which remains to be distilled. None the less, some bold initiatives have been attempted.

Initiatives

Peter Hall and Paul Cheshire (1987) provide a challenging agenda for urban regeneration which, they argue, must embrace both economic and social objectives. First, urban regeneration should enhance the image of the inner city as a place to live and work by (a) improving its built environment and infrastructure of roads and public transport, and (b) developing the viable skills of local people. Second, it should incorporate measures to help offset the costs of change, especially as some of those fall on the poorer members of local society. It is likely that the city authority must operate in partnership with private developers in order to achieve these ends and to bring in new and viable activities rather than attempt to prolong the existence of moribund forms of employment and land use.

The transformation of London's Docklands from an abandoned port area to a renewed part of the city provides a good example. The state-funded London Docklands Development Corporation (LDDC) has certainly changed the 'image' of the area (both through its publicity campaign and literally on the ground). The 'enterprise zone' on the Isle of Dogs, with its tax advantages and simplified planning procedures, has contributed to this transformation and Docklands has been perceived as an investor's bonanza (Church and Hall 1986; Church 1988). New and converted housing (very largely for sale and often at high prices), a range of modern employment opportunities, a new light railway, a business airport, a vast office complex and new shopping facilities add up to a remarkable – and in many ways successful – change (Marmot and Worthington 1986). But as house prices escalated and new jobs were occupied by incoming white-collar home-owners investing in the Docklands, one was left wondering what had been gained by ordinary, working-class local people (Townsend 1987). The demand for expensive office space and high-cost housing in Docklands has fallen during the recent recession and asking prices have tumbled as developers seek to fill their properties, and the financial collapse of developers Olympia and York has plunged Canary Wharf into crisis. The LDDC has paid more attention to social and environmental issues in recent years, but fundamental questions remain. Have the interests of the poorer members of local society been defended? Or is that an impossible dream in a planning scenario that

seeks to encourage the fastest possible transition to the economy and society of the twenty-first century?

In central Scotland, the Glasgow Eastern Area Renewal (GEAR) project has operated since the mid-1970s to effect a comprehensive social, economic and environmental regeneration of 1400 ha from which population and industry had retreated since 1960. Objectives included promoting job creation, training for employment, overcoming social disadvantages, diversifying the social structure, and improving and maintaining the environment of buildings and open spaces. Pacione (1985) reports that important improvements have occurred in Glasgow's east end since the mid-1970s and that many of these are the outworking of the GEAR project. Housing has been improved, overcrowding reduced and buildings restored (Wannop 1990). Now there is less privately rented accommodation but more housing provided by the local authority, housing associations and private developers. Landscaping has enhanced the visual environment. By contrast, unemployment remains high and the GEAR project has not been particularly successful in achieving its social goals (Donnison and Middleton 1987). Both 'east end' experiments reinforce the message that planning is not just about buildings and landscapes but rather about promoting the quality of life in all its aspects. We have seen, all too disturbingly, that it is much easier to change the urban environment than to improve people's lives.

The challenge of the post-war estates

A quarter of a century ago, West Europeans used to think that the inner-city crisis that was so evident in the USA would never happen in their great cities. Their hope was sadly misplaced. Nor could they have imagined that the big estates of social housing that they were building at that very time would soon become the second focus of desperation and deprivation. Hundreds of thousands of blocks of relatively low-cost housing were built on tens of thousands of estates (Castells 1983). Construction standards were often poor and features of internal and external design were questionable. Vast numbers of inner-city dwellers and migrants from the provinces and from overseas were housed in this way. Roughly the same social housing product can be found throughout Western Europe, from London or Paris to the rapidly growing sprawls around Mediterranean cities. For example, in France no less than one-fifth of all families live in *grands ensembles* and two-fifths of all households headed by immigrants. Provision of adequate schools, shops and other community facilities sometimes came as an afterthought. Even so, the estates continued to grow, with Glasgow's Easterhouse housing nearly 60 000 people at its peak, Sarcelles and Massy-Antony (near Paris) 50 000 and 40 000 respectively, Gropiusstadt (West Berlin) 45 000 and Les Minguettes (Lyons) 35 000 people (Barrère and Cassou-Mounat 1980; Tuppen 1988).

Whole estates as well as individual blocks acquired particular roles. Some were 'respectable' and reasonably well maintained; others were used by housing authorities as 'dumping grounds' for the homeless, for immigrants, for large or one-parent families, for those on social security. Many estate dwellers who could afford to move on to 'better' estates or to individual houses did just that, thereby reinforcing the socioeconomic deprivation of the people and the places that they left behind. At the same time, unemployment rates were rising throughout Western Europe and the great estates became problematical in environmental as well as human terms, with internal damp and structural defects becoming widespread in very many blocks (Wynn 1984).

By the early 1980s, vandalism, crime, racial attacks, deprivation and despair characterized many post-war estates as well as the inner cities. The violence at Broadwater Farm estate (Tottenham, north London) in 1985 was particularly tragic but less devastating incidents were far from uncommon. Disturbances around Lyons in the hot summer of 1981 and around Paris in 1983 encouraged the French government to promote a national scheme for neighbourhood improvement that would support and to some extent finance 'bottom-up' projects for improving environmental, social and economic conditions on problem estates (Tuppen and Mingret 1986). Local politicians, administrators, community leaders and residents were responsible for drawing up ideas and implementing them on the ground (Clout 1987c). Over 120 schemes were launched and, not surprisingly, have had greatest success in enhancing housing, open spaces and community buildings. As part of these projects, some particularly problematic blocks of housing have been dynamited and surrounding neighbourhoods remodelled. The social and visual 'image' of some estates has been improved and interesting experiments have been started to involve residents in managing and maintaining their estates, to convert 'no-man's land' open spaces into private gardens and, indeed, to privatize some blocks and their local services. But whether this kind of neighbourhood intervention can ever reach the roots of social inequality, unemployment and racism remains open to debate. Further riots in Brussels, around Lyons and several British cities at the start of the 1990s demonstrated that the challenge of wise management remained. The distribution of wage-earners on income support payments in greater Paris evokes this array of social and environmental issues in both the inner city and the new estates in working-class neighbourhoods of northern and eastern parts of the city-region (Fig. 8.4).

Privatization of blocks of the city-region or sale to property companies has been implemented by some local authorities in the UK, with the London borough of Wandsworth being a prime example. The appearance, quality and monetary value of what were formerly 'problem' blocks have certainly changed, as different residents with much higher incomes (and very different political persuasions) have moved in. The negative side of this policy is that other blocks and estates have been used more emphatically as 'dumping grounds' for the deprived. Privatization and allied changes were recommended in 1986 by an independent enquiry into Glasgow's post-war housing estates, where many apartments are damp, open decks invite crime and violence, and the financially starved housing authority has left repairs undone. The idea was for between one-quarter and one-half of Glasgow's corporation housing to be sold off to housing associations, cooperatives, development companies and private individuals.

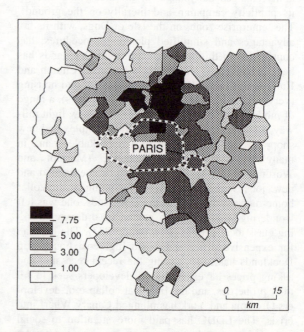

Fig. 8.4
Paris: number of persons receiving income support payments per thousand inhabitants 1989 (after Informations RECLUS, Montpellier).

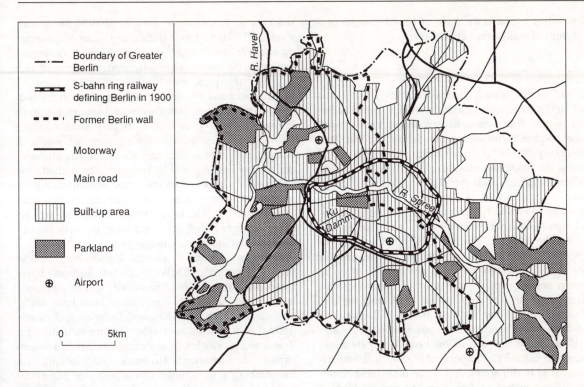

Boundary of Greater Berlin

S-bahn ring railway defining Berlin in 1900

Former Berlin wall

Motorway

Main road

Built-up area

Parkland

Airport

0   5km

R. Havel

R. Spree

Ku Damm

Fig. 8.5
The urban structure of Berlin.

United Berlin: from city to city-region

Two into one

The Berlin Wall was erected in 1961; on 9 November 1989 it was breached and the way opened for reuniting the two Germanies (see Ch. 13). After almost three decades of largely separate existence, the two 'cities' of Berlin came together again to form a single metropolitan area of 3 335 000 people, with a further million in the immediate surroundings (Krätke 1992). Occupying the territory of Greater Berlin of the 1920s, united Berlin now has the same status as Germany's other *Länder* (Fig. 8.5). For decades West Berlin had functioned as a subsidized showcase of capitalism, while East Berlin had performed the role of East Germany's capital (Elkins 1988). Despite profound differences in regime, the two parts of Berlin had certain features in common (Ellger 1992). Their physical extent was contained by the Wall, in the case of the West, and by policies which favoured social housing (as well as a severe shortage of funds and building materials), in the East. The net result was large estates of apartment blocks in the newer neighbourhoods of each part of the city. Both East and West Berlin performed important routine manufacturing functions but neither was at the forefront of industrial innovation. Western firms had long moved their headquarters to other parts of the Federal Republic, while Eastern enterprises, starved of investment, were old-fashioned compared with those in the West. Not surprisingly, neither part of Berlin displayed well-developed service-sector activities when judged by West European standards.

Political unification, the decision of the German parliament to transfer federal government activities from Bonn to Berlin, and the city's location as a potential bridgehead in east-central Europe have transformed Berlin from a geographical curiosity on the European map to a magnet for migrants (rich and poor), investors, developers and construction firms, bringing not only the hope of redevelopment but also management problems in their wake. As a place of new opportunities, Berlin is without rival in Europe. The future of its central area, the organization of transport and communications, and the management of the city's surrounding region in the *Land* of Brandenburg represent three of many urban challenges.

Unique opportunities

As well as two central business districts, freshly unified Berlin inherited extensive stretches of land adjacent to the Wall which had remained undeveloped since bombardment during the Second World War (Heineberg 1977). A major controversy has raged around the future use of land surrounding the former Potsdamer Platz and the Leipziger Platz. Some observers have envisaged this site as the ideal location for office towers to house multi-national corporations and high-level service-sector activities that Berlin will need if the city is to generate substantial numbers of new jobs. These will be required to compensate for jobs that will disappear through deindustrialization in eastern and western districts alike. Eastern factories have already lost the protection of the former regime; in similar vein, firms in West Berlin will lose their subsidies and be exposed to the powerful blast of competition from other parts of the Federal Republic and elsewhere in the EC. Prospects of high-tech reindustrialization in Berlin are modest, since innovative activities are entrenched so firmly in and around Munich and Stuttgart. Other observers, paying more attention to principles of urban conservation, have argued that Berlin has remained largely a low-rise city and that its future development should respect this characteristic, rather than covering its inner areas with high towers. This critical central part of the united city will also have to accommodate three clusters of federal government offices (only partly in existing buildings to either side of the former Wall) and the Kulturforum which will receive collections from other museums.

There is a dual dimension to the challenge of urban transportation in Berlin. Without doubt, efficient connections by road, surface railway and underground lines must be ensured between the two parts of the city and stations be renovated where necessary. More problematic is the desirable relationship between public transport and the private car. A strong case may be made for taking advantage of Berlin's unique circumstances to provide the city with an effective network of public transport for the next century.

The third challenge involves planning Berlin and its region in an appropriate way to cope with future pres-

sures for development. At the start of 1990 Berlin was short of 100 000–150 000 dwelling units, depending on how generously one viewed dilapidated accommodation in the inner (especially eastern) parts of the city. In subsequent months that figure has risen as Berlin continued to receive migrants from rural and small-town eastern Germany and from east-central Europe (see Ch. 4) as well as from western *Länder*. Berlin anticipates accommodating new national and international roles in the years ahead and large quantities of new housing will be required. The unified city authority has launched an important rehabilitation programme in inner eastern districts and has started new construction on the urban fringe within its tightly defined *Land*. Faced with escalating property prices, land shortages and tight planning regulations in Berlin, investors are turning to sites in Brandenburg. Major firms (for example, BMW, Mercedes, Siemens) have moved into what has become the city's outer commuting zone and numerous applications have been made for building houses, offices and factories and establishing golf courses and other leisure activities in Brandenburg, which will accommodate Berlin's future airport. In addition, Berliners are looking to Brandenburg's lake-dotted countryside for places to build or buy weekend cottages. Berlin is fast outgrowing the limits of its own *Land* and is posing questions that will have to be resolved at a truly regional scale (Schmoll 1990).

Further reading

Burtenshaw D, Bateman M, Ashworth G J 1991 *The European City: a Western perspective*. Fulton, London

Castells M 1978 *City, Class and Power*. Macmillan, London

Castells M 1983 *The City and the Grassroots*. Arnold, London

Fielding A J 1982 *Counterurbanization in Western Europe. Progress in Planning* 17: 1–52

Harrison P 1983 *Inside the Inner City*. Penguin, Harmondsworth

Herbert D T, Smith D M (eds) 1989 *Social Problems and the City: new perspectives*. Oxford University Press, Oxford

Merlin P 1971 *New Towns*. Methuen, London

White P E 1984 *The West European City*. Longman, London

Wynn M (ed) 1983 *Housing in Europe*. Croom Helm, London

Wynn M (ed) 1984 *Planning and Urban Growth in Southern Europe*. Mansell, London

9

Agriculture and rural change

Agricultural populations

Farming represents by far the most extensive land use in Western Europe, and any view of geography which emphasizes the study of the landscape must recognize the importance of this 'rural space'. Large parts of the European rural landscape have been almost totally humanized by 2000 years of continuous cultivation – even longer in the more spatially restricted oecumene of ancient Greece. For the rural geographer with a historical bent, Western Europe still offers a fascinating mosaic of agrarian regions whose 'personalities', particularly in France, Portugal, Spain and Italy, still reflect the *genres de vie* so dear to the heart of Vidal de la Blache. With such a prolonged backdrop of settled agriculture, it is small wonder that farming systems are so complex and varied. Primeval landscapes are rare.

By the late 1980s, land in agricultural production in Western Europe was approximately 150 million ha, or 44 per cent of the continent's land area. France and Spain, each with over 30 million ha, together contributed over two-fifths of total farmland. Yet rural areas are not monopolized by farming. As transport improvements, particularly the diffusion of the private motor car in the post-war period, allow the urban and rural world to interpenetrate more and more, so rural areas are seen as attractive environments for recreation, holidays and first and second homes. At worst, such pressures may eventually negate that very rurality that once attracted them. At best, planning future uses of Western Europe's countryside demands that a delicate balance be established between conserving valuable resources and permitting sufficient change and economic development for the living standards of country dwellers to be improved – by no means an easy task (Clout 1984, 1991).

In 1985, 14 308 000 people were employed in farming in Western Europe – 8.6 per cent of the total work-

force. This proportion has been falling rapidly in recent decades (It was 19.1 per cent in 1965, when farming employed 27 975 000.) Table 9.1 shows that the decline of the agricultural workforce has been especially dramatic in Spain, Italy and Finland. Only Greece and Portugal retained more than a fifth of their economically active population in farming in 1985; while Greece still retained one-quarter of its workforce in agriculture in 1989, the Portuguese figure had dropped to 16.9 per cent. The figure for the UK, on the other hand, was already down to 4 per cent by 1960; this country had honed down its farm population in earlier decades. Agriculture in the UK is, however, more important than its manpower figure suggests, for British farmers produce more than two-thirds of the food consumed in the county, including virtually all the milk, eggs, barley, pork and poultry.

Rural depopulation

The remarkable loss of workers from agriculture since the war implies a dramatic reshaping of rural demography. The nature and effects of this exodus have already been described in Chapter 4, notably the selectivity of the out-movement, unbalancing the residual population and threatening the viability of rural communities. Not all those who quit farming necessarily leave the area, however: there is a distinction to be drawn between rural exodus and agricultural exodus.

Rural depopulation has affected not only farmers but also a diverse assortment of rural labour moving in search of industrial and tertiary sector work. This movement also has a social dimension as rural migrants seek the better recreational and social conditions of urban areas – aspirations which are, however, not always realized. A good example of this social component is the widespread movement of rural workers out of agriculture in the Po Valley, despite the fact that farm labour here is better paid than unskilled

Table 9.1 Economically active population in agriculture, 1950–89

	Working population in agriculture (%)						Numbers in agriculture
	1950	1965	1975	1980	1985	1989	1989 ('000)
Austria	32.3	19.3	11.9	9.0	7.0	6.0	217
Belgium/Lux.	12.5	6.6	3.9	2.9	2.2	1.9	80
Denmark	24.5	14.4	9.2	7.3	5.9	4.9	139
Finland	46.0	23.5	15.8	12.0	9.3	8.4	213
France	28.7	17.9	11.1	8.6	6.7	5.5	1398
Germany (West)	24.0	10.8	6.6	5.8	5.1	3.8	1123
Greece	52.6	47.2	36.6	30.9	25.9	24.8	964
Ireland (Rep.)	48.7	31.4	22.4	18.6	15.4	13.9	189
Italy	42.8	24.8	15.4	12.0	9.5	7.5	1737
Netherlands	14.3	8.7	6.2	5.5	5.0	3.8	236
Norway	25.9	15.5	10.0	8.3	7.0	5.5	117
Portugal	44.3	37.7	28.1	25.6	23.0	16.9	788
Spain	49.7	34.0	21.5	17.1	13.7	11.3	1623
Sweden	20.3	11.1	6.9	5.7	4.7	4.0	175
Switzerland	11.0	9.4	7.0	6.2	5.5	4.2	141
UK	6.1	3.4	2.7	2.6	2.5	2.0	581

Source: FAO Production Yearbooks. Rome.

factory work (Romanos 1979). Village studies by social anthropologists have also shown that beyond the macrostructural economic explanation of rural depopulation lie powerful cultural factors. At the local level, *who* moves out of agriculture, and out of the region, is often determined by land inheritance patterns and types of tenure (Douglass 1971).

Outmigration has had devastating economic and social consequences for the exporting rural areas. One critical effect is the draining of the young segment of the labour force, dramatically distorting the residual population and leading to what Barberis (1968) has called the feminization and senescence of the agricultural labour force. In Italy, half the farmers are now over 55 years of age – a staggering obstacle to the technical and capital improvement of that country's farm sector. Beyond a certain threshold, depopulation makes it impossible for villages to sustain a viable existence. Schools, shops and other services close down, and this only serves to increase the unattractiveness of rural life (Clout 1984). Declining public transport poses a particular problem for those without cars, such as poor people and the elderly. In Western Europe deprivation thus has a rural dimension which is often overlooked.

Another result of depopulation is the abandonment of farmland and the consequent decline in agricultural production. Food must be imported to compensate these production losses. Nor does outmigration necessarily lead to farm rationalization. When rural migrants leave, they do not sell their land, but hold on

to it as insurance; it appears as patches of fallow in the rural landscape. In a study of farming in the south Italian region of Basilicata, Lane (1980) notes that small abandoned plots are a frequent sight. Their owners now live in Argentina, the USA or former West Germany. In contrast to the situation in the 1950s when this region was chronically overpopulated and the agricultural ecosystem severely overstretched by overgrazing and the erosive ploughing of steep slopes, now the area could sustain a much more intensive agriculture, but lack of labour and technical expertise among the remaining, mostly old, farmers prevents this happening (King and Killingbeck 1990). In Mediterranean Europe, land abandonment is most serious in upland regions: as terraces crumble, further erosion takes place and topsoil, once carefully conserved, is washed away into silting valleys (De Reparaz 1990).

To this evidence from southern Europe can be added corroborative material from northern Europe, where rural depopulation has been in existence in many areas since the early nineteenth century. In industrial countries it was primarily the availability of urban and factory jobs which pulled workers out of farming. In one of the most detailed British studies, Saville (1957) found that rural depopulation continued unabated throughout the 1851–1951 period, and was especially marked in remoter areas and in smaller villages. Drudy (1978) then pointed out that the phenomenon was not confined to marginal agricultural regions of Britain but was present in prosperous farming regions as well. His study in north Norfolk showed a 7 per cent drop in population between 1951 and 1971, with many parishes losing a third or more of their population. In northern Europe, the chief causes of rural depopulation appear to be agricultural processes (mechanization, concentration on fewer crops, etc.) which lead to high rates of redundancy in farming and the lack of alternative employment opportunities in rural areas (Gasson 1974; Verrips 1975). Very recent trends are ambiguous: in some areas rural depopulation appears to be continuing; in others it is slowing down or even moving into a phase of rural repopulation.

Along with rural depopulation and the exodus from agriculture goes the demise of peasants and indeed farmers as a social class. This is signalled in the titles of a number of relevant books; *Peasants No More* (Lopreato 1967), *The Vanishing Peasant* (Mendras 1970) and *The European Peasantry: the final phase* (Franklin 1969). What is happening is not just a simple change of occupation: it is the upheaval of a whole way of life. This occurs even where the agricultural exodus is mollified by the growth of some alternative employment *in situ*. In some parts of Europe, such as the Alps and the Pyrenees, farmers have switched to tourism as an alternative existence; they earn more money for less work in the hotels or on the ski-lifts and, as a result, the whole ethos of farming is undermined (Greenwood 1976; Vincent 1980). In other areas, such as north-eastern Italy and Swabia, the diffusion of small-scale industry into rural areas enables family farms to survive as social units by providing off-farm work for some members of the household. In the wilder corners of Europe, like Calabria and Corsica, there is no alternative: here, whole villages have been evacuated. These ghost settlements stand in the landscape as crumbling tombstones to a society that is no more.

Part-time farming

After the exodus from agriculture, the second major socio-structural characteristic of West European farming has been the expansion of part-time farming in recent decades. The EC labour force survey data for seven countries show that second-job holding is widespread in agriculture, especially in the Republic of Ireland, Italy and the former West Germany (Alden and Saha 1978). While agriculture in the seven countries surveyed employs only 7 per cent of the working population, it accounts for 31 per cent of those holding second jobs. In the former West Germany the phenomenon of 'worker-peasants' has been well established for generations, whereas in Italy and the Republic of Ireland a substantial part of the new manufacturing labour force created since the 1960s has retained links with family farming.

Part-time farming occurs when one or more members of a farm-based household are gainfully employed in work other than, or in addition to, farming the family's holding. The phenomenon may be examined under either the labour or the income share, although both can lead to complex problems of classification and definition given the diversity of experiences in part-time farming in Western Europe. If the farmer himself works off the farm, in a local factory for instance, the situation is straightforward. If the spouse of the farmer works off the farm, then again the situation is fairly clear since such off-farm earnings will almost certainly be pooled with farm income. However, if children residing on the farm have off-farm employment, contribute only casually to farm labour and do not pool their incomes with that of the holding, they have relatively little to do with the farm as a production unit and it is therefore doubtful as to whether this constitutes part-time farming.

Part-time farming is often assumed to be a kind of transitional phase of employment on the route from a predominantly agrarian society to a modern, industrial-based economy. In fact, it can be shown that part-time farming has deep historical and structural roots in the regional specialization of production and in the spread of domestic rural industry which occurred after the demise of the feudal system (Sivini 1976). In a more modern context, it can be seen as a functional adaptation to capitalistic development, providing a social base from which an individual can pick up a range of possible income supplements, either from local industry, services and public works, or as a temporary labour migrant elsewhere (Pugliese 1985).

At the risk of over-simplifying, two main models of part-time farming in western Europe can be identified. The distinction is based on entrepreneurial roles and on the location of the non-agricultural work. The first category consists of farmers who have a small business operated from, and perhaps also located on, the farm. Examples are the farm and the guesthouse, the farm and the camp site, the farm and the transport business, etc. The second model is one of 'pluriactivity' and consists of farmers who work outside the farm in factories. Here, to use Franklin's terminology, the farmer becomes a 'five o'clock farmer', an *Arbeiter Bauer* (worker-peasant), performing the heavy farm-work with the aid of a tractor in the evenings and at the weekend (Franklin 1971). In the first model, the farmer retains entrepreneurial functions in his second, non-agricultural occupation. The worker-peasant, on the other hand, abandons his, for he accepts the orders of others within the strict discipline of the factory system.

The German experience has been well described by Franklin (1969) who forecasts that worker-peasants will eventually become workers and that the agricultural element will disappear, ultimately leading to the disposal of farm property. This may be true for the former West Germany but not necessarily elsewhere. Part-time farming is increasing in many West European countries and there is ample evidence of the investment of factory earnings in agricultural holdings, which is an indication of the part-time system's stability.

The worker-peasant structure also offers a cushioning effect in times of industrial slump, and helps to maintain population and services in rural areas. But it has important 'backwash' effects on local agriculture. One of these is the decline in milk cows, especially in areas where women can find off-farm employment easily. Second, part-time holdings are usually small and excessively fragmented. Relative to their size, they are frequently over-mechanized, since labour time is limited. Lastly, there is the appearance of *Sozialbrache,* portions of farmland left fallow for social rather than agronomic reasons, simply because worker-peasant families lack the time or the inclination to cultivate it. In some German rural districts, up to half the land may be under 'social fallow' (Franklin 1969).

Recent changes in the German rural economy have altered the context of part-time farming (Thieme 1983). While little remains of the traditional augmentation of farm incomes from work in mining and cottage industries, some continuation of the worker-peasantry may be observed through farmers' participation in local manufacturing and tourism. This continuity is most marked in upland areas.

Somewhat different perspectives on part-time farming are introduced by the Italian and British cases. The importance of part-time farming in Italy was brought dramatically to the fore by the publication of the 1970/71 round of censuses. According to the 1971 population census there were 3 195 000 employed persons working in agriculture. According to the 1970 agricultural census the figure was 6 518 000! The difference is largely explained by the censuses' differential treatment of part-time employment in agriculture (Barberis 1977). While part-time farming in Italy cannot be divorced from that country's deep-seated agrarian problems, it is possible to view the phenomenon in quite a favourable light. Surveys in the Modena district of Emilia and in the region of Marche have revealed that part-time farms are more dynamic than full-time holdings. On average, part-time farmers were found to be younger, more educated, more mobile and more innovative; they had higher overall incomes and

a stronger work and managerial ethic (Cavazzani 1976; Pieroni 1982).

Not only does part-time farming characterize marginal upland regions such as Marche, it also appears to be endemic on the rural–urban fringe. Farmers located in this area are subject to strong employment influences from the nearby town. Furthermore, inconveniences may result from vandalism, pilferage, pollution and the shortage of farm labour. The farmer may be encouraged or forced to change his land-use regime, perhaps by abandoning livestock production and switching to market gardening. Anticipation of urban expansion may lead to land speculation and de-investment in farming, producing a ragged, semi-derelict landscape at the urban periphery. Around large British cities some farmers are switching from livestock farming and horticulture to 'horsiculture', providing stabling and riding facilities for the 'suburban cavalry' – an army of mostly young female horse-riders. Also to be found particularly in Britain are two other types of part-time farmer: those who have businesses or professional jobs but want to live in the country and farm as a hobby; and those who have chosen to 'opt out' of the urban life style and are prepared to accept a more simple way of life, often with greatly reduced income (Jolliffe 1977).

For a long time there were no policies for part-time farming because it was regarded as a transitory phenomenon. Only in recent years has it begun to be recognized as a permanent structural feature of the agricultural scene. Now part-timers can qualify for some assistance under EC schemes. Yet policy-makers are still inhibited, because they do not know enough about it: with the exception of Germany, no national government in Western Europe directly encourages part-time farming.

Land tenure

Agrarian structures do not necessarily coincide exactly with rural social structures or with economic modes of production in farming, but an analysis of land tenure adds an important institutional dimension to an understanding of agriculture in Western Europe. Clout (1971) sees land tenure, farm size and farm layout as a complex web of structural constraints interposed between the farming population and the land it cultivates, holding back agricultural progress in many areas. These features tend to reflect the requirements of agriculture in the past; they undergo only slow natural change. Less than 1 per cent of Western Europe's farmland comes on to the market each year, making structural changes such as farm enlargement excruciatingly slow. This situation may change in the fairly

Table 9.2 Structure of agriculture in the European Community, 1987

	Total no of holdings ('000)	% holdings			% agricultural area owner-farmed
		< 5 ha	5–50 ha	< 50 ha	
Belgium	93	38.5	56.7	4.9	31.7
Denmark	87	2.9	80.2	17.0	81.7
France	982	24.0	59.2	16.8	46.7
Germany (West)	705	32.8	61.4	5.8	63.6
Greece	953	77.4	22.2	0.4	77.1
Ireland (Rep.)	217	16.0	74.9	9.0	96.0
Italy	2784	77.2	21.4	1.4	80.0
Luxembourg	4	26.2	52.3	23.8	51.6
Netherlands	132	33.3	62.8	3.9	64.5
Portugal	636	83.4	15.5	1.2	66.3
Spain	1791	59.9	34.9	5.2	69.8
UK	260	19.3	49.6	31.1	62.6
Total/average	8644	59.3	35.2	5.5	64.9

Source: Eurostat (1991b).

near future given that so many of Western Europe's farmers are elderly and approaching retirement. Farming is an unusual occupation in that a large proportion of the new entrants come from within the sector. In France in the 1970s, only one-tenth of the people who took over farms were not farmers' sons or sons-in-law (Blanc 1987). Yet it is clear that a high proportion of farmers have no direct heir, either because they are childless or because they do not expect their existing heirs to succeed them on the farm. Only about one-quarter to one-third of farmers who are at least middle-aged have on-farm successors. The evidence on total successors (i.e. including heirs taking over the farm from outside) is unclear. Fennell (1982) suggests that around one-half of farms will have no direct heir. Partly, the proportion depends on the off-farm economic climate and job opportunities and, of course, on the viability of the farms themselves. Even where there is a successor there is no guarantee that the holding will be farmed full-time, or even at all.

Every West European state has a wealth of legislation affecting landownership and land tenure but little of it is aimed at accelerating structural change. Measures tend to favour owner-occupiers and to enhance the rights of tenants against landlords. Hired workers are fast disappearing in most countries although they remain important in British and Iberian farming. Throughout Western Europe the sanctity of the 'family farm' is preserved with an almost religious zeal – indeed, the 'yeoman farmer' is close to the social ideology of both the Catholic and Protestant churches. Yet the practical and theoretical grounds for the superiority of the family farm are not well established.

Three main parameters may be recognized under the umbrella of land tenure: types of tenure (owner-farming, tenancy, sharecropping, large estates etc.); the distribution of ownership of farmland; and land fragmentation or the splitting of farm holdings into scattered parcels. Each constitutes a distinct and important axis of analysis and will be considered in turn. Comparative data on farm structure are good for EC countries and Table 9.2 provides a résumé of farm size and tenure data.

Types of tenure and modes of production

The matrix of land-tenure types of Western Europe can be dissected in various ways. One fundamental distinction is between family farms and large capitalistic enterprises. The difference here lies in the use of labour of members of the farm family; capitalistic farms rely mainly or entirely on hired workers. Another basic difference can be drawn, on the basis of the destination of the product, between peasant and commercial farming. Peasant farms, which can also be regarded as a sub-group of family farms, involve a high degree of consumption of farm produce by the farm family; relatively little may be destined for the market and then only for local outlets. Commercial farming, on the other hand, is totally market-orientated. Thirdly, tenure differences may be defined according to the legal relationship between the farmer and the land: owner-farmers, tenant farmers, share-croppers, etc. These three typologies interrelate to produce a complex pattern of tenure types, the subtleties of which there is no space to examine here. However, some general trends are very clear, particularly for EC countries.

The most prominent feature of West European land tenure is the dominance of small owner-operated farms. Table 9.2 shows that 65 per cent of the EC's agricultural land is owner-farmed. The idea persists that peasant proprietorship lends backbone to the body politic; it is part of the 'man–family–business–community' ethic of rural society and reflects the social function of land as a source of work as well as of wealth. As a particularly labour-intensive form of enterprise, the peasant farm possesses a socially useful residual function that best manifests itself in times of unemployment and economic stagnation. Independence and hard work are two attributes epitomized in the peasant farmer who owns and controls his own assets. He also sets his own objectives: he can work for as long as he likes or for as little reward as he likes, for he is his own boss. Whether he can impose such behaviour on other members of the farm family is another issue, and undoubtedly gives rise to much intra-family conflict, particularly between father and son.

As an agricultural mode of production, family farming can also involve tenanted farms and rented land. There is a wide variation in the importance of tenanted land in Western Europe and in the forms that tenancy can take. Within the EC the proportion of farmland that is tenanted varies from 4 per cent in the Republic of Ireland to nearly 70 per cent in Belgium. Among other EC countries, France, the Netherlands, Luxembourg and the UK all have around 40–50 per cent of their land under rental. The tenancy situation is complicated, however, by the existence of mixed tenures (farms partly owned and partly rented) which are common in western Germany, France, Belgium, Austria and Spain. Farms under mixed tenure tend to be larger than average in those countries where they are important, which also tend to be the countries where tenancy is important, and smaller than average in countries such as Norway and Austria where tenancy is unimportant (Dovring 1965).

National attitudes towards tenancy vary enormously. Generally it is on the wane, not so much for economic reasons as for the outdated exploitative disadvantages which are assumed to accompany landlordism. Events in agrarian history still wield considerable influence. The high Irish figure for owner-occupancy must be seen in the light of nineteenth-century Irish opposition to the British occupation and its attendant landlordism. Even though the benefits of tenancy are understood in Ireland today, the expansion of renting is politically unacceptable in a land where the words 'landlord' and 'tenant' are repugnant. The Danes look back with great pride on the emancipation of their copyholder peasantry in the eighteenth century; since that time, Danish agrarian thought has stressed that the farming community's best interests are served if agriculture is based on self-ownership. In southern Italy, landlord–tenant relationships, based on social class divisions of an almost feudal nature, left little doubt as to who was the beneficiary and who the oppressed: rents were extortionate and contracts insecure. Such conditions lasted well into the post-war period (King 1970). Now, however, rents are controlled and are of only nominal proportions. Tenancies are inheritable and tenants of four years' standing have pre-emption rights. Indeed, the Italian landlord–tenant system has virtually ceased to function, at least officially. Instead it is replaced by informal 'seasonal tenancies' and under-the-counter payments by landlords to outgoing tenants in order to get possession of the land.

On the other hand, in many north-west European countries tenanted farms are larger and more efficient than owner-occupied holdings. The efficiency of Dutch, Belgian and British agriculture has more than a little to do with the flexibility that a structure of rented farms permits. In France, tenancy predominates in the north which, by and large, also contains the richer agricultural land. Many of the rich farms of the Dutch polders are rented from the state. In Britain, conventional wisdom has it that the landlord–tenant arrangement has operated well, but it is really only in this country that an independent and strictly commercial landlord class exists which fulfils any management and capital-formation function (Harrison 1982).

Share-tenancy or *métayage,* where the 'rent' is paid in kind rather than cash, seems destined for imminent oblivion. Its main areas of importance are, or rather were, southern France and central Italy, although it is also to be found in parts of Iberia. In the classical Tuscan *mezzadria,* the crop was divided equally between the landlord and the share-tenant or *mezzadro* but, in practice, cases existed where the division was much more in favour of the landlord. Nevertheless, *mezzadria* holdings were larger than other peasant farms and security of tenure was assured for life and could even be inherited. The *mezzadro* not only had his land but also a farmhouse and most other capital equipment supplied by the landlord. Some economies of scale were available through the *fattoria,* the landlord's estate headquarters for his *mezzadria* tenancies. Encouraged by Mussolini, *mezzadria* expanded to embrace 22 per cent of Italian farmland by 1947, since when it has pitched into rapid decline, covering 12 per cent of land in 1961 and 3.2 per cent in 1982 (King and Took 1983; Venzi 1988).

More degraded forms of sharecropping are also fast disappearing. Typical is the south Italian *comparteci-pazione* contract, where tenants are paid in kind for little more than their labour on the landlord's land: security is limited to one year, there is no attempt to form a properly coherent holding and the entrepreneurial functions of the tenant are virtually zero. Such contracts have no part to play in modern agriculture.

Farm labourers, unlike owners, tenants and sharecroppers, have no formal tenure ties. The use of hired labour is the main distinction between capitalist and peasant farming, the former inevitably larger scale than the latter. In former West Germany, for instance, 93 per cent of labour input on farms of 10–20 ha is supplied by the family, while on farms of over 50 ha, 82 per cent of the labour is hired, most of it on a full-time basis. In northern Italy the capitalist and peasant systems are sharply divided by topography. The break of slope that marks the edge of the Po Plain is also the agrarian division between two socio-economic systems: large capitalist farms run with wage labour in the plain; small peasant-owned or rented holdings in the hills and mountains (Fuller 1975).

When compared to family farms, capitalist farms nearly always have higher returns per unit of labour, better technical performance, greater specialization and more sophisticated market links. It is, therefore, rather strange to learn that hired workers are declining faster than the farming population as a whole. In six countries of the EC – the UK, the Republic of Ireland, the Netherlands, Denmark, West Germany and Luxembourg – the decline in hired workers during 1960–80 (50–80 per cent) was roughly twice as fast as for farmers (25–45 per cent). Clearly they have moved rapidly out of agriculture because of their lack of formal tenure status, low wages and the seasonality of much farm labour. Southern Italy's *braccianti* and Spain's *braceros* – unskilled labourers hired by the day – have all but vanished from the rural scene: they have been in the vanguard of the labour migrants streaming abroad. Fixed-wage workers, often skilled, are a more stable element of agriculture in certain regions, notably the UK, the Paris Basin, Languedoc and the North Italian Plain.

Distribution of landownership

The wealth of data provided by the EC farm surveys reveal remarkable contrasts in the pattern of land ownership in different countries. Some of these data are summarized in the middle columns of Table 9.2. In 1983 the average farm size in the Ten was 13.6 ha, but national averages ranged from 3.6 ha (Greece) to 64.5 ha (UK). Regional variation within national boundaries is also considerable for the larger countries of the EC. The average French farm size is 25.5 ha, yet most farms in the north-east are over 30 ha while most in Brittany, Aquitaine and the Midi are under 10 ha (Winchester and Ilbery 1988). In Italy, Greece, and Portugal around four-fifths of farms are below 5 ha. Such holdings are only viable economically if they are devoted to factory farming or market gardening. This is largely the case in the Netherlands and Belgium where farming is efficient and yet about half the holdings are below 20 ha, but not in Southern Europe where most of the small farms are semi-subsistence peasant plots. Denmark and Ireland are notable for their predominance of medium-sized farms (Table 9.2). Spain has a polarized landholding regime, with large farms accounting for most of the land and small-holdings accounting for most of the farm population. Farms are also mostly small in Western Europe's other major non-EC countries: Austria, Switzerland, Norway, Sweden and Finland. To take the Finnish example, in 1979, farms in the 2–10 ha class covered 59 per cent of all arable land, those over 25 ha only 9 per cent. The average Finnish farm has 11 ha of cropland and 35 ha of forest. The entire country contains only 367 farms with more than 100 ha of cropland (Cabouret 1982).

Of the three countries which have inherited polarized landholding structures, two have had significant land reforms in the post-war period – Italy in 1950 and Portugal in 1974. Pressures for land redistribution in Spain have been resisted; here, structural policy concentrates instead on the consolidation of fragmented holdings, discussed in the next section.

The Portuguese land reform was one of the more significant products of the 1974 revolution. Regionally, it was concentrated in the Alentejo, a rolling landscape of unirrigated wheat, sheep and cork estates worked by underpaid landless labourers with a long tradition of political radicalism. Here, Communist-inspired land seizures anticipated the land reform law of mid-1975. Latifundian capitalism was defined as the main target for expropriation and more than 1 million ha were taken over in a matter of months. New 'collective production units' united the various seized estates within the framework of local administrative boundaries, substantially concentrating the land and the labour force. The collectives were given names like 'Che Guevara' and 'In Lenin's Footsteps'. The fact that the reform increased the size of the operating unit, and did not split the estates into smallholdings, is related partly to political ideology but also reflected a belief in the economic superiority of large-scale farming, at least under the Alentejo's prevailing agrarian regime. It was also suggested that the rural labourers, lacking a tradition of peasant landownership, did not want to become titleholders of their own small plots (Cabral 1978), but this view has been criticized (Routledge 1977).

When the Communists lost their position of strength in the Portuguese government after November 1975, the impetus of the land reform weakened and went into reverse. Some usurped land was given back to landlords, a process often accompanied by violence in the countryside (King 1978). Support to the collectives was diluted and shifted instead to the peasant and capitalist sectors, in line with Portugal's integration with the EC.

Portuguese land reform was, therefore, politically significant but its economic impact has been relatively modest. The 1 million ha held by the new production units represent 33 per cent of the cultivated area in the reform region of Alentejo and 14 per cent of the national farmed area. Some 59 000 rural workers participate in the collectives; this is 35 per cent of the regional and 7 per cent of the national active rural population. Economic performance is difficult to monitor because of the withdrawal of government support. Agriculturally, the main change has been the development of livestock production. Outside the strictly economic sphere there is no doubt that Alentejo's rural workers now enjoy more employment, much better social welfare and, indeed, a complete transformation in the social relations of production (de Barros 1980).

Much less need be said of the Italian land reform which forty years after the event, appears as a dated and expensive irrelevance in the overall strategy of Italian agricultural development. Operating mainly in the latifundian areas of the Mezzogiorno, it moved in the opposite direction to the Portuguese reform, splitting up some 680 000 ha of large estates into 113 000 small peasant plots of around 2–10 ha. A large proportion of these plots were too small to support a family and were subsequently abandoned or used merely as part-time bases in the countryside. Only in certain rich irrigated areas such as the Sele Plain south of Naples and the Metaponto Plain stretching west and south of Taranto have the reform holdings emerged as viable units (King 1973).

Many countries have measures aimed at farm enlargement and amalgamation. Legislation in France, Germany, the Netherlands, Norway and Sweden encourages the use of land falling vacant for the enlargement of adjacent holdings. But the results of this kind of natural change are slow: it takes a decade to raise the EC's mean farm size by just 1 ha.

The French policies for farm enlargement are well known. In 1960, twenty-nine SAFERs (*Sociétés d'Aménagement Foncier et d'Établissement Rural*) were set up; their objective was to buy land offered on the open market and use it to enlarge other farms that were too small. In addition, SAFERs can acquire abandoned land, improve it and use it for the same end. Recently additional SAFERs have been set up in northern France and in Corsica, bringing the total to thirty-one. The SAFER policy has had most impact in the south-west, centre and south of France, especially in Aquitaine and the Massif Central; in the north-east and Brittany high land prices have limited the scope for action (Jones 1989a). Since they started the SAFERs have bought up about 1.5 million ha of land and sold over 1.3 million ha. These figures represent almost 5 per cent of France's agricultural area and mean that SAFERs have acted as a stabilizing influence in the French land market (Winchester and Ilbery 1988). The main problem SAFERs face is deciding which farms should receive land: should it go to large or to small farms? Considerable variation has occurred in the way SAFERs operate: they have, in places, been accused of nepotism (allocating land to acquaintances of board members) and of helping uneconomically small farms to persist (Hirsch and Maunder 1978). However, SAFERs are unable to carry out as much structural improvement as they would wish because of shortage of funds. Although most land still goes to farm enlargement, an increasing proportion, now about 25 per cent, is used to establish young farmers. Stocks of land may also be sold off in connection with *remembrement* (consolidation) schemes, or used to compensate owners who have lost land through public schemes, e.g. road widening (Winchester and Ilbery 1988).

The British experience has seen most vacated land bought up by the larger farms; farm amalgamation among smallholdings has been rare (Clark 1979). Where the land market is tight, farms may only be able to increase their size by piecemeal addition of blocks located at some distance from the holding, thereby increasing fragmentation. Renting additional farmland is not a policy actively encouraged anywhere in Western Europe; indeed, in Denmark, institutional landlords are forbidden. Denmark has also consistently legislated to limit the extent of landholdings. In 1978 the maximum limit on Danish farms was reduced from 100 to 75 ha, and residence by owner-farmers on their properties has been obligatory since 1973.

Fragmentation and consolidation

Land fragmentation refers to the internal spatial structure of the farm; it is the condition whereby a holding is split into many small non-contiguous plots. In France, for example, fragmentation is a legacy of the Napoleonic code of equal land inheritance; with each passing generation, land is divided up among all direct heirs. Large family sizes, common in many parts of rural Europe until recently, thus provoked rapid fragmentation. The phenomenon is also compounded by the frequent insistence of each heir on a share of each part of the patrimony: a holding of four plots divided among four heirs yields sixteen fragments.

The disadvantages of fragmentation are largely economic. Time is wasted by the farmer in travelling to his scattered plots and in moving his implements, animals and machinery back and forth; expenses on fencing, water supplies and buildings are greater. Some of these economic costs have an ecological basis, such as problems of soil conservation, drainage and irrigation. Other disadvantages arise in the social and administrative fields of rural life. Land planning becomes very difficult, for fragmentation increases the number of people involved in making a decision about a particular area. Cadastral authorities have difficulty in monitoring precise patterns of ownership, particularly where plots become fractioned into pocket-handerchief size or where, as sometimes happens in Mediterranean areas, trees are owned separately from the land. Social tension is caused by disputes over ownership, access and damage. Some of these problems are well documented in Greece where property division is often marked by acrimony and even violence (Herzfeld 1980; Thompson 1963).

Fragmentation can, however, have a rational basis, as the work of ecological anthropologists in Alpine and Mediterranean regions shows. In these areas, microvariations in altitude, aspect, soil, precipitation, temperature and wind provide an ecological setting in which scattered farms are the logical outcome. Especially where farming is subsistence-oriented polyculture, the multiplicity of ecological niches provides ideal conditions for a range of crops and, moreover, spreads the risks of destruction by frost, hail or wind. In the Alps the key to farming success is the seasonal oscillation of man and beast which unites mountain pastures and forests with the valley-floor meadows (Netting 1981). Galt's (1979) study of viticulture on the Italian island of Pantellaria shows how fragmentation can be rational even in a monocultural setting. Here, possession of plots in different parts of the island causes variations in the grape crop in terms of sugar content and harvest time; grape-picking labour can thus be staggered and the risk of complete failure is minimized.

These exceptions apart, there is no doubt that fragmentation is a serious and widespread problem in European farming. Levels of fragmentation were, and still are, particularly high in southern European regions like Galicia, northern Portugal and Crete. In Denmark and Sweden, on the other hand, hardly any serious fragmentation exists, for these countries had their farms consolidated in the eighteenth and nineteenth centuries.

The progress of land consolidation in West European countries since the last war has been highly variable (Jacoby 1959; Lambert 1963). In Portugal, Italy and Greece almost nothing has been done. France has pioneered *remembrement,* parcel exchange, since a law of 1941 when 15 million ha were estimated to need consolidation. Forty years later, half this target has been achieved. Most progress has been in the north and east where fragmentation was acute and yet prospects for the mechanization and improvement of agriculture good (Clout 1968). In the south and in the mountains, costs are greater, benefits probably less and peasant antipathy stronger. In these areas the future must be envisaged in terms of farm enlargement, livestock rearing, afforestation and recreation (Clout 1974).

In contrast to the voluntaristic nature of French *remembrement,* land consolidation in former West Germany, Switzerland and the Netherlands is more highly organized and integrated with regional and settlement planning. Resettlement of farm families is an expensive process, requiring new farmhouses, roads and other rural services, but is necessary to decentralize farmers from overcrowded villages to their newly consolidated holdings. In the Netherlands, important consolidation and resettlement schemes have transformed the landscape of the Rhine and Meuse valleys. The German policy of *Flurbereinigung* has also seen a shift from simple plot amalgamation to settlement relocation. Following laws in 1937 and 1953, rapid extension of consolidation took place, especially in Bavaria and Swabia where fragmentation was most severe (Mayhew 1970). However, much of this early work is now considered obsolete and must be redone. A new Farm Consolidation Act was passed in 1976; now consolidation embraces other aspects of rural planning such as recreation and environmental preservation. By 1980, about a third of the target still remained to be achieved. At the current rate of progress this should be attained by the end of the century – by which time the technical side of farming may have changed so much as to demand further reorganization programmes!

The country where land consolidation has made most progress in recent years is Spain. In the early post-war years much Spanish farmland was impossibly fragmented, especially in the northern half of the country. In the village of Santa Maria de Ordax in León province each farmer had 6–7 ha divided into 80–120 tracts dispersed over a radius of 5 km. Spain's first agricultural census of 1963 showed that the average number of plots per holding was 13.7. Fragmentation was particularly acute in Galicia where the average number of plots per farm was 32. It was less severe in Catalonia and the Basque provinces which have a tradition of *mayorazgo,* passing land to the eldest son, and in the south where large estates predominated. In 1957 the Spanish Secretary of State for Agriculture estimated that 8 million ha of farmland needed consolidation.

Spain's programme of land consolidation started in 1952 with the creation of the *Servicio de Concentración Parcelaria*. The programme has three main aims. Firstly, it tries to assign to each farmer a single plot equal in area and quality to the several parcels previously possessed. Second, the new holding should be sited within easy reach of the farmer's dwelling (where necessary, new country roads are built to facilitate access). Third, the titles of the new holdings are confirmed in the property register. Progress was slow throughout the 1950s (less than 50 000 ha consolidated per year), peaked in the late 1960s and early 1970s (350 000–400 000 ha annually), and then fell back in the late 1970s to around 200 000 ha per year (Guedes 1981). Most progress has been made in the central and northern provinces of Spain, Galicia and Catalonia excepted. After consolidation it is possible to practise such techniques as contour ploughing, soil improvement and irrigation; mechanization is more feasible on larger, more compact plots. Productive effects do, however, depend on the type of agriculture. In dry zones, output only increases marginally. In irrigated areas, where consolidation has the additional payoff of facilitating better use of irrigation water, increases of the order of 70 per cent have been reported. Consolidation also encourages the replacement of peasant individualism by an appreciation of the value of collective action, and groups of consolidated smallholdings are now being farmed cooperatively. There are, however, some problems. One is the lack of legislation to prevent continuing subdivision. Another is the small size of most consolidated farms: at 5 ha on average, most are still *minifundios*. In Galicia, a difficult area, the policy is fuzzily conceived in relation to the local agrarian structures; it proceeds only slowly and on an *ad hoc* basis (O'Flanagan 1982).

While initially the cost of Spanish consolidation was low compared to French, West German and Swiss schemes, costs escalated after 1972 to roughly four times their level in the 1960s. Nevertheless, the overall statistics of the policy are illuminating and encouraging. By 1978, 5 million ha of Spanish farmland had been consolidated, leaving 3 million ha still to be treated. Nearly 1 million proprietors had been affected, their total number of parcels being reduced from 13 688 688 to 1 790 971. Mean plot size rose from 0.35 ha to 2.52 ha and the average number of plots per holding for those affected was cut from 14.6 to 1.9 (Guedes 1981).

Land use and production

From the human and social side of agriculture we pass to the economic. Statistically, rural land use has been well documented in most West European nations since the introductory work of Stamp (1965). The tradition of detailed land-use mapping has become firmly established in many countries although none can match the clarity and beauty of the British Land Use Survey sheets published at 1 : 25 000.

It is, therefore, rather strange that almost no attempt has been made at an overall treatment of Western Europe's agricultural geography. Van Valkenburg's work constitutes a partial exception but his analysis is based on a simple distinction between high and low 'standards of land use' (van Valkenburg 1959, 1960). In the absence of a truly integrative study, this section proceeds on two bases: a broad and brief description of Western Europe's land-use structure; and a nation-based tabulation of FAO area and production data for the late 1980s.

Climate largely dictates the latitudinal range of agricultural types. Given the importance of cereals in Europe's mid-latitude farming regimes, a series of east–west grain belts can be recognized (Fig. 9.1a). This east–west pattern is rotated along the north-west margins by the equitable influence of the North Atlantic Drift. At the meso-scale, topography interrupts this pattern, preventing crop growth in northern Europe in the Scottish and Scandinavian mountains, and introducing cooler climates in the uplands of southern Europe, notably the Spanish Meseta, the Pyrenees, the Massif Central, the Alps, the Apennines and the Pindus Mountains. Even more localized farming types result from micro-scale variations in soil and aspect. For instance, the vine, a typically Mediterranean crop, is found in scattered locations much further north, from the valley of the Loire to Koblenz on the Rhine: here, warm soils, sunny slopes and sheltered gorges enable it to flourish well outside its natural geographic habitat. In Brittany, the mildness of the Atlantic winters and the early spring allow the

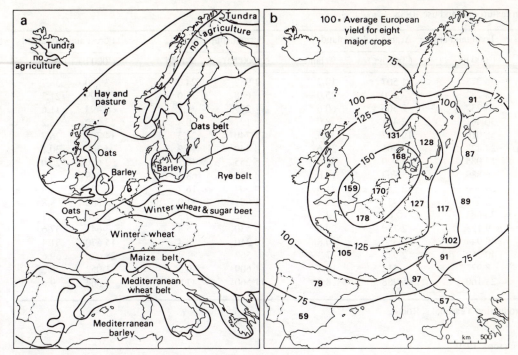

Fig 9.1
Agricultural features of Western Europe: (a) main crop belts; (b) patterns of yield intensity (after van Valkenburg and Held 1952).

cultivation of early vegetables along certain valleys and sheltered coasts. Transport may also play a role: the cultivation of *primeurs* around Avignon is related not only to the environmental conditions of the southern Rhône valley but also to the development of fast access routes to Paris (Minshull 1978).

The influence of urban markets is more fully evident in Fig. 9.1b which, in spite of some uncertainty over the validity of the figures, can be interpreted as a macro-Thünian portrayal of western Europe's cropping intensity, based on spatial variations in the yields of eight major European crops. A zone of high production centres on the North European Plain, the Rhine delta and south-east England. In this zone are to be found most of Western Europe's main centres of urban population. Away from this core of high intensity, yields diminish in all directions. Again, climatic factors are prominent: exposure to the west, frost to the north, continentality to the east, drought to the south. Topography too plays a role, for the core is girt by hill and mountain areas where yields are inevitably lower. Particularly to the south, differences in the technical standards and intellectual abilities of farmers are important, for it is in Iberia, southern Italy and Greece where most of the really backward areas of European

agriculture are found. Of course, there *are* areas of high-intensity production in these southern countries – such as the Po Plain (which can be regarded as a southern outlier of the North European Plain), the citrus orchards of Valencia and Sicily, and the specialized vineyards of the Douro Valley (producing port) and Jerez de la Frontiera (producing sherry) – but these are spatially less continuous than the zone of high-intensity production to the north.

Tables 9.3, 9.4 and 9.5 set out national statistics on land use, livestock and agricultural production. It should straightaway be emphasized that comparison between countries can be hazardous because the definitions of categories of land use vary. Particular problems arise over little-used agricultural land like mountain pastures and communal land. The area figures are, therefore, indicative rather than intrinsically accurate.

Climate, topography and soils are the main factors determining the national proportions of land under farming. Population density has also played a role, especially in the past. In the UK, 77 per cent of land is in agricultural use, despite extensive urbanization and the fact that only 2 per cent of the working population earn their living from farming. High proportions of

Table 9.3   Main features of land use, 1988

	Total land area ('000 ha)	Arable and cropland ('000 ha.)	% land	Permanent pasture ('000 ha)	% land	Forest and woodland ('000 ha.)	% land
Austria	8 273	1 507	18.2	1 982	24.0	3 200	38.7
Belgium/Lux.	3 282	820	25.0	688	21.0	699	21.3
Denmark	4 237	2 566	60.6	217	5.1	493	11.6
Finland	30 461	2 441	8.0	127	0.4	23 222	76.2
France	55 010	19 547	35.5	11 740	21.3	14 700	26.7
Germany (West)	24 428	7 466	30.6	4 449	18.2	4 360	30.1
Greece	13 085	3 929	30.0	5 255	40.2	2 620	20.0
Ireland (Rep.)	6 889	963	14.0	4 688	68.1	339	4.9
Italy	29 406	12 149	41.3	4 907	16.7	6 735	22.9
Netherlands	3 392	931	27.4	1 081	31.9	300	8.8
Norway	30 683	862	2.8	102	0.3	8 330	27.1
Portugal	9 195	2 750	29.9	530	5.8	3 641	37.6
Spain	49 944	20 380	40.8	10 210	20.4	15 670	31.4
Sweden	40 260	2 935	7.3	562	1.4	28 020	69.6
Switzerland	3 977	412	10.4	1 609	40.5	1 052	26.5
UK	24 160	6 988	28.9	11 560	47.8	2 364	9.8

Source: FAO Production Yearbooks. Rome.

Table 9.4   Main crops, 1989

	All cereals '000 ha	'000 m.t.	kg/ha	Main types of cereals (% area) Wheat	Barley	Maize	Oats	Rye	Other products (1985) Sugar beet ('000 ha)	Pulses ('000 ha)	Fruits ('000 m.t.)	Vegetables ('000 m.t.)
Austria	947	5 009	5 290	30.4	31.7	19.7	7.1	8.3	43	—	755	463
Belgium/Lux.	376	2 324	6 189	50.9	34.6	3.4	7.7	1.6	120	2	370	1 172
Denmark	1 572	8 808	5 603	21.3	68.4	—	2.3	7.8	73	105	88	260
Finland	1 193	3 791	3 177	12.5	51.3	—	32.6	2.5	31	4	113	169
France	9 417	57 140	6 068	49.8	23.2	19.1	4.4	0.9	490	244	13 788	7 391
Germany (West)	4 658	26 212	5 627	33.2	39.9	3.7	11.9	8.7	403	23	3 517	2 198
Greece	1 385	4 403	3 180	60.6	20.9	13.7	2.9	0.6	43	54	3 885	4 429
Ireland (Rep.)	340	2 072	6 086	20.5	73.6	—	5.9	—	34	—	21	289
Italy	4 639	16 935	3 650	63.0	9.7	18.9	3.8	0.2	233	205	18 749	14 677
Netherlands	202	1 368	6 762	69.9	21.3	—	6.0	2.7	131	25	444	2 820
Norway	347	1 252	3 608	11.4	50.0	—	37.7	0.3	—	—	134	217
Portugal	1 066	1 825	1 713	27.6	8.0	32.0	14.7	14.5	3	298	1 585	1 896
Spain	7 792	19 440	2 495	27.0	55.5	6.9	6.2	3.0	178	444	12 117	9 104
Sweden	1 276	5 530	4 334	18.2	44.1	—	31.1	3.0	52	46	185	295
Switzerland	198	1 203	6 093	53.3	27.7	10.3	5.4	2.7	15	—	739	340
UK	3 903	22 526	5 771	47.3	49.0	—	3.4	0.1	200	90	510	3 920

Source: FAO Production Yearbooks. Rome.

land devoted to farming are also to be found in the Republic of Ireland (82 per cent), Greece (70 per cent) and Denmark (66 per cent). The harsher physical conditions in northern Europe mean that countries here have only small proportions of their land devoted to farming: Norway 3 per cent, Finland 8 per cent and Sweden 9 per cent. Forest becomes the dominant land use instead, covering more than three-quarters of the surface of Finland. In Switzerland and Austria the extent of cultivated land is also limited. In Alpine Europe and in the Repubic of Ireland, permanent pasture dominates over arable. In Mediterranean countries

Table 9.5 Livestock numbers, 1989 ('000)

	Cattle	Pigs	Sheep	Goats
Austria	2 541	3 874	256	32
Belgium/Lux.	2 967	5 950	190	8
Denmark	2 226	9 105	86	—
Finland	1 379	1 327	59	3
France	21 780	12 480	12 001	1 103
Germany (West)	14 659	22 589	1 464	52
Greece	731	1 226	10 376	5 970
Ireland (Rep.)	5 637	961	4 991	9
Italy	4 606	13 820	1 405	34
Norway	932	750	2 248	92
Portugal	1 359	2 326	5 354	745
Spain	5 050	16 100	23 797	3 100
Sweden	1 662	2 320	396	—
Switzerland	1 850	1 869	340	70
UK	11 902	7 626	29 046	61

Source: FAO (1990).

a significant proportion of cropland is contributed by permanent crops like vines and orchards of olives and citrus fruits: this proportion is 25.4 per cent in Greece, 23.9 per cent in Italy, 24.2 per cent in Spain and 16.5 per cent in Portugal. It is also in the Mediterranean that irrigation becomes such a vital component of farming. Greece has 24 per cent of cropland irrigated, Italy 24 per cent, Portugal 18 per cent and Spain 15 per cent. Interestingly 26 per cent of Dutch farmland is irrigated – a reflection of the intensity of agriculture in the Netherlands. Elsewhere, figures are much lower: France 6 per cent, West Germany 4 per cent, the UK 2 per cent.

A comparison of agricultural production indices for the triennium 1987–89 (1979–81 = 100) shows how Western Europe's farmers are increasing their yields. The biggest increases were recorded by Denmark (120), Belgium (116), Spain (116) and the Netherlands (114). These figures illustrate firstly the dynamism of the Spanish agricultural economy (Spain also achieved the biggest output increase in the 1970s – 36 per cent between 1969–71 and 1979–81) and secondly the capacity of Europe's most intensive farming nations to go on achieving high productivity gains.

Table 9.4 gives a breakdown of the main crops grown in the countries of Western Europe. Cereals dominate in terms of area, but fruit and vegetable harvests are quantitatively heavier in some countries – the Netherlands and the familiar group of Mediterranean nations. France and West Germany produce between them 45 per cent of Western Europe's cereals, with France producing twice as much as West Germany. There are considerable differences in the type of cereals grown: in Denmark, 68.4 per cent of cereal land is

barley for livestock feed; wheat is important in all countries except Scandinavia and the Republic of Ireland; Norway, Sweden and Finland have around a third of their cereal land under oats; maize is the most widely grown cereal in Portugal. Cereal yields also reveal an interesting hierarchy. The Netherlands leads by a considerable margin, with a mean yield four times that of Portugal. Spain is the leading producer of pulses, while Italy heads both fruit and vegetable output. Vines, an intensive and remunerative crop, cover large areas in Spain (1 380 000 ha), Italy (1 046 000 ha) and France (982 000 ha), and are also significant in Portugal, Greece, West Germany, Austria and Switzerland. France and Italy are the main producers of wine (5 891 000 and 5 890,000 metric tonnes respectively in 1989), followed by Spain (3 030 000 metric tonnes). These area and production figures for wine have been falling quite rapidly. Wine, especially low-grade 'table-wine', has been in a crisis of overproduction, due in part to falling consumption levels in France and Italy. Ten years ago, taking the average for 1979–81, the three big producers generated 8 075 000 metric tonnes (Italy), 7 063 000 (France) and 4 142 000 (Spain).

Regarding livestock (Table 9.5) France has most cattle, West Germany most pigs, the UK most sheep and Greece most goats. With the exception of dairy cattle, these numbers have shown a consistent tendency to increase in recent years. Livestock distribution, like that of many crops, is often highly regionalized within countries. In Italy, for example, most cattle are in the North Italian Plain and most sheep in Sardinia, Lazio and Abruzzo.

Nearly all Western European countries are experiencing substantial losses from the agricultural area, while woodland and urban areas show a persistent increase. Most of the agricultural loss is from cropland, although in the UK the decline is mainly in pasture. Yet the downward trend in farmed area is not paralleled by declining total output. The spectre of insufficient food supplies by the end of the century is a figment of the imagination. Indeed, the opposite circumstance applies, and is a major dilemma of EC agricultural policy.

The Common Agricultural Policy (CAP)

In spite of trenchant criticism, the CAP has undoubtedly acted as a potent force in the creation of the Common Market. Absorbing three-quarters of the Community budget over the years (but generally two-thirds during the 1980s), the CAP represents something of a success story when compared to relative non-achievement in the regional and industrial policy fields.

Before the 1960s, each member state operated its

own farm policy to ensure an adequate national food supply, the siege mentality of war and scarcity lingering on (Clout 1984). Cost considerations took second place to an emphasis placed firmly on production. Trade policies reflected this fixation. Differing prices and support levels ensured protection against imports for each country; there was no way of getting farm produce to flow freely between members of the Six. A community policy for prices, production levels and marketing was thought essential. Agriculture needed to be strengthened: in 1958, it employed more than a fifth of the Six's working population but it accounted for only 8 per cent of the EC's gross product.

Several stages in the rather slow evolution of the CAP can be recognized (Bowler 1985). The first stage was the evolution of the principles of a farm policy as they were enshrined in the 1957 Treaty of Rome, specifically Articles 38 to 47. The key article was 39.1, which laid down the following objectives of the CAP: to increase agricultural productivity by promoting technical progress and the optimum deployment of labour; to ensure a fair standard of living for members of the agricultural community; to stabilize markets, assure the availability of supplies and ensure that supplies reach consumers at reasonable prices. Article 39.2 stipulated that, in the working out of the CAP, account was to be taken of the social structures of agriculture and of the disparities between farming regions. In Article 40 the scope of the proposed common legislation was spelled out: it was to include regulation of prices, aids for the production, marketing and storage of various products and common machinery for stabilizing imports and exports. It also provided for the establishment of common funds to finance the policy – what came to be known as the Agricultural Guidance and Guarantee Fund, or FEOGA.

The second stage started with harmonization of national marketing systems in the early 1960s and of product prices in the late 1960s. Initially, the marketing arrangements were harmonized for cereals, pigmeat and eggs in 1962: other products followed. The basic principles of policy were the establishment of a single market area, free international movement of produce within the Six, uniform external tariffs, Community preference in agricultural trade and the sharing of the CAP's financial burden. Product price harmonization started with common cereal prices in 1967; their role as a human staple and livestock feed made cereals a key sector. In 1968, common prices were set for milk, milk products, beef, rice sugar and olive oil; in 1970 for tobacco and wine. The support prices contained a bewildering variety of types – there were intervention, base, exclusion, indicative, minimum, reference, withdrawal and threshold prices – but all amounted to assuring farmers a guaranteed minimum income.

It soon became obvious that a product-based policy did not fit the realities of the EC's agricultural geography. Price levels and price differentials between products were not of an order to avoid surplus output and reorientate production in desired directions. There was no mechanism for relating overall supply to overall demand. Surpluses were drawn forth in cereals, milk and sugar. Later, other products bulked into 'mountains' or pooled into 'lakes'.

The problem lay in the duality of the EC's agricultural systems between small-scale peasant farming and the commercial sector based on large holdings. Prices set in order to guarantee a reasonable income to the former were so favourable that they encouraged large production increases in the latter. It was discovered that pricing policy could not simultaneously meet farm income and farm production objectives: the 'guarantee' section of FEOGA was sabotaging the 'guidance' element. The cost of supporting the production of agricultural surpluses through intervention buying became the main stimulus to a realignment of the CAP in the late 1960s. The struggle to achieve this realignment is still going on.

This ushered in the third phase, which separated price policies, which were to orientate production in the commercial sector, from a new set of structural measures designed to deal with the problems of the small-farm sector. This dual approach, initially set out in the Mansholt Plan of 1968, recognized that a successful agricultural policy could not treat all farmers equally.

The Mansholt Plan and after

This was the work of Sicco Mansholt, former Dutch Agriculture Minister and Vice-President of the EEC Commission on Agriculture. Mansholt's diagnosis of the ills of European farming could be summed up as too many farms, too many people working in farming and too much agricultural land. He proposed a three-pronged attack to rescue farming from its crisis of overproduction and structural dualism. Firstly, he wanted to see the EC's farm population down to 5 million by 1980 (it was 20 million in 1950). Secondly, small peasant holdings were to be replaced by larger agricultural enterprises practising modern techniques. Mansholt laid down that a suitable farm for the 1980s might comprise 80–100 ha of cereals or raise 40–80 dairy cows, 150–200 head of beef cattle, 450–600 pigs, or 100 000 head of poultry per year. These were ambitious targets given that, at that time of Mansholt's memorandum, two-thirds of the EC's farms were fewer than 10 ha and two-thirds of dairy farmers had less than five cows! Thirdly, he prescribed a 7 per cent reduction in the farmed area, so that by 1980 the culti-

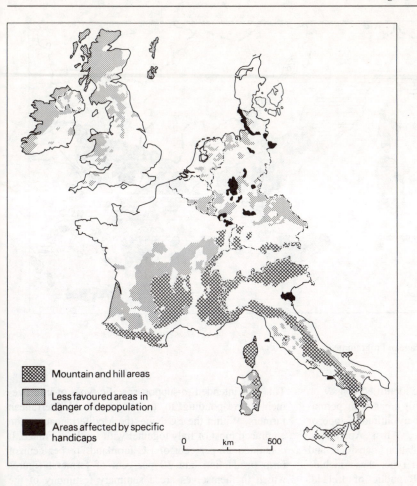

Mountain and hill areas

Less favoured areas in danger of depopulation

Areas affected by specific handicaps

0   km   500

Fig. 9.2
Less favoured areas in the European Community, according to Directive 75/268.

vated surface of the Six should be down to about 70 million ha. Land taken out of agriculture was to be used for recreation and afforestation. The areas affected by this geographical shrinkage of farming were not spelled out but clearly would include regions of limited potential such as the Massif Central, Provence, Corsica, Sardinia, Sicily and southern Italy.

Mansholt's memorandum engendered considerable hostility. His use of terms like 'business-oriented production unit' sent shivers of apprehension through elderly farmers and farming associations devoted to the ideal of the family farm: they feared that this was a prelude to some kind of collectivization. A revised and moderated report was presented in 1970, but the basic strategy was unchanged. Mansholt's shock treatment was certainly effective in promoting discussion based on a new realization of the predicament of the Common Market's farm sector.

In 1972 the Community approved three measures which moved in the direction indicated by Mansholt, shifting the thrust of the CAP more towards structural change. These measures provided investment aids for approved farm modernization plans (Directive 72/159), gave lumps sums or annuities to retiring farmers (72/160) and promoted 'socio-economic guidance and training' for those leaving agriculture (72/161).

At about the same time, however, the CAP was unhinged by new developments reflecting the uneven development of member states. Between 1969 and 1973 most Community countries either floated, revalued or devalued their currencies in a succession of changing parities which wrought havoc with the concept of common agricultural prices. As a result, common prices became a fiction between uncommon partners. Further ructions were caused by the first round of EC enlargement in 1973.

From 1975 the CAP became geographically more sensitive. Directive 75/268 provided income aids or 'compensatory allowances' – based usually on the number of livestock units per farm – to farmers in mountain, hill and other less favoured areas. This honoured the commitment obtained by the UK during its entry negotiations, but was of benefit to other countries too, especially to the Republic of Ireland, France and Italy. Three

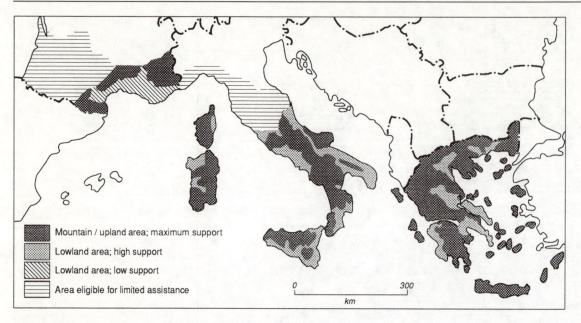

Fig. 9.3
The European Community's Mediterranean Programme.

main types of 'problem area' were identified (Fig. 9.2). Firstly, there are mountain and hill areas with permanent handicaps to farming caused by altitude, slope or poor soils (Pyrenees, Auvergne, Alps, Apennines, Corsica). Areas with low population densities and severe depopulation are a second type; these include upland areas in Scotland, the Republic of Ireland, Wales, Belgium, southern Germany and Sardinia. Thirdly, there are more restricted areas with special handicaps such as poor drainage, lack of infrastructure, the need for environmental conservation or the need to preserve a tourist potential. The 'less favoured regions' policy does not necessarily cover all areas with low farm incomes but it does mark a very important change in the CAP, recognizing the inability of farmers in the areas concerned to produce on competitive terms.

Under the original 1975 classification, 31 per cent of the utilized agricultural area of the Community was listed as 'less favoured'. In 1986 a new classification was introduced, raising the proportion to 48 per cent. This extension was partly linked to enlargement: Greece, for instance, has 78 per cent of its agricultural area designated as 'less-favoured', higher than any other Community country except Luxembourg. But partly it was also due to extension of the less favoured surface in countries like Ireland and the UK.

Another significant new development, which has also furthered the spatial sensitivity of the CAP, was the 'Mediterranean package' introduced from 1978 onwards.

This was intended to support specific types of improvement corresponding to the needs of Mediterranean farmers. Within the EC Nine, the Mediterranean region embraced most of Italy together with southern France. It contained 17 per cent of EC farmland, 18 per cent of farm production and 30 per cent of farmers: figures which in themselves are a summary testimony of the region's low labour productivity. Measures adopted by 1981 included irrigation in the Mezzogiorno and Corsica, the development of a farm advisory service in Italy, reafforestation and the improvement of rural infrastructures in upland areas of Italy and southern France, flood protection in the Hérault Valley of southern France, and support for cooperative ventures in Italy. Such schemes qualified for 25 and occasionally 50 per cent contributions from FEOGA.

In 1979 'integrated rural development programmes' were introduced in the Western Isles of Scotland, the French *département* of Lozère and the Belgian province of Luxembourg. Under these schemes the CAP combined with the regional and social funds to promote a mix of agricultural and non-farming projects, belatedly acknowledging the inability of the CAP alone to solve farmers' problems in difficult environments. This strategy was taken a stage further in 1983 with the integrated plan for the Mediterranean. Expenditure on this six-year programme started in 1985 and is designed to improve job opportunities and raise income levels in the still strongly rural areas on the southern margins of the EC where farmers face poor and hazardous ecological conditions and outdated land tenure structures. The Integrated Mediterranean Plan operates over the five most southerly planning regions of France, most of Italy and virtually all of Greece (Fig. 9.3). It leaves out highly developed urban areas, such as Athens, and major touristic zones. Maximum intensity of support is given to highland and lowland areas in Greece, southern Italy, Sicily, Sardinia and Corsica; limited assistance is applied to central Italy and south-western France. Sixty per cent of the Plan budget comes from central EC funds and 40 per cent from the three beneficiary countries, with France the main donor. The benefits, however, are distributed more in favour of Italy (45 per cent) and Greece (38 per cent). The money is being spent in three main directions: first, improving agricultural production and making it more market-oriented (40 per cent of expenditure); second, creating alternative rural employment opportunities, for instance in tourism and craft industry (33 per cent); and third, funding afforestation, fishing and training schemes (27 per cent) (Bertelsmann Foundation 1991).

While the Mediterranean orientation of the CAP in the late 1970s and early 1980s rectified to a certain extent the northern bias of the earlier phases of the CAP, there were more fundamental aspects of reappraisal to be noted. Some of these are as follows.

Inequities, lakes and mountains: some criticisms of the CAP

A cynic would maintain that the CAP is the very negation of what its name means. It is not Common because it is a source of endless bickering between member countries; it is not Agricultural because it is basically a prices and income strategy; and it is not a Policy because it reflects a miasma of interest groups. Stuart Holland (1980) has argued that the CAP, the EC's 'agricultural albatross', is both economically efficient and socially unjust. He maintains that its support prices are too high, its consumer prices are excessive, it functions as a disincentive to structural reform in agriculture and its subsidies, through price support, do not adequately distinguish relatively rich from relatively impoverished farmers. His criticisms still ring true today.

Inequity appears to have been built into the CAP from the start. One of the main reasons commonly given for the original Six developing the CAP was that this was a requirement by France as a *quid pro quo* for West German manufacturers' access to the French market. Thus the fact that the CAP benefits some states a lot more than others was accepted at the out-

set. Since it involves a transfer of income to agriculture from the rest of the economy it is clear that this will involve net transfers of income between states, mainly from food-importing nations like West Germany, the UK and Italy to food exporters such as the Republic of Ireland and Denmark.

So far so good. But then two clear *social* injustices appear. First, it is the lower-income groups which suffer relatively more through food price increases than do the better off, because a much higher proportion of expenditure by the former goes on food. So, if food prices rise, this has the effect of making the distribution of income less equal (unless low-income families happen to be farmers, but most are the urban poor). Second, there is inequity within the agricultural sector. There is a positive correlation between public subsidies and value-added per head of the farming population. In other words, the CAP spends more money supporting the more productive and better-off farmers than it does on the poorer ones. The latter, handicapped by harsh environments, exploitative tenures and antiquated farming techniques, get a subsistence living; the former make large profits. This is a regressive policy by any criterion of social justice (Ritson 1982).

The CAP's poor defence of consumer interests is understandable to the extent that the individual farmer has a much bigger stake in it than the individual consumer: a 10 per cent rise in CAP prices is likely to yield an increase in farm incomes of around 20–30 per cent and yet decrease consumers' real income by no more than 1 per cent. Consumers' arguments have in fact latched on to the paradox of food surpluses and high prices rather than the general price-raising impact of the CAP *per se*.

Surpluses have undoubtedly become the most publicized and most criticized aspect of the CAP. At worst, they are a moral outrage in a hungry world: at best, they are a clumsy method of guaranteeing farm incomes at the expense of the consumer. More than 60 per cent of the CAP's budget has been used to dispose of farm surpluses. Some foods which are perishable have to be destroyed, chiefly fruits and vegetables; others, such as sugar, can be stored or converted to other uses. Cereals grown for human consumption can be diverted to animal feed. Some surplus wine has been converted to industrial alcohol. Export subsidies enable sale on world markets. Part of the butter mountain – estimated at one time to be equal in weight to the entire population of Austria – has been sold to the former USSR at a fraction of its real production cost. New surpluses may come with enlargement. Already Greece has started a currant 'hill'. Greater problems lie in store with Spanish olive oil: the EC Twelve contains 2 million families involved in olive production.

With some products it is, in theory, relatively easy to encourage farmers to switch to other crops. With dairying, however, the situation is much more difficult because of the large numbers of farmers involved, many of whom are small and have little alternative to milk production. France is the Community's largest milk producer, yet the structure of its dairy industry is fragmented and inefficient, in contrast to the situation in the UK and the Netherlands (Barton and Young 1975). With attempts to despecialize out of wine in the vine-monocultural areas of Languedoc, policy runs up against the cultural importance of the vineyards to this region's rural society (Jones 1989b).

Reforming the CAP

Towards the end of the 1970s, the Community's institutions began to realize that the unlimited nature of price guarantees had to be reconsidered. Expenditure on market-price support from the Guarantee section of FEOGA had grown by an annual average of 23 per cent during 1975–79. In 1977 a 'co-responsibility levy' of 1.5 per cent (later raised to 2.5 per cent) was introduced on all milk destined for some form of processing. Proceeds from the levy were to be used for the expansion of markets and the search for new outlets and improved products. However, at this stage the levy was not explicitly linked to a defined target for Community milk production: and it was not effective in checking the increase in the milk surplus.

In 1981 the notion of 'guarantee threshold' was introduced, and this played an important role in the operation of the CAP during the 1980s. The threshold idea is simply that producers must share in the responsibility for generating any production over and above a certain level. This policy of 'punishment' for overproduction found its first major expression in the introduction of milk quotas for individual producers in 1984. Milk quotas have had mixed results: successful in some countries in some years but insufficient to match the generally declining pattern of demand, which has been quite sharp in some products such as butter. Overall, the quotas led to a 12.5 per cent fall in EC milk production between 1984 and 1988, while butter stocks were reduced from 1.3 million tonnes in store in 1986 to 860 000 tonnes at the end of 1989 (Robinson 1991).

The mid-1980s were a watershed for the CAP, comparable in significance with the Mansholt period of 1968–70 (Harvey and Thomson 1985). Against a background of failed dairy policy, difficult problems in the cereals, beef and wine sectors, UK refunds and

the impending accession of Spain and Portugal, a new package of measures was agreed upon which set the hitherto spiralling cost of FEOGA within the budgetary discipline of the average growth of EC revenues. Following the Dublin summit of 1984 and the Green Paper of 1985, the Council agreed to the extension of guarantee thresholds to a wide range of commodities and pegged product price increases to modest proportions (Commission of the European Communities 1985b).

In the late 1980s increasing attention was paid to the strategy of reducing the cereals surplus by programmes of 'set-aside' – incentives to take arable land out of production for a number of years. One of the first measures was the British Ministry of Agriculture's ALURE scheme (Alternative Land Use and Rural Economy) to encourage farm diversification through alternative uses of farmland, notably woodland and farm businesses. This £25 million package was launched in 1987; preliminary indications are that only a relatively small group of entrepreneurial farmers have taken up the scheme (Cloke and McLaughlin 1989).

The EC set-aside scheme, launched in 1988, covers arable land. To qualify for aid, at least 20 per cent of a farmer's arable area must be set aside for a period of five years. The arable land withdrawn from production must be left fallow, converted to woodland or used for permitted non-agricultural purposes such as farm tourism or keeping horses. The level of compensation, which varies between countries up to a maximum of 600 ECU per hectare in the Netherlands, should ideally be high enough to provide farmers with the incentive to withdraw land but not higher than the income they would have earned from cropping the same land. Compensation may also vary according to the type of conversion carried out: in the UK, for instance, the rates are £200 per hectare for permanent fallow or woodland, £180 for rotational fallow, £150 for non-agricultural use, with lower rates applying to land within the less favoured areas (Ilbery 1990).

Table 9.6 shows the pattern of set-aside results in its first year of operation, 1988–89. The overall figures, 38 164 applications and 434 300 ha of arable land represent tiny fractions, however – less than 1 per cent of total arable land for instance. Response has varied greatly between EC countries: West Germany recorded the biggest response, with two-thirds of all applications and 39 per cent of the set-aside land, while in Denmark and Luxembourg the scheme was not applied, and Portugal is exempt. Controversy surrounds the Italian figures, a suspiciously high proportion of which originate from Sicily.

The relatively poor uptake of the set-aside scheme

Table 9.6   Results of the EC's set-aside scheme, 1988/89

	No. of farms	Area set aside (ha)	Average per farm	Area set aside as % of:	
				Arable	Cereals
Germany (West)	25 289	169 729	6.7	2.4	3.6
Italy	9 301	155 603	16.7	1.8	3.1
UK	1 750	54 779	31.3	0.9	1.3
France	1 002	15 707	15.6	«0.1	0.1
Spain	518	34 229	66.1	0.3	0.4
Netherlands	195	2 621	13.4	0.3	1.3
Ireland (Rep.)	77	1 310	17.0	0.1	0.3
Belgium	32	329	10.2	«0.1	«0.1
Total	38 164	434 310	11.3	0.9	1.3

Note: The scheme was not applied in Denmark and Luxembourg; figures are not available for Greece; Portugal is excluded from the scheme. *Source:* Jones and Munton (1990).

partly reflects national governments' lack of real enthusiasm. For instance, France only offers a maximum premium of 350 ECU per hectare, much less than the EC maximum of 600. According to Jones and Munton (1990) the French government has two reasons for not pushing the scheme harder. Firstly it is reluctant to forgo the export revenue derived from the EC budget support for exportable surpluses. Secondly, there is a fear that high rates of set-aside could occur in fragile regions where any loss of economic activity might result in rural depopulation.

It is now realized that the set-aside policy is unlikely to reduce cereal surpluses to any significant degree. Even in Germany, where the scheme has made most progress, the net reduction in output due to land withdrawal is less than the annual increase in production due to improving yields. Evidence from German farmers also reveals that it is often the least productive land which is set aside, and more intensive farming practices are then carried out on the remaining cereal land (Jones 1991). Even within the EC, support for set-aside has waned following the low commitment from farmers and national governments.

Whatever the future of the set-aside programme, it will increasingly be seen as part of a wider package of measures designed to promote the interests of rural dwellers. The change of emphasis away from the CAP illustrates the growing awareness that farming is no longer the sole, or even the major, channel through which the future of the rural regions of the Community can be planned. The key document which enunciates this shift of policy is the 1988 Green Paper on 'The future of rural society' (Commission of the European Communities 1988a), whose main points will be summarized below.

Before we do this, reference must be made to the other major influence on changing the context of the CAP in the 1980s, the EC's Mediterranean enlargement. This enlargement (Greece joining in 1981, Spain and Portugal in 1986) had much greater agricultural significance than the 1973 enlargement when the UK, Ireland and Denmark joined. Enlargement from the Nine to the Twelve entailed a 55 per cent increase in the farming population, a 57 per cent increase in the number of farms, a 49 per cent increase in the agricultural area, and a 24 per cent rise in agricultural production.

Greece, Spain and Portugal share common features of the Mediterranean farming regime, producing much the same range of commodities (sheep products, hard wheat, olives, vine products, citrus fruits, etc.), many of which are also in competition with produce from Italy and southern France. In general the CAP's market support mechanisms for Mediterranean products are less generous than they are for 'northern' products like beef, cereals and dairy produce. Because of this the main Community approach has concentrated on structural measures such as irrigation and cooperatives, as noted in the earlier discussion on Mediterranean integrated development.

Despite their similarity, the three Mediterranean countries under discussion pose different problems for the CAP. In Greece agriculture retains more importance, relatively speaking, than any other Community country: one-quarter of the working population is engaged in farming, and agriculture (especially fruit and vegetables) accounts for 30 per cent of total exports. Yet no other EC country has such a hostile environment for agriculture: 40 per cent of the agricultural area is mountainous terrain, and the farm structure is impossibly fragmented. Because of its small size and its earlier entry than Spain or Portugal, Greece has been integrated into the CAP relatively successfully, absorbing only 3 per cent of FEOGA. Trading patterns for Greek farm produce have not be dislocated since they were already EC-oriented (Marsh 1979).

Portugal, like Greece, has a small agricultural economy in European terms but, unlike Greece, it is a net importer of agricultural produce. The farm sector is inefficient with crop yields and farm incomes which are the lowest in Western Europe. According to Monke (1986), any problems the Community has in accommodating Portuguese agriculture are minimal in comparison to the stresses EC membership is creating for the country's predominantly small-farm sector, while the freeing of trade with the rest of the Community is worsening the trade deficit in agricultural products. The main exports are wine, tomato concentrate, cork, fruit and vegetables; they account for about one-third of the value of agricultural imports, which are mainly beef, cereals, rice and animal feed. As a substantial net importer of foodstuffs, Portugal can eventually expect to become a net contributor to the CAP budget; this inevitable situation will, however, be compensated by a high level of funding for structural improvements – notably 70 million ECU over ten years (Unwin 1989). Unfortunately the restructuring of Portuguese agriculture to target levels specified by the EC can only result in massive outflows of farmers from the rural sector, with little indication of where these people will go or what they will do.

The impact of Spain on the EC's agricultural economy is much greater, for its agricultural area is two and a half times that of Portugal and Greece combined. Like the rest of the economy of Western Europe's fastest-growing country, Spanish agriculture is in a dynamic expansionist phase. Output has increased faster than most other European countries over the past twenty years, and plans exist for a 300 000 ha increase in the irrigated area. Considerable exportable surpluses of stoned fruit, citrus, olives and olive oil, wine and grapes, and dried fruit are generated. Spanish farm production costs, although rising sharply, are still significantly below those of its main rivals in Europe, France and Italy. France's worries over the impact of Spain's accession to the EC are perhaps too forcibly stated (witness the 'sadla war' fought in Roussillon in the 1980s between French farmers and lorry loads of

cheap Spanish produce) but the fierceness of competition cannot be denied. Italian agriculture, too, is compromised by the scale of Spanish production. Other countries with serious misgivings about the EC's southern enlargement are those Mediterranean countries (Morocco, Algeria, Tunisia, Cyprus, Israel, etc.) which have preferential trade agreements with the EC for their agricultural produce; these agreements have been eroded as a consequence of Iberia's entry to the EC. In fact Morocco's 1989 application to join the EC (which was refused) was related to the loss of trading privileges.

The CAP and the future

Digesting Iberia has not been the only, nor indeed necessarily the most important, problem facing the CAP over the past half-dozen years. Nor is the CAP the major issue in a Community stricken by high unemployment and a variety of social problems and still coming to terms with the events in Eastern Europe. For a while in the 1980s the CAP seemed to lose direction. Many adjustments were made to try to take account of new developments, with the result that detailed provisions became more and more involved, and less and less consistent with the policy's original objectives, which themselves were far from clear-cut and not without internal contradictions. For a while cutting the cost of the CAP seemed to be the only key objective, indeed almost an obsession, but this was hardly the basis for rational and well-planned policy (Fennell 1985).

The rationale for the CAP, and for the whole field of rural planning in the Community in the 1990s, was set out in the aforementioned statement on 'The future of rural society' (Commission of the European Communities 1988). This made several important points. The first was to stress the diminishing importance of agriculture in rural regions, both as an employer and in terms of its contribution to the regional product. Thus, out of 166 regions in the EC there are only 10 where the share of agricultural employment is more than 30 per cent and only 17 where farming accounts for more than 10 per cent of the regional product: these regions are nearly all in Mediterranean Europe. Farming's contribution to GDP in the Twelve is now down to a little over 3 per cent, the EC's farming workforce has halved since 1965, and the utilized agricultural area has reduced by 8 per cent over the same period. In spite of these shrinking parameters, agricultural production has increased steadily by about 2 per cent per year, significantly faster than the average annual growth of internal demand, 0.5 per cent.

The second key point of the 1988 Green Paper was

to acknowledge the importance, in policy terms, of diversification in the economic structure of rural areas, and to recognize that this process has been accompanied by a diversification of the social fabric (partly through repopulation) and a reassertion of local political autonomy. In this shift of emphasis from *agricultural* policy to *rural* policy special attention is given to rural tourism and rural industrialization as providers of employment, either on a part-time or full-time basis. The paper noted, for instance, that 60 per cent of new employment created in Italy since the 1970s has been located in rural areas, and also drew attention to the buoyant rural economies of parts of France and Germany. Much of this rural employment buoyancy has been based on a structure of part-time farming. Within the farm sector, particular stress was laid on forestry, both to take out some of the surplus land and to satisfy the timber deficit.

The third novelty was to recognize not only the geographical specificity of different rural regions (this had already been taken up in previous rounds of policy) but to express this regional dimension in terms of different rural socio-economic and physical processes. Three so-called 'standard problems' were identified, each corresponding to a different geographical *milieu* and each with its own set of policy options. The *first* standard problem is designated as that of the pressure of modern development, encountered in rural districts near to big conurbations, particularly in the northern part of the EC (south-east England, the Netherlands, the Paris–Brussels–Bonn triangle, the Po Valley), and in many coastal areas of touristic development (Mediterranean coasts of Spain, Italy and France, the Algarve, Balearic Islands and the south coast of England). The problem is above all one of planning land use in the face of competing interests which threaten to transform the countryside and upset environmental stability. Here the strategy is less one of pushing economic development than strengthening measures to protect the rural environment. Careful and quite restrictive planning is necessary. The *second* problem is that of rural decline which, to varying extents, blights many outlying rural areas, especially in the south of the Community (Mezzogiorno, inland Spain and Portugal, Greece, Ireland). The strategy here is development through economic diversification. Life can be breathed back into these areas, the Green Paper maintains, by harnessing indigenous potential in rural tourism and small firms to create economically justified jobs, The *third* standard problem is encountered in the most remote areas of the Community – mountain areas and islands (Alps, Pyrenees, Scottish Highlands and Islands, Corsica, Crete, etc.). Here rural decline, depopulation and land abandonment are already marked, and the scope for economic diversification may be extremely limited. How to sustain a minimum threshold population by some combination of economic and social activity is the key task, otherwise erosion, desertification, and social isolation for the few remaining old people will be the inevitable results. Some of these areas may be too remote or sparsely populated for tourism to have much impact. Development may be slow, constant efforts are needed to give a *raison d'être* to such areas. Development of a forestry–timber activity may be a viable option.

The 1988 Green Paper's fourth key point – already signalled some years earlier – is to allow a greater play of market forces in order to curb the upward trend of farm production and spending. The EC has now begun the task of dismantling price support systems to help reduce surplus production. The impact of such measures is strongly felt, not just by the big farmers who have made fortunes out of price subsidies but more particularly by farmers in areas where small holdings predominate or where crop specialization is at the vulnerable end of the product range. Thus the low-income Mediterranean regions of the EC involved in the production of olive oil and table wine – products relatively well supported by the CAP – stand to suffer greatly from reduced price support. Structural measures offer some compensation for this, but they are not necessarily applicable to all farmers. Moreover the effectiveness of such structural policies, whether they are aimed at enlarging holdings or at diversifying and upgrading enterprise systems, is often dependent on the efficiency of local institutions such as the SAFER or, in the case of crop diversification in the vine-dominated Languedoc, the *Compagnie Bas-Rhône* for irrigation water (Jones 1986, 1989a). In short the CAP is now less monolithic than in the past. Agricultural production is no longer seen purely in quantitative terms; instead there is increasing emphasis on quality (e.g. of wine) and local distinctiveness.

Finally, and perhaps most crucially, agricultural policy now has an explicit environmental content, which has been present in most EC publications since the mid-1980s (Symes 1992). The background to this has been the increasing realization during the 1980s of the 'environmental disbenefits' of the CAP's 'engine of destruction', driven by the growth of industrial-type production based on economic efficiency, profit (and subsidy) maximization and maximum use of technological and chemical inputs (Robinson 1991). The list of environmental problems associated with modern intensive farming methods is long and includes destruction of precious ecosystems such as wetlands and downlands, loss of landscape amenity (e.g. by hedgerow removal for field enlargement), eutrophication of water bodies, groundwater pollution by agro-chemicals, straw-burning, soil erosion and many more. Recently the environmental backlash of industrial farming methods has extended to the products themselves with scares over salmonella in chicken, eggs and cheese, and the spread of bovine spongiform encephalopathy ('mad cow disease').

While the specification of an environmental dimension to future policy is welcome, there are strong grounds for arguing that the measures introduced in the late 1980s to alleviate environmental stress resulting from agricultural activity are essentially tinkering with the results of price support policy rather than addressing fundamental problems of ecology and long-term sustainability of agricultural systems. Price support reduction, quotas and set-aside schemes are largely economic rather than ecological measures (Robinson 1991).

The most recent expression of the thrust of the 1988 Green Paper is the July 1991 MacSharry set of proposals for the future of the CAP. These represent the most fundamental reshaping of the CAP since its inception thirty years ago. Once approved, they will be introduced in 1993 and be fully operational by 1996. The proposals are specifically designed to redress the problems of declining farm incomes, unstable markets, surplus food stocks, increasing budgetary costs, and damage to the environment caused by intensive production. The main features are: major reductions in support prices (by 5–35 per cent according to product); supply control measures (lower milk quotas, set-aside schemes, calf disposal premia, etc.); improved direct support towards small and medium-sized farmers (e.g. exemption from quotas and set-aside); special agro-environmental programmes supporting extensive production methods and the reduction in environmental damage; an accelerated programme of afforestation; and new measures to facilitate early retirement.

Competing pressures for rural space

The predominantly agricultural community has become the exception in large parts of rural Western Europe. Although disguised by the unchanging character of the rural landscape, this process is revealed by employment statistics, the pattern of commuter flows and the steady transformation of consumption patterns, clothing and standards of living. Public and, even more, private transport have broken down the physical isolation of rural existence, while education, the television and other mass media have comprehensively eroded its cultural isolation.

Above all, rural industrialization has regenerated many countryside districts. The 'urban–rural manufac-

turing shift' (Keeble, Owens and Thompson 1983) has taken industry, mainly small and medium firms, out of congested urban and metropolitan areas into more spacious surroundings, a process aided by the changing natures of Europe's industrial mix from heavy to light, technologically advanced and therefore more footloose enterprises. This 'productive decentralization', which also involves changes in the structure of industry, has been studied in many parts of Europe, but the classic area is the north Italian region of Emilia (Brusco 1982; Cooke and da Rosa Pires 1985). The 'Emilian model' blends small industries and a range of tertiary activities with a modernized, and mainly part-time, agriculture to create a prosperous equilibrium which, moreover, does not disturb the existing framework of village and small-town settlement.

Many other influences are also at work to change and repopulate the countryside (Clout 1984). The countryside has become fashionable for city dwellers who have responded to the attractions of open space, pleasant landscapes and the wholesomeness of nature. The phenomenal post-war growth in private car ownership (25 cars per thousand population in 1950, 324 in 1981) has both enabled natives of rural areas to travel to work in nearby towns, and encouraged urban dwellers to move into the countryside without changing their employment. Motives for moving into the countryside vary. Some of the new ruralites are affluent purchasers of large houses set in their own grounds; other incomers have more limited finances and are attracted by the availability of cheaper types of housing which are more pleasant than comparably priced accommodation in the city; yet others are retired people who are either returning to their villages of origin or seeking out new settings to live out their remaining years.

Further pressures on the countryside are brought by second homes and rural tourism, both of which bring people into rural areas on a temporary basis. Clout (1984) estimated a total of 3 million rural second homes in the EC in the early 1980s, with nearly half of these in France. Roughly 20 per cent of all French and Scandinavian households are possessors of a second home; in these countries, and in the UK, there are now more second homes than farms in the countryside. In some districts, second homes outnumber primary residences: this can happen where the demand for weekend cottages is particularly strong, as on the fringes of the Paris Basin, or where depopulation has been particularly intense, as in some parts of the Provençal Alps. Rural tourism generally creates more beneficial impacts in the field of rural employment than do second homes, although the pressures on the landscape and the environment may be greater, at least at certain times of the year.

These changes bring not only economic revolution to the countryside, but also strong sociological and even psychological impacts. Europe's current agricultural problems are rooted in the ambiguity of farming as both an economic activity and a way of life. The contrast is typified by setting alongside each other the elderly Greek peasant with 10 ha of hills and rocks, his baggy trousers a uniform of the past, and the young East Anglian farmer with 1000 ha of fertile loam, a barnful of huge machines, and computerized accounts. This dualism will persist for some time, but is becoming more blurred. As Henri Mendras wrote in *The Vanishing Peasant*, there will always be men who till the soil and manage their farms, but the farmer who is content to run a small subsistence holding in conformity with traditional routines, seldom going beyond the limited horizons of his own soil, is becoming rarer and rarer, and his peasant soul is disappearing with him (Mendras 1970). The current age structure of Europe's farmers makes this statement particularly appropriate at the present time. Currently about half of Europe's farmers are over 55 years of age, and half of these have no successors. Most of the older farmers run smallholdings in the southern European countries; almost two-thirds manage 5 ha or less. Within the next decade or so many of these elderly farmers will die or become pensioned off, at which point the age structure of the farm population will rejuvenate (Clout 1991).

To return, in final conclusion, to the land: the art of development in a crowded countryside involves formulating not just a policy for farming but rather a land policy to resolve land-use conflicts, as well as creating an efficient socio-economic planning mechanism to cater for the various interests and social groups in rural areas. Agriculture can no longer be considered only in terms of productivity or yield, nor is it purely the primary sector. It has become part of the secondary sector through the intensity of its industrial inputs and transformation into capitalistic agribusiness, and it is currently moving into the service sector. The rural world should be considered as a mosaic: semi-natural habitats; settlements; areas for agriculture; and areas for sport, recreation and leisure pursuits (Montanari 1991). Farmers, instead of being viewed as a destructive force, sweeping away much-loved landscapes as they strive to maximize profits, should be made managers and protectors of the countryside. For, as Newby (1979) has stressed, the land may be privately owned, but the landscape is publicly consumed.

Further reading

Bowler I R 1985 *Agriculture under the Common Agricultural Policy: a geography*. Manchester University Press, Manchester

Clout H D 1971 *Agriculture*. Macmillan Studies in Contemporary Europe, London

Clout H 1984 *A Rural Policy for the EEC?* Methuen, London

Dovring F 1965 *Land and Labour in Europe in the Twentieth Century*. Nijhoff, The Hague

Duchêne F, Szczepanik E, Legg W 1985 *New Limits on European Agriculture*. Croom Helm, London

Franklin S H 1969 *The European Peasantry: the final phase*. Methuen, London

Franklin S H 1971 *Rural Societies*. Macmillan Studies in Contemporary Europe, London

Hill B E 1984 *The Common Agricultural Policy: past, present and future*. Methuen, London

Jones G, Robinson G M (eds) 1991 *Land Use and the Environment in the European Community*. Biogeographical Monograph No 4, University of Edinburgh

Robinson G M 1990 *Conflict and Change in the Countryside*. Belhaven, London

Tracy M 1989 *Government and Agriculture in Western Europe*. Harvester, London

10

Recreation and conservation

Dimensions of demand

Both the demand for recreation and the scale and variety of provision have grown dramatically in Western Europe, as in all Western industrial societies, in the past half century. People generally now have more leisure time, and more money than ever before to spend on it. Combined with the extensions to personal mobility conferred by the motor car and cheap air travel, these factors have enabled them to diversify their interests into a bewilderingly large number of different recreational activities. This in turn has led to an enormous expansion in the spatial scope of recreation: there is a growing predilection for travelling away from the immediate locality of the home, neighbourhood or town to satisfy recreational needs. There has also been a marked change of focus: rather than specific sites or locations being the main object, whole areas with a number of different attractions are often taking their place. The change of emphasis is important for it goes some considerable way to explaining the growing concern for environmental conservation. Rural and, to some extent, urban landscapes are seen as the backdrop against which much recreational activity takes place and they are being viewed increasingly as a rather static part of the cultural heritage. It is well known that people tend to harbour idealized and often sentimentalized images of far-away places, at odds with the reality of constant evolution and change; a desire to preserve what is best from the past for the enjoyment of present and future generations is an important stimulant to the burgeoning demands for the enactment of conservation measures in both town and country (Lowenthal 1968). This chapter examines the nature of the growth in demand for recreation and the effect that it has had on specific parts of Western Europe through the growth of tourism. It also analyses the nature of the response by both governments and private agencies to the calls for conservation measures and the effect that these have had on rural and urban environments.

In the past generation a number of influences have acted in concert to expand the amount of leisure – time free from work and other obligations – most people have at their disposal (Parker 1976). It is not simply that the number of hours worked per week has been reduced, although this has contributed to the change, but more importantly that the actual organization of work in society, the general state of affluence of the majority, and the overall demographic trends have all combined to alter radically recreational aspirations and opportunities.

Holidays with pay are now the norm in industrial societies and the majority of West Europeans only work a five-day week. People therefore not only have more spare time, it tends to be organized into predictable blocks, be it at weekends or annual holidays, thus allowing for the long-term planning of more ambitious recreation projects. There has also been a marked change in the actual composition of the populations of all the West European countries, the most significant feature of which has been the general ageing that has occurred. In the region as a whole, 13 per cent of the population is above 65 years of age, compared with only 10 per cent in 1950 and 6 per cent in 1910 (Table 10.1) and, just because this age group has so much leisure time, it has a disproportionate impact on patterns of recreation and leisure.

Other recent changes in society have also had profound effects upon recreation. The length of time spent in full-time education has steadily increased and more young people than ever before are taking higher education training at universities. Among the twelve members of the EC alone, their numbers rose from 3.9 million in 1970 to 6.2 million in 1990. It is not being argued that those in full-time education necessarily have more leisure time, but it certainly is true that young people are more likely to participate in sports and other more active recreations, and education services frequently provide them with the physical facilities to do so.

Another, more distressing, social change that has had a marked impact on the amount of leisure time for

Table 10.1 Population over 65 years of age in Western Europe in millions (%)

1910	1950	1970	1990
14	26	38	47
(6.1%)	(9.3%)	(13.0%)	(13.2%)

many people is unemployment. In Western Europe as a whole its rate more than doubled after the shock rise in the price of crude oil in 1973 and in 1991 stood at 11.4 per cent of the population of working age. It would, of course, be ridiculous to imply that the first, or even the major, concern of unemployed people is to seek out new recreational opportunities, but none the less many of them do suddenly find that they have time to devote to hobbies and other activities that were previously denied to them.

Important as these changes in the socio-economic structure of society are, in terms of the impact on recreation all pale beside the explosive growth in private car ownership (Table 10.2). In 1950 there were only 6.1 million cars on the roads of Western Europe compared with some 142 million today. Put another way, one person in forty-eight owned a car in 1950, compared with approaching one person in three in 1990. The impact of this transformation in personal circumstances for recreation was prophetically highlighted by Michael Dower in the early 1960s and his predictions have proved uncannily accurate (Dower 1965). Most people are now able to choose where they want to travel within a wide radius of their homes, without the time and spatial constraints imposed by public transport. They are also spared the inconvenience of making their way to central collecting points, such as bus and railway stations, thus encouraging them to act much more easily on impulse and significantly increasing the amount of time spent on recreation. The major impact of widespread car ownership has been felt in the countryside, previously largely inaccessible to the mass of the population, where informal recreation is now a major land use (Simmons 1975). One of the greatest difficulties facing those trying to cater for these new recreational demands has been the fact that the planners of the 1940s and 1950s, who are still largely responsible for the framework of public access to the countryside, failed to foresee the revolutionary increase in demand that the private car would bring to rural areas.

Despite their importance, the new developments ushered in by the improved access to transportation for the mass of the population need to be kept in perspective. For most people, the home and the immediate locality still remain the focus of their leisure-

time activities. A detailed report on the leisure needs of the population of one London borough showed that, while holidays and shorter outings were considered important and desirable by people at all levels, the bulk of their recreation was locally based and was going to remain so (Dower et al 1981). The report urged that those in both government and private industry recognize this fact when formulating policies for invest-ment in recreation facilities. Indeed, it went further, arguing that the lives of many significant groups, such as young married women, the elderly and ethnic minorities, were so constrained in terms of recreational choice that sports and other facilities needed to be arranged so that the particular difficulties faced by these, and other, groups could be effectively countered. These findings from London are in no way exceptional and have been corroborated by many other studies (see, for example, Glyptis 1981), for while there is some debate as to precisely which social characteristics are likely to determine the types of recreation people choose, there is no doubt about the continuing paramount importance of opportunities at or close to home. The systematic information available for Western Europe also indicates that while the nature of provision by local authorities is enormously varied, it is directed consistently towards those people with a better than average education and with middle to high ranking occupations. This was one of the main conclusions of an enquiry, sponsored by the Council of Europe, into recreation policy in fourteen Western European towns in thirteen countries (Mennell 1976). The picture that emerges confirms the evidence of a lack of recreational opportunities for most disadvantaged groups, already highlighted in more detailed individual studies, and indicates that the same situation is widespread throughout the region.

Tourism

Economic significance

Tourism is defined as the temporary, short-term movement of people to destinations outside the places where they normally live and work, for other than business or vocational reasons, and their activities during the stay

Table 10.2 Number of private cars in use in Western Europe, in millions

1950	1960	1970	1980	1990
6.1	22.9	64.8	95.3	122.4

at these destinations (Burkart and Medlik 1974). As an aspect of recreation, it has grown in line with the improvements in, and expansion of, communications that have been going on steadily since the mid-eighteenth century. The building of the turnpikes, the spread of the railway network, widespread ownership of private cars and the growth of mass air travel have all in their turn stimulated tourism to a point where it is now a significant part of every West European economy. Of the estimated 300 million international tourist arrivals world-wide in 1986, more than 50 per cent were from Western Europe.

In fact, this figure is a gross underestimate of the total amount of tourism in the region, for it only reflects international movements and takes no account of the very large volumes of domestic tourism in most West European countries. By the same token, the contribution of Western Europe to international tourism is highly emphasized when compared with other industrial nations, in particular the USA, because the very large number of relatively small states increases the number of international as opposed to domestic movements (Hudman 1978).

Within Western Europe there is very considerable variation in both the absolute and the relative net importance of tourism among the different countries (Table 10.3). In 1986 Spain earned most, followed by Italy, France, Austria, Greece, Portugal, Turkey and Switzerland. All these countries had a positive balance in excess of $US 100 million and they represent the southern margin of Western Europe. The distribution reflects an understandable desire on the part of the majority going on holiday for sun and warm seas and, in the case of winter sports, the longer winter days of the Alps in comparison with the mountains of the north in the Highlands of Scotland and Scandinavia. It is apparent from Table 10.3 that, in general, southern Europe's gain is northern Europe's loss in financial terms, with some countries, notably Germany, running large deficits on their tourism accounts .

Obviously, absolute figures take no account of the differences between states in respect of either population or the overall size of the individual economies. If international tourist receipts are expressed as a proportion of total exports (Table 10.4), then a somewhat different picture emerges. Four countries – Spain, Greece, Austria and Portugal – dominate, with tourism accounting for a significant proportion of their exports. In no other country does the proportion exceed 10 per cent and for most it is less than 5 per cent. This agrees closely with the contention that the greater a state's economic potential the less will be the significance of tourism in its economy (White 1976). It also supports the thesis that tourism is well suited to the kind of eco-

Table 10.3 International tourist receipts less international tourist expenditure, 1986 (million US$)

Spain	10 442
Italy	7 095
France	3 197
Austria	2 722
Greece	1 338
Portugal	1 250
Turkey	916
Switzerland	862
Ireland (Rep.)	−24
Iceland	−35
Denmark	−354
Finland	−473
Belgium/Luxembourg	−618
UK	−765
Sweden	−1 267
Norway	−1 437
Netherlands	−2 524
Germany (West)	−12 838

Source: OECD (1988).

nomic circumstances found in states at the periphery rather than the urban–industrial core of Western Europe. The level of capital investment needed by the tourist industry is relatively low and thus does not make undue demands on a country's financial resources; it is labour-intensive and can therefore make a substantial contribution to the employment situation; and returns are rapid, making an immediate impact on national finances (Christaller 1964). Nevertheless, such economic determinism always needs to be tempered by not losing sight of the critical importance for success of physical attributes, such as sun and scenery, explaining why it is the southern periphery rather than the northern one that has benefited most from tourism in Western Europe. In common with other industries that offer a rapid return on investment, tourism is intensely competitive and very susceptible to the vagaries of the economic climate. The upheaval experienced by all the Western industrial nations in the wake of the doubling of oil prices in 1973 interrupted the sustained growth in tourism enjoyed throughout Western Europe in the preceding decade. Although the downturn was quite quickly reversed, the renewed growth rates have been much more modest and the competition to attract visitors much more fierce (Williams and Shaw 1992). Superficially the rankings of the importance of national tourist industries in 1970 and 1986, shown in Table 10.4, reveal only minor alterations over the intervening years, but the shifts

Table 10.4 International tourist receipts, 1986, in rank order of the proportion they form of total exports

		Amount (million US$)	Proportion of total exports (%)	1970 ranking
1.	Spain	11 945	21.1	1
2.	Greece	1 835	17.9	3
3.	Austria	6 928	17.7	2
4.	Portugal	1 583	13.5	5
5.	Switzerland	4 240	8.9	6
6.	Italy	9 853	8.6	7
7.	Turkey	1 228	5.9	11
8.	Denmark	1 759	5.8	8
9.	France	9 580	5.2	9
10.	Ireland	659	4.3	4
11=.	Iceland	42	3.2	17
	UK	7 921	3.2	12=
13=.	Finland	597	3.0	10
	Sweden	1 543	3.0	18
15.	Germany (W)	7 826	2.6	15=
16.	Norway	992	2.4	12=
17.	Belgium/ Luxembourg	2 269	2.1	15=
18.	Netherlands	1 906	1.8	14

Source: OECD (1988).

that have occurred have been quite fundamental. Turkey has significantly improved its position, as have most of the other countries of southern Europe, especially Greece, Italy and Portugal. At least some of the growth of tourism in these countries would appear to be at the expense of Spain, though it still heads the ranking. In 1970 tourism accounted for 33 per cent of the country's total exports, but by 1986 this proportion had fallen to 21.1 per cent. In part this reflects a general diversification that has taken place in the economy, but it also illustrates how the Spanish tourist industry has lost out to its rivals in the competition for trade.

Local impact

The impact of tourism on the receiving areas, like the previously remote Alpine valleys of Austria and Switzerland and the Mediterranean coast of Spain, is dramatic, not least because many of the more popular destinations have gone from being among the most isolated and poor areas of Europe to centres of mass tourism in the space of less than a generation (Barker 1982). The changes that occur in the local economy, the society and the landscape as the scale of tourism grows are complex and varied, but Fig. 10.1 (redrawn from Kariel and Kariel 1982) provides an effective

summary of the developments that occur over time. Initially, the small number of tourists coming to remote areas expect little except the opportunity to enjoy the landscape and to share vicariously in the lives of the local population. For their part, the residents see the tourists as a minor diversion from their primary task of eking a living from the land. If the tourist demand grows, this relationship begins to change, and with increasing rapidity. Tourism establishes a niche in the local economy and challenges the total dependence on agriculture; facilities – like improved and expanded accommodation – are built using private capital, followed by public investment in communications and other services. Quickly the local economy changes from being mainly dependent on agriculture to being mainly dependent on tourism and, far from accepting the visitors as an optional extra, the local community has to begin to compete through advertising and more ambitious investment in facilities for their custom. Investment on such a scale inevitably presupposes borrowing money, so that the populations of the receiving areas eventually find themselves in a position where they are as wedded to tourism as they previously were to agriculture, with the added danger of being even less in control of the means of production.

The Kariels (1982) go on to make the very important point that the changes that occur are not simply economic but fundamentally alter the whole basis of the local society. Initially, the influences on the self-sufficient, subsistence-level, agrarian society are relatively modest: knowledge of the outside world increases; more time is spent with guests and less with family and friends; working hours become more rigid; children are less involved, because they tend to help less around the home and the farm; the wife earns an increasing share of the family income independently of her husband. After a time these changes in the economic circumstances of the family are followed by changes in social structure of the community as a whole: there tends to be a decreased emphasis on religion; family size falls; levels of education improve; outmigration is slowed; the spirit of cooperation in the community is replaced by a growing sense of competition between individuals; medical care improves; there is a growing dependence on services, such as shops, outside the immediate local community.

Finally, a new equilibrium is established: paid labour replaces volunteer–cooperative community work; children participate in sport; the concepts of employment and unemployment become established; employment and housing for children when they grow up is routinely found outside the confines of the family; the wife earns a growing proportion of the family

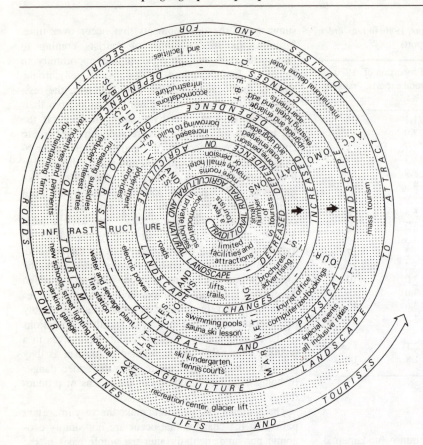

Fig. 10.1
Spiral of economic and infrastructural growth and landscape
changes influenced by tourism
(after Kariel and Kariel 1982).

income on her own account and consequently alters
fundamentally her role in the family; immigration of
workers occurs; there is an increased incidence of marriage breakdown; holidays are taken away from home;
a revival of interest occurs in traditional customs and
local language, but mainly for their curiosity value. In
other words, a previously remote society has been
sucked into the mainstream of the technological,
urban-orientated, credit and monetary exchange society from which it was previously largely insulated.
There is no intention of implying that this is in any
way a disaster, or even undesirable, but tourism does
impose a whole new series of constraints on the societies to which it reaches out. Also, the speed of change
frequently means that local people had little time to
adapt to their new circumstances and are in danger of
being overwhelmed by them.

Examples of the above transformation can be found
in many of the previously more remote and backward
parts of Western Europe, but nowhere are they more
starkly illustrated than in Spain. Between 1959 and
1964 the country's foreign currency earnings from
tourism leapt more than tenfold, from under £20 mil-

lion to well over £200 million, with the numbers of
foreign tourists growing from 4 to 14 million in the
same period (Naylon 1967). By 1964, tourism was
earning nearly as much foreign currency as all Spain's
other exports put together and, although its relative
significance has subsequently declined, it has none the
less had a profound and irreversible effect on the economy, the society and the landscape.

Yet the effects have been highly localized and, even
by 1980, 60 per cent of Spanish hotel capacity was concentrated in only seven of the country's fifty provinces.
Five of these – Gerona, Barcelona, Alicante, Malaga
and the Balaeric Islands – were on the Mediterranean,
so that the rest of Spain, except for the capital Madrid
and Guipuzcoa on the Atlantic coast bordering France,
was left largely untouched by the direct impact.
Equally, the tourist season is highly concentrated with
nearly 50 per cent of foreign visits still occurring in the
months of July, August and September. Such uneven
spatial and temporal patterns raise considerable difficulties for the economy as a whole, for not only are the
benefits of tourism very unevenly divided nationally,
there are also long periods during the year when much

of the labour force even in the tourist areas is either unemployed or underemployed.

A graphic illustration of the way in which tourism can radically alter the environments affected by it is provided by the two most southerly of the Balaeric Islands, Ibiza and Formentera (Pacione 1977). Until the mid-1960s both were relatively untouched by the flocks of visitors from all over Europe who flew into Minorca and Majorca, but in the last twenty years they have rapidly begun to catch up. Existing roads have been metalled and new ones constructed in an attempt to produce a better integrated route network and, as a result, the number of cars registered in San Antonio alone – one of ten small towns on Ibiza – rose from 960 in 1967 to 2500 in 1975. Between 1950 and 1975 the number of hotel beds in the town rose from 500 to more than 12 000, and the local architecture has come to be dominated by multi-storey buildings, at the expense of the traditional, single-storey, stone-built houses. In the countryside, many of the scattered farmsteads have been abandoned as the inhabitants have found better-paid jobs in the towns. The decline in the size of the rural population has also encouraged the mechanization of agriculture and forced farmers to allow the more inaccessible parts of their holdings to revert to a less ordered and manicured state. Within the towns themselves, workers from mainland Spain have come on a seasonal basis to cater for the every need of the tourists and, like them, they cannot be accommodated in a traditional agricultural society. Shops and other services have had to be provided on a large scale, thus further altering the infrastructure and layout of the settlements.

Faced with such fundamental changes at all the major tourist locations, the Spanish government has had to modify its initial attitude of uncritical encouragement for expansion in the industry. In the 1950s and 1960s, when the economy was already weak, it was difficult to justify infrastructural investment in roads, water supplies, sewers and other services, especially as there was no guarantee that the expansion would continue. Now this kind of investment and measures to protect the environment are being forced on the government just to prevent the most popular resorts on the Balaeric Islands, the Costa Brava and the Costa del Sol from deteriorating to such an extent that tourists are deterred rather than attracted. In 1977, the Secretariat of State for Tourism drafted a new law to curb the environmental impact of tourist development, and also took positive action to spread demand away from the Mediterranean coast to less developed parts of the country, thus reducing the environmental pressures in the most crowded areas and spreading the benefits of the tourist industry more widely (OECD 1980b).

The problems caused by a few popular, or even over-popular, locations and a very restricted season are common to all tourist countries and, like Spain, most have made strenuous efforts to diversify their attractions and increase the number of resorts. The greatest success in this search has undoubtedly attended the Alpine countries, which have succeeded in turning the winter inconvenience of snow and the difficulties of having to travel on skis into a multi-million dollar industry. In the winter of 1979–80, for example, foreign visitors spent $US 3400 million in Switzerland alone; and France, Germany, Austria and Italy also enjoyed large inflows of foreign currency (EFTA 1981). Nearly half of the world total of 3000 ski resorts is found in Western Europe, the success rooted in the fact that high, snow-covered mountains are found close to large and relatively affluent industrial populations (Barbier 1978). People in the region can travel quickly and relatively inexpensively from their homes to the ski resorts. It is noticeable that the more inaccessible mountains, like those of the Scandinavian countries and the Sierra Nevada in Spain, are markedly less popular as ski resorts, though in the case of Scandinavia the northerly location and consequently short winter days is certainly an added deterrent.

Skiing and all winter sports are a developer's dream, for they utilize resources of both land and people at a time when they would otherwise be severely underemployed. Most of the sports also require large quantities of equipment and specialist clothing and this has given a considerable boost to manufacturing industry in Austria, Switzerland, Norway and Sweden. Nevertheless, even such specialist equipment manufacturing can become the victim of its own success. In all four of the countries mentioned above, the market for Alpine skis has recently begun to decline, despite the continuing growth in the popularity of skiing, as manufacturers elsewhere have begun to find it worth their while to compete in the lucrative market. The obvious response of course is to diversify and all the four countries have begun to promote strongly cross-country skiing, which requires different equipment from its more popular downhill cousin. So far they appear to have tapped yet another buoyant market, but there is always the danger that it too could eventually suffer from outside competition should the market grow too large.

Nature conservation and landscape protection

National parks

Concern for the future of the natural environment and a desire not to see the rural landscape swamped by urban development and twentieth-century technology are common popular causes throughout the nations of Western Europe. Nevertheless, despite the unanimity that exists on the general ideals, there is great variation in the way in which the individual governments have gone about translating them into practical policies. Throughout Europe the primeval landscape, untouched by civilization, has long since disappeared, so that conservation is really about conserving semi-natural ecosystems and the remnants of the rural economies that formed them. As the experience of every country in this regard is somewhat different and the way in which each assess the value of what remains varies, so the conservation policies themselves all have a distinctive national flavour. In the UK, for example, much of the emphasis has been on public access to open country and the need to preserve the illusion in the landscape of a pre-industrial rural scene. By way of contrast, in Germany, where there are more than 1100 nature reserves, the focus has been much more narrowly directed towards nature conservation. Short of detailed examination of each individual policy, however, it is often hard to say precisely what even their general aims and objectives are, because apparently synonymous designations have quite widely differing meanings in the various West European countries. The most blatant example of this is the confusion that still surrounds the term 'national park'. It can mean anything from a state-owned wilderness, where public access is strictly controlled and limited so that the environment can evolve with a minimum of human interference, to privately owned upland farmland, where access for recreation is actively encouraged and the only controls over development are planning regulations covering new buildings and mining operations.

A degree of comparability has been introduced through the efforts of the International Union for Conservation of Nature and Natural Resources (IUCN), an offshoot of the United Nations, which in the past twenty years has published three world lists of what it defines as national parks. To merit inclusion, an area must be relatively large, with one or several ecosystems not materially altered by human exploitation and occupation. It must also either be of special scientific, educational or recreational interest for its wildlife, geomorphology or general habitat, or possess a natural landscape of great beauty. In addition, the national government of the country concerned must have taken steps to protect the special features for which the area is to be designated and visitor access has to be restricted to those whose purposes are 'inspirational, educative, cultural or recreative' (IUCN 1974). In order to ease the confusion, the IUCN also requested governments not to designate as national parks any of the following:

- a scientific reserve which can be entered only by special permission;
- a natural reserve managed by a private institution or a lower authority without some type of recognition and control by the national government of the country;
- a special reserve, such as a fauna or flora reserve, a game reserve, a bird sanctuary, or a geological or forest reserve;
- an inhabited and exploited area where landscape planning and measures taken for the development of tourism have led to the setting up of recreation areas, where industrialization and urbanization are controlled and where public outdoor recreation takes priority over the conservation of ecosystems. It is further recommended that any area wrongly designated as a national park according to the above definition should be renamed as soon as possible.

Under the terms of the IUCN definition, all but the very smallest West European states have national parks (Fig. 10.2), although their number, the proportion of the total land area designated, and the national park area per capita vary widely. On most counts the Scandinavian countries emerge with the most active records (Table 10.5), with Sweden having no less than twenty-six separate national parks, 1.042 per cent of its land area designated as parkland, and 6.111×10^{-4} km^2 of parkland per head of population. Elsewhere the picture is very variable and reflects more the enthusiasm of individual governments than the intrinsic value of the habitats and landscapes. For instance, the heavily populated Netherlands with its almost total lack of spectacular scenery has eighteen national parks, covering 1.436 per cent of its land area, while Switzerland has but one, covering only 0.4 per cent. On the other hand, the UK has nineteen designated areas, but they are all very small and encompass only 0.365 per cent of the land area, while Portugal's one national park incorporates 0.5 per cent of its territory.

The great value of such international comparisons is that it fosters a recognition of common problems and helps to promote joint and compatible solutions. This is extremely important in Western Europe where scenery and valuable natural habitats are not respecters of the closely packed national boundaries. There are now a number of examples of jointly administered national parks spanning international frontiers, such as the Hohes Venn–Eifel park on the German–Belgian border and the Maas–Schwahn–Nette park on the German–Dutch border. The scope for further similar cooperation, especially among the Alpine and the Scandinavian countries, is considerable, but so far practical political considerations have outweighed the advantages of joint action.

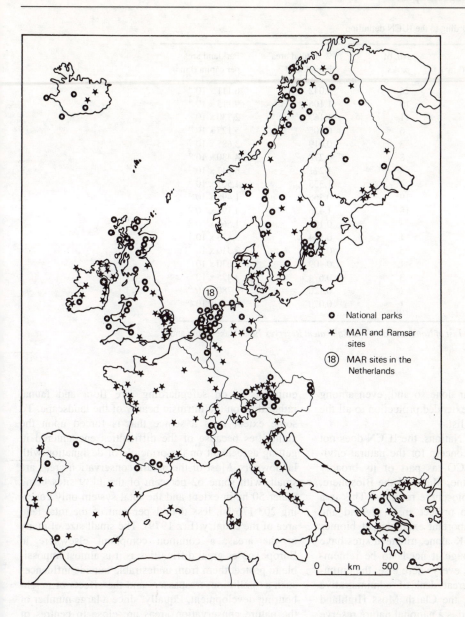

National parks
MAR and Ramsar
sites
(18) MAR sites in the
Netherlands

Fig. 10.2
National parks (according
to the IUCN definition,
1975), MAR and Ramsar
sites.

Wetlands and other habitats

National parks are not the only international designations sponsored by the IUCN; since 1965 it has published a list of temperate marshes, bogs and other wetlands in Europe and North Africa which are in need of permanent conservation and management if they are not to disappear under the pressure of agricultural improvement or other development. The sites are known as the MAR list, the name being chosen because M, A and R are the first three letters of the word for wetland in so many European languages (*Marscii*, marsh, marshland, *marécage, marius, maris-*

ma). As can be seen from Fig. 10.2, there are 163 wetland sites on the list, spread among eighteen different countries, and one of the principal aims of the scheme has been to draw international attention to the dangers facing these habitats, in the hope that a convention can be agreed to ensure their long-term protection. This goal was achieved in 1971 when the Convention on Wetlands, of International Importance, especially as Waterfowl Habitat was signed at Ramsar in Iran. However, the painful slowness with which any international agreement can be actually implemented is illustrated by the rate at which the Ramsar convention has been ratified by individual European governments.

Table 10.5 National parks according to the IUCN definition

Country	Area of parkland (km^2)	No. of parks	% of total area in parkland	Parkland area per capita (km^2)
Sweden	5 072	26	1.042	6.111×10^{-4}
Norway	3 585	10	1.106	9.115×10^{-4}
Germany (West)	2 902	11	1.167	2.730×10^{-5}
Finland	2 375	6	0.705	5.125×10^{-4}
Italy	1 865	3	0.619	3.685×10^{-5}
France	1 520	8	0.278	4.430×10^{-5}
Iceland	1 310	4	1.271	6.047×10^{-3}
Austria	1 119	4	1.335	1.501×10^{-4}
UK	839	19	0.365	1.498×10^{-5}
Netherlands	591	18	1.436	4.191×10^{-5}
Spain	540	3	0.107	1.560×10^{-5}
Portugal	501	1	0.544	5.836×10^{-5}
Greece	281	5	0.213	3.310×10^{-5}
Switzerland	167	1	0.409	2.630×10^{-5}
Ireland (Rep.)	43	3	0.674	3.425×10^{-5}
Belgium	37	1	0.120	3.780×10^{-6}
Denmark	30	1	0.070	6.085×10^{-6}

Sources: IUCN *United Nations List of National Parks and Equivalent Reserves 1986.* Nake-Mann and Nake (1979); Wilkinson (1978).

Less than half have so far done so and, even among those that have, none has extended protection to all the sites on the original MAR list.

Even within the United Nations, the IUCN does not have a monopoly of the concern for the natural environment. In 1970 UNESCO, as part of its broadly based ecological programme, Man and the Biosphere, launched the concept of biosphere reserves. This is a world-wide programme to protect areas of land and coast, which represent important examples of biomes at a global scale. In the UK alone, nineteen sites have been so designated, although it needs to be remembered that here, as well as everywhere else, the enthusiastic activity hides a great deal of administrative duplication. For example, the Claish Moss Highland Region in western Scotland is a national nature reserve in the UK, is on the MAR list, has been designated under the Ramsar convention, and is also a biosphere reserve.

The designations made by the IUCN and other international bodies represent only a very small proportion of the protection afforded under conservation legislation in the individual West European countries (Orme 1990). However, as was pointed out earlier, the differing priorities and legislation mean that the practice varies in detail from country to country. In Germany the national government, spurred on by the 1976 Federal Nature Conservation Law, has devoted much of its attention to preserving natural habitats. The emphasis is on safeguarding rare flora and fauna, rather than on the intrinsic beauty of the landscape. To some extent it is a stance that is forced upon the authorities because of the difficulties encountered in getting agreement on the principle of designation with landowners. Most of the nature conservation areas are small, with some 62 per cent of the 1139 sites being under 50 ha in extent and the total system only covering 203 176 ha, less than 1 per cent of the total land area of the country (Erz 1979). The small size of many of the areas, a common complaint elsewhere in Europe, poses great difficulties as it is almost impossible to protect them from undesirable outside influences such as pollution, trespassers and the effects of neighbouring development. Equally, since a large number of the nature conservation areas are close to centres of population, they are under constant pressure from people wishing to use them for more general recreation. This, in turn, leads to tension between the state and the largely private landowners, who resent the intrusions by the general public. They are deeply suspicious of the limitations that designation as a nature conservation area imposes on the freedom to do what they want with the land that they own. Finally, there are quite large variations between the various *Länder* in the federation in the amount of land set aside for nature conservation purposes. In part, of course, this reflects differences in the intrinsic quality of the natural environment, but much more it is evidence of the degree of commitment. The amount of land that is protected and the resources in both manpower and money devoted to managing it are vital ingredients, which ultimately determine the effectiveness of nature conservation programmes, not only in Germany but anywhere in Western Europe (Council of Europe 1990).

Experience in France and the United Kingdom

In France, the most important rural conservation initiative has been the system of *parcs naturels régionaux*, made possible by the central government decree of 1 March 1967. It enabled local and regional authorities to combine to set up joint administrative machinery for running and financing such parks. There are thirty-two of them spread right across the country, ranging in size from a massive 206 000 ha in the Landes area of south-west France to only 10 300 ha in the Saint Aman-Raismes park on the Franco-Belgian border. The purposes of the parks are to preserve and enhance both the natural environment and the cultural heritage of the areas protected, to provide facilities for visitors and to ensure the flourishing survival or rural life and traditions (Blacksell 1976). The concept is a grand one, but the success with which it is realized in each instance depends entirely on the level of local commitment; the central government contributes nothing directly to either the finance of the administration of this park system. Control, in most cases, is vested in a joint authority comprising the *département* and the rural *communes,* with the *département* providing the bulk of the money and the *communes* implementing most of the decisions. Obviously such a system is fine so long as it works, but if any of the constituent authorities decides not to cooperate then there is little that can be done about it. The great advantage of local, as opposed to national, control is that it ensures that the people actually living in the parks have a direct say in the way in which they are developed and managed. One of the main objections to national and international conservation strategies is that they ride roughshod over the interests of the local population; the administrative system devised for the *parcs naturels régionaux* overcomes this problem, albeit at the cost of clear central direction and policy control (Woodruffe 1990).

The West European country with the most varied and extensive system of nature and landscape designations is the UK. Ever since the late 1940s, a comprehensive network of national nature reserves has gradually been designated and it now provides a measure of protection to scientifically important habitats throughout England, Northern Ireland, Scotland and Wales. At the end of 1982 there were 219 national nature reserves, many owned or leased by the Nature

Conservancy Council on behalf of the government (Nature Conservancy Council 1982). Others are established under Nature Reserve Agreements by which the owner and occupier retain their rights but agree, in the interests of nature conservation, to certain conditions affecting land use and management.

This kind of control over development and change is obviously the ideal if one wants to be certain that nature conservation objectives should always be paramount, but it is clearly impracticable in a densely populated country to tie large tracts of land up for such single-purpose use. In the majority of the protected areas in the UK, the demands of nature conservation have to compete with other government objectives, notably the need to provide access to open country for recreation, as well as with the legitimate interests of the private landlords, who actually own the land and usually depend on it for their livelihood. In England, Scotland and Wales there are more than fifty areas set aside, because of their exceptionally fine upland scenery, in the belief that they require special protection as a part of the national heritage. The ten areas in England and Wales are actually called national parks, although in most respects they are much closer to the French *parcs naturels régionaux* than national parks in the sense that the term is normally used; significantly, they do not appear on the IUCN list. The national scenic areas in Scotland are purely landscape designations, as their name implies, without any pretensions otherwise. With both the national parks and the national scenic areas, however, the crucial factor affecting their public management is that not only is the majority of the land privately owned, but also that administrative responsibility is primarily with the local authorities. Apart from making the initial designation, the main concern of the national government agencies is to exert a watching brief and to channel a certain amount of finance, so as to encourage the execution of the national conservation and recreation objectives for these areas.

The principle of designating land in the UK as an expression of concern about its value for the national heritage has gone well beyond the concept of the so-called national parks. There are also forty-one areas of outstanding natural beauty covering nearly 10 per cent of the land area of England, Northern Ireland and Wales, many of them incorporating lowland farming landscapes which are otherwise unprotected. In addition, thirty-five lengths of coastline extending over 1161 km have been dubbed heritage coasts, in the hope that the significance of the sea in the development of England and Wales will receive due weight in conservation policies.

As if this were not enough, the Broads in East Anglia, one of the nation's most interesting man-made landscapes, was considered so important that a special form of designation, the Broads Authority, was worked out to ensure its long-term protection. A similar plan has also been agreed in principle for the New Forest in Hampshire, an ancient royal forest and one of the best preserved areas of broad-leaved woodland in the UK (Edwards 1991). On their own, these responses are of only local significance, but they are symptomatic of a more general problem facing those concerned with nature and landscape conservation. In the UK and, indeed, throughout most of Western Europe, the natural landscape is only part of what needs to be preserved; of equal importance is the economic and social system that has modified and moulded the land surface. If this is to survive, it requires more than the simple isolation of the land in question that is at the heart of the concept of a nature reserve. The essential elements of the economic system have to be fixed in time as well and this involves people and the way in which they live, as well as land, flora and fauna. Managing farmland so that it retains elements of pre-industrial agricultural practice is an extremely difficult exercise and not one that any single, simple solution is ever likely to do successfully. It is, therefore, hardly surprising that in the UK so many different approaches have been adopted to what is essentially a single problem.

There is one other rather different difficulty facing those wanting to see a greater measure of protection given to the countryside for the purposes of nature conservation in the UK. Large tracts of rural land are owned and managed either directly by government, or by its agencies, such as the Forestry Commission, for specific purposes not connected with conservation. There has been increasing pressure on the government to broaden the remit of these bodies so that their land is managed for purposes other than the primary function, which in the case of the Forestry Commission is timber production. It is a campaign that has met with considerable success and now the Commission has to include provision for public access and recreation in its management plans. The Forestry Commission administers sixteen national forest parks where it mixes its responsibility for tending trees with providing facilities for informal recreation. Such strategies for multiple use of land are the key to rural conservation in an area like Western Europe, where competition for land is always likely to be great.

Conserving the built environment

Context

Europe has an urban heritage stretching back to medieval times, and towns and cities are at the very heart of its culture. It is therefore hardly surprising that there has been a widespread propensity to view the city as an artefact and to ensure that adequate planning and other controls exist for conserving the urban fabric (Evenson 1981). Indeed, even when cities have been struck by a calamity, such as the wholesale destruction that resulted from the bombing of Germany during the Second World War, there has always been a strong tendency to reconstruct painstakingly what was there before, so as to emphasize the thread of continuity with the past, rather than introduce new shapes and novel layouts (see Ch. 8).

The seeds of organized concern about the preservation of urban areas in Europe can be traced back to the eighteenth century in some countries like France, Germany and the UK, but it is in the last fifty years that the conservation ethic has come to be so firmly embedded in the planning process. The main reason for this, of course, is the greatly increased pace of growth and change in the twentieth century. Not only have people continued to flock into urban areas, new technologies, particularly those related to transportation and building, have opened up previously undreamt of development options. On the one hand, land may now be used very much more intensively than ever before, because of the advent of steel-framed high-rise buildings rising to twenty storeys and more. Before the end of the nineteenth century, anything higher than three or four storeys was beyond the scope of mass-construction techniques. On the other hand, new forms of urban transportation, starting with the railway and progressing through trains and buses to the motor car, have dramatically extended the built-up area of cities. They have produced a much greater segregation of land uses, as people generally are more easily able to travel within the urban area, and have also created new land-use demands in the form of railways and roads, which have had to be carved out of the existing urban fabric. Conscious of the rapid change going on all around them, citizens throughout Europe have begun to demand that not all the old be sacrificed to the demands of the new (Laborde 1989).

A second reason why conservation has recently grown in significance is the much greater strength and effectiveness of urban government. The growth of planning systems and the proliferation of regulations governing new development have enabled the desire on the part of individuals to oppose new development and change to be translated into fact. Urban conservation in Western Europe has always drawn its strength from grass-roots support, with the small pressure groups formed to save particular townscapes, such as Georgian Bath or medieval Paris, subsequently amalgamating to create a national movement like the

Fig. 10.3
Demonstration projects
selected for the European
Architectural Heritage Year,
1975.

British Civic Trust (Burtenshaw, Bateman and
Ashworth 1981). To some degree the movement has
also managed to transcend national boundaries. The
Council of Europe has taken a number of initiatives in
the past three decades to protect the built environment,
the most successful being European Architectural
Heritage Year in 1975. Under this scheme, forty-nine
sites throughout Western Europe were selected with a
view to protecting a representative selection of urban
landscapes (Fig. 10.3). The list is in no way comprehensive, nor is it intended to be; the object of the
whole exercise has been to choose sites where the
cooperation of the local authorities is assured, so that
an integrated approach to the conservation task can be
seen to be working. This explains the inclusion of
Bologna, Taranto and Verona in Italy, three towns
where coordination has been successfully achieved,
and the exclusion of Venice where, despite international
recognition, the constituent local authorities have
failed to agree on a joint plan to save the city from

ever-more serious flooding (Council of Europe 1987).

As a policy for the future of the city, urban conservation presents a number of difficulties. One of the
main reasons for the rapid cycle of demolition and
reconstruction is the changing pattern of functions
within the urban area, the pressure of new demands on
land leading to a constant process of redistribution in
the mosaic of land use. The most spectacular manifestation of this process is the suburbanization of residential development as business and commerce price it out
of central locations. Unfortunately, the purpose-built
buildings for one form of land use can prove less than
ideal for another. Urban conservation planners have
always been faced with the delicate task of accommodating new functions into the existing fabric, often
in the knowledge that adapting buildings will be
more expensive than complete demolition and redevelopment. Even when the financial pressures
can be resisted, it is not always possible to find a
demand for floorspace in a renovated neighbourhood.

Conservation studies undertaken by the Ministry of Housing and Local Government in Bath, Chester and York showed that in the historic centres of these cities as much as 40 per cent of the available floor area, especially in the upper storeys, was vacant (HMSO 1969; Insall and Associates 1969; Esher 1969).

Tourism has frequently provided a convenient, though rather too easy, solution. Historic buildings, either singly or in large groups, are a considerable attraction to visitors and many can be relatively easily adapted to house the service activities, like hotels, restaurants and gift shops, which tourism demands. However, the scale of tourist demand is unlikely to be sufficient, except in a few isolated cases, to fill all the available space. In any case, much of the attraction of towns and cities is the vibrance that the concentration of activities creates and this is all but lost if buildings are preserved as in a museum (Ashworth and Tunbridge 1990).

A large part of conservation in urban areas is concerned with residential building, very often the houses of ordinary people that have become rundown and fallen into decay. In such cases, one of the consequences of renovation can be a process of gentrification, whereby low-income families are replaced by others that are better off (Hamnett and Williams 1980). Such a change is obviously a danger with any kind of improvement to the urban fabric, but is particularly likely in conservation areas because of the way in which the middle classes have dominated the conservation movement in Western Europe. Local groups, such as the Bath Preservation Trust, constantly monitor changes and campaign for conservation policies; they also invariably come to believe their own propaganda and are keen to move to the refurbished neighbourhoods. In the process, rents are driven up and the original residents tend to drift away. Conservation is not necessarily a panacea for urban problems; pursuing it too enthusiastically can turn towns into monuments to the past, where nostalgia is substituted for drive and a sense of purpose. This kind of atmosphere will be all too familiar to those who have visited Georgian Bath or baroque Salzburg.

Government action: national and local

Throughout Western Europe there is a commitment by governments to the principles of urban conservation and to the importance of the built environment in the cultural heritage, but there is considerable variation in the legislation and policies. As a result, there is little consistency in the scale and quality of the conservation work that has taken place (see Ch. 8). In France, systematic protection of individual buildings goes back to the 1830s, and over the years some 29 000 have been

nationally listed, thus affording them the protection of the central government. Since 1962, when the Malraux Act came into operation, it has also been possible to designate conservation areas ('safeguarded sectors') in towns so that whole districts may be preserved. An initial list of 400 towns was drawn up where such conservation areas were thought to be appropriate, but only just over 60 have been approved and in only a handful of these have the actual conservation plans been agreed. The main problems have been the length of time it took to survey the areas involved and the lack of adequate finance, which has caused the scope of even agreed schemes to be repeatedly reduced. Even where work has gone ahead, as in the Marais district in central Paris, there has been considerable criticism because of the way in which many of the original residents were forced out after restoration as a result of the increased rents, and because social networks, far from being preserved, were totally disrupted (Kain 1981, 1982).

In most countries in Western Europe, urban conservation is much more in the hands of local rather than central government. The UK has the huge total of over 170 000 nationally listed buildings, but the responsibility for their conservation is largely in the hands of local planning authorities, as is the choice of conservation areas. Some central government money is available to help with the cost of preserving particularly outstanding buildings and townscapes, but the initiative for obtaining it lies almost wholly with the local authority. There is therefore considerable regional variation in the enthusiasm with which conservation policies have been pursued. It was no accident that two of the first town schemes, conservation plans for a whole settlement, were for Totnes and Ashburton, both towns in the county of Devon, an authority with an unusually strong commitment to preserving the historic urban fabric. Nevertheless, despite the rather weak central direction, the principle of urban, as of rural, conservation is probably more firmly accepted in the UK than anywhere else in Western Europe, not least because it is a policy that can be relatively easily understood and implemented by local planning authorities.

Elsewhere, urban conservation policies are similar, though with marked variations in the degree of control by the different levels of government and the extent to which the policies have been implemented. Many of the oldest and most valuable historic urban landscapes are in the Mediterranean countries where the record is generally poor. In Italy, a few towns such as Urbino and Arezzo in the Marches, Bergano in Lombardy and Bologna have been almost completely preserved, but against this must be set the fact that effective comprehensive plans for Venice, Florence, Rome and Naples have yet to be agreed (Burtenshaw, Bateman and Ashworth 1991). In Greece, it was not until the 1970s that comprehensive planning legislation came on to the statute book and, consequently, the task of conservation has hardly begun in the country that was one of the first centres of urban civilization (Papageorgiou-Venetas 1981; Leontidou 1990). Nevertheless, in comparison with most other parts of the world, Western Europe is extremely fortunate in the extent to which its urban heritage is known and recorded and in the effectiveness of the policies that have been developed to conserve it.

Western Europe is unique in the diversity of cultures found within a small area and in the high level of appreciation that exists about their importance. Despite the political fragmentation, there is a strong feeling of cultural unity and a sense of belonging to a single coherent tradition. This sense of identity has been fostered by the generally high level of economic development, which has helped to foster tourism and made travel, both at home and abroad, an accepted form of recreation. Actually visiting the relics of their cultural history has made people deeply aware of both their value and their fragility and stimulated a general determination to see them conserved. The aims of recreation and conservation policies are by no means always identical and can sometimes conflict, but it is in Western Europe that the greatest synthesis has been achieved.

Further reading

Burtenshaw D, Bateman M, Ashworth G J 1991 *The European City: a western perspective*. Fulton, London

Council of Europe 1987 *Management of Europe's Natural Heritage – Twenty-five Years of Activity*. Council of Europe, Strasbourg

Kariel H G, Kariel P E 1982 Socio-cultural impacts of tourism; an example from the Austrian Alps. *Geografiska Annaler (Series B)* **64**: 1–16

Williams A, Shaw G 1992 *Tourism and Economic Development: West European experiences* 2nd edn. Belhaven Press, London

11

The encircling seas

Marine resources

The sea has always played an important role in the life of Western Europe; there are sixteen states with coastal waters in either the Atlantic, the North Sea, the Baltic or the Mediterranean and some countries, like France, Spain Sweden and the UK, abut more than one of these major bodies of water (Alexander 1963). Traditionally, access to the sea has been the key to participation in intercontinental trade and has been exploited by many European states since the Middle Ages. Even today, with the widespread importance of air travel, access to the sea is a much-prized asset, offering scope for greatly enhanced flexibility and independence in international affairs at all levels.

For very much longer, the sea has also been a reliable source of food. Even in modern times, using quite primitive equipment and more or less random hunting techniques, people have usually found fish for the taking in quantities sufficient for their needs. Only in the past generation has pressure on these and other marine resources begun to mount seriously, necessitating formal agreements about the division of the ocean's bounty.

Although recent, the change has none the less been very rapid, as the technological advances that have transformed all aspects of twentieth-century life have been applied to exploiting the seas. Not only have fish been caught with great efficiency in unparalleled quantities, thus endangering stocks, but other resources, in particular oil and natural gas, are now being exploited on a large scale. Rather than acting as a gateway to the rest of the world and providing a marginal supplement to food supplies in Western Europe, the resources of the sea are now an integral part of the industrial economy of countries such as Norway, the UK and the Netherlands, providing the bulk of their fuel supplies (see Ch. 5). Naturally, other coastal states are eager to share in this new-found wealth. They have begun to

take an unprecedentedly keen interest in the boundaries and extent of their coastal waters and have inaugurated frantic explorations to ascertain whether or not they too are in a position to exploit hitherto undreamt of mineral wealth. The fact that so far most have been disappointed has done little to dampen their enthusiasm since, at a time when mineral resources of all kinds are an expensive drain on most West European industrial economies, the benefits of success are potentially enormous.

Territorial waters

From the middle of the seventeenth century there was a general acceptance that the territorial waters of coastal states should extend for 5.5 km (3 nautical miles) out to sea from the high-water mark. They were defined in this way for reasons of defence rather than economics, as the distance represented the theoretical range of cannon fire and was therefore the zone that could be defended effectively from the land. During the last century there has been mounting dissatisfaction with this definition and today the tacit agreement has all but collapsed but, in the absence of any international agency governing the exploitation of the oceans, extensions to territorial waters have been *ad hoc* and opportunist (Hodgson and Smith 1979). The USA claimed 19.3 km (12 miles) over a century ago, and all but a minority of coastal nations have now followed suit, but for most this proved only a very temporary solution, given the growing demands being made on coastal waters.

The first concerted progress towards a resolution of the problem was provided by the United Nations Convention on the Law of the Sea, which was finally signed in 1964 (Johnston 1976; Glassner 1990). It gave states the right to exploit the mineral resources of their coastal waters to a depth of 200 m (656 ft) or beyond if exploitation was technically feasible. In the furtherance of such exploitation, states were permitted to build permanent structures, such as oil wells, so long as they were not accorded the status of islands with territorial waters of their own and providing they did not pose an undue hazard for shipping. The Convention also gave states rights to sedentary species of living things on the seabed, such as shellfish and crustacea, but not to fish. The latter exclusion was very significant, for at the time it removed the major item of economic interest in coastal waters. In general the Convention has worked well, although some difficulties have emerged. In shallow, semi-enclosed seas, like the North Sea, the Baltic and the Mediterranean, the claims of neighbouring states overlap and a dividing line has had to be agreed. Also, the outer territorial limit of the shelf, as defined in the Convention, depends on the technological ability to exploit its resources, rather than any uniquely defined boundary, which does not make for any permanent definitions in the present climate of rapid progress (Smith 1980).

Faced with such shortcomings, further negotiations were inevitable and these took place within the context of the third phase of the United Nations Convention on the Law of the Sea which began in 1973 and was finally agreed in 1982, though it has still not been finally ratified by the governments of all the signatory countries (Blake 1987). The agenda was extremely broad, but the main concern was to agree on the establishment of a 322 km (200 mile) exclusive economic zone for states with coastal waters, embracing both the resources of the seabed and the coastal waters themselves. By 1975 there was broad agreement on the principle of such a zone and most West European states have moved independently to enact appropriate legislation. Iceland first unilaterally imposed an 80.5 km (50 mile) zone from which foreign fishing boats were banned around its coasts in 1972 and then three years later extended it to 322 km. The initial reaction was one of fury and righteous indignation on the part of the UK and the other West European countries, whose fishing fleets were suddenly excluded from waters they had fished for generations, but the so-called Cod Wars that ensued were totally ineffective (Mitchell 1976). Iceland showed itself as being quite capable of enforcing its new limits and very quickly the disgruntled governments turned their attention to laying claim to similar 322 km exclusive economic zones around their own coasts.

All the major West European countries have now accepted the principle of a 322 km exclusive economic zone and, where possible, laid claim to the areas in question, but the extensions have raised difficult problems. In some instances the distance separating neighbouring states is much less than 644 km and an alternative definition has had to be employed. In the majority of cases the solution has been a median line drawn equidistant from the nearest points on the base lines of the territorial waters of the countries concerned, but this was not universally acceptable. The median-line solution would have given West Germany only a very small area of the North Sea in relation to the length of its coastline and this caused a protracted dispute with Denmark and the Netherlands. The issue was eventually resolved by these two countries ceding 7000 and 5000 km^2 respectively of their initial territorial claims to Germany, thus giving it the all-impotant north-westward extension of its sector into the more promising oil and gas exploration areas of the North Sea. Elsewhere, as between the UK and Ireland in the Irish Sea, the median line has been accepted provisionally but is still subject to ratification (Blacksell 1979). This is significant, because the uncertainty has so far effectively precluded any exploration for oil and gas in the disputed zone (Fig. 11.1). Further complications are created by isolated islands. For instance, the exclusive economic zone of the UK has been much extended by the existence of several islands around its shores, but its claim to a 322 km territory around the barren, rocky outcrop of Rockall (way out in the Atlantic and annexed for security reasons in 1955) is strongly disputed by both Ireland and Denmark and still remains to be settled (Brown 1978).

Further difficulties have arisen with the definition and implementation of exclusive economic zones as a result of political changes in Europe, notably the establishment and subsequent enlargement of the EC. In 1970, when the organization still only consisted of the six original members, a Common Fisheries Policy was agreed which, among other things, gave Community fishermen equality of access to all fishing grounds in the maritime waters under the sovereignty or jurisdiction of any member state. At the time it was a relatively uncontentious decision, since the bulk of fish were caught outside the narrow territorial zone as then defined. Now, of course, the situation has altered dramatically with the newly imposed 322 km territorial limits and the consequent pressure on domestic waters around the coasts of Western Europe from fishermen excluded from traditional fishing grounds elsewhere in the world. Within the EC too the matter became considerably more important after 1973 when Denmark, Ireland and the UK all joined, as Denmark and the UK both had large fishing industries. What for most member states had initially been a matter of somewhat peripheral concern was now an issue of vital national importance. Agreement was hampered further, because of uncertainty as to whether the EC or individual member governments should resolve the

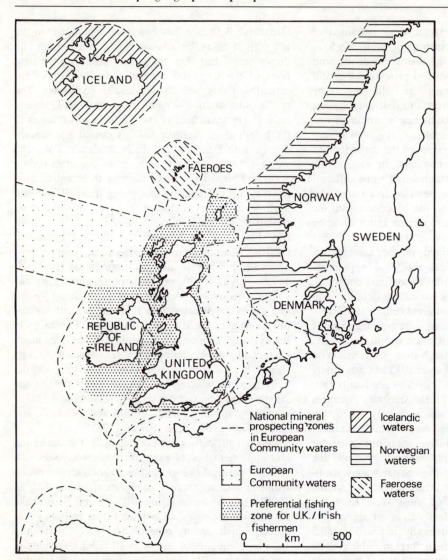

National mineral
prospecting zones
in European
Community waters

European
Community waters

Preferential fishing
zone for U.K./ Irish
fishermen

Icelandic
waters

Norwegian
waters

Faeroese
waters

0   km   500

Fig. 11.1
West European territorial
waters in the North Sea
and North Atlantic.

issue of fishing rights in Community waters (Allen 1980).

When the Community was first enlarged in 1973, the principle of a Common Fisheries Policy was retained, but in a substantially modified form. Rather than there being unrestricted access, during a ten-year transition period up to the end of 1982 all members were entitled to claim a 9.6 km (6 mile) exclusive zone off their coasts for vessels which traditionally fished in those waters and which operated from ports in that geographical coastal area. This zone was extended to 19.2 km (12 miles) off parts of the coasts of Denmark (including the whole of the Faeroes and Greenland), France, Ireland and the UK. It was assumed that a permanent solution would be worked out during the ten-year transition period but, inevitably, the discussions

then taking place within the United Nations Conference on the Law of the Sea and the unilateral extensions of territorial waters to 322 km made some interim revision essential. The members decided that, rather than attempting to enact Community legislation, all of them would extend their fishing limits around their North Sea and North Atlantic coasts from 1 January 1977 to 322 km without prejudice to the access already agreed for the fishermen of any member state under the terms of the temporary Common Fisheries Policy. At the same time, the Commission promised to put forward proposals in due course for similar extensions in the Mediterranean, but so far these have not materialized.

Even with the promise of this future action over the Mediterranean, the European Commission's proposals still fell short of the needs of member states, as a number also had territorial waters in the Baltic, not to mention outside Europe around overseas dependencies. In reframing their national legislation, most wished to include all their territorial waters and not just those in the North Sea and the North Atlantic. Denmark extended its limits in the Baltic, the Kattegat and the Skagerrak, as well as claiming 322 km around Greenland and the Faeroes. Germany extended its territorial waters in the Baltic as well as in the North Sea. Both France and the UK have also claimed extended limits off the coasts of a number of overseas territories, or territories for whose external relations they are responsible. In the case of France these include Guyana, New Caledonia, French Polynesia, French Antarctica, the Wallis and Futuna Islands, the islands of Tromelin, Glorieuses, Juan de Nova, Europa and Brassas du India, Clipperton Island, Réunion and Mayotte. The UK has claimed 322 km off the coasts of Bermuda and the British Virgin Islands; and France and the UK together have made a joint claim off their condominium of the New Hebrides (Churchill 1980).

In the seas off the shores of the EC itself the extensions of territorial waters have raised serious problems, most of them relating not so much to the principle of extension as to how best to manage the available resources so as to maximize the benefit for all the members. Given the unequal distribution of both hydrocarbons and fish stocks, friction was almost inevitable, particularly where the allocation of fisheries was concerned. The fishing industry is not necessarily dependent on nearby, shore-based facilities for its successful operation, unlike the development of oil and gas fields. Therefore the exploitation of fisheries can take place more or less independently of the particular coastal state in whose waters they happen to be located and competition for available resources is correspondingly greater. The existence of the EC has, if anything, hindered rather than helped, for it introduced a new legal dimension and imposed a very short time-scale because of the need to negotiate a revision to the original Common Fisheries Policy by the beginning of 1983.

The fishing industry

The broad continental shelf and the shallow seas around the west coast of Europe have been exploited as a fishery since time immemorial. Indeed, until the present century there was nowhere else in the world where fishing was carried on to a comparable scale, and even in 1980 some 20 per cent of the total world catch of 56 million tonnes was landed in Western Europe. As recently as 1969, however, the proportion

Table 11.1 West European fishing catches by region 1980 and 1989 in '000 tonnes live weight

Country	1980	1989	Increase/decrease 1980 to 1989 in %
Belgium	46	39	-15.22
Germany (W)	622	408	-34.41
Denmark	1845	1922	+4.17
Greece	104	129	+24.03
Spain	1336	1182	-11.53
France	789	818	+3.67
Ireland (Rep.)	107	252	+135.51
Italy	455	553	+21.53
Netherlands	317	337	+6.30
Portugal	283	322	+13.78
UK	988	836	-15.39
EC	6897	5576	-19.16
Faeroe Islands	302	304	+0.66
Iceland	1425	1519	+6.59
Norway	2928	1780	-39.21
Austria	3	5	+66.66
Switzerland	4	4	—
Finland	129	97	-24.81
Sweden	208	243	+16.82
Non-EC	4999	3952	-20.95
Total	11896	9528	-19.91

Source: Commission of the European Communities (1992d).

was 31 per cent and the sharp decline reflects the growing interest in the sea as a reliable source of food in other parts of the world, rather than any absolute decline in European fisheries.

The importance of the industry to the individual countries of Western Europe varies considerably, as can be seen from Table 11.1. In general, the largest fleets and catches are found on the northern and western peripheries of the continent, with Norway, Iceland and Denmark dominating the picture. The combined population of these three countries is only 9.65 million and the large fisheries are central to their economies and obviously, therefore, of great political significance to them. All have welcomed and, in the case of Iceland, pressed for the extensions to territorial waters; they have also viewed with great suspicion the proposals for equal access to all the territorial waters of the EC. Concern about the effect that this would have was one of the main reasons for Norway withdrawing its application for membership in 1972 and Denmark lagged some months behind the other members of the EC in accepting the terms of the Common Fisheries Policy.

In order of importance, the other nations with large fisheries are Spain, the UK and France but, in terms of their economies as a whole, there are very much less important than those in the Scandinavian countries discussed above. Also, in both the UK and France the size of the fishing industries has declined sharply in recent years as access to traditional deepwater fishing grounds on both sides of the North Atlantic has been progressively barred. Elsewhere the trends are variable, though the size of the industry everywhere is small (Coull 1972). Nevertheless, a number of countries within the EC, in particular Ireland, have seen their industries grow substantially against the general background of decline, evidence of the benefits that have accrued to some from having easier access to Community waters.

The main fishing grounds in Western Europe are found in the shallow waters that stretch north from the coasts of France and the Low Countries across the Irish, the North and the Norwegian Seas and on over the Atlantic to the coastal waters of Iceland and Greenland (Ministry of Agriculture, Fisheries and Food 1981). A wide variety of different species are to be found and the North Sea in particular is also important as a spawning ground. Traditionally, cod and herring have formed the bulk of the catch with smaller amounts of haddock, plaice and sole also being taken where they are locally plentiful. Almost inevitably, however, the increased scale and efficiency of fishing in the past generation have put enormous pressure on existing fish stocks, with the herring being particularly badly affected. Throughout the 1960s the catch of this fish in the North Sea alone averaged nearly 900 000 tonnes, but over-exploitation led to a rapid decline and from March 1977 until 1982 fishing for North Sea herring was prohibited entirely by mutual agreement. A similar decline caused a ban to be imposed to the west of Scotland in the Celtic Sea in 1978 and the catch for the Irish Sea was limited to 10 000 tonnes a year. The stocks of other species too have varied, though it is not clear whether this is due to over-fishing or long-term periodic changes in migration patterns. Inexplicably, the annual cod catch in the North Sea nearly doubled in the decade 1965–75 to 216 000 tonnes while, at the same time, there was an almost parallel decline in the sprat fishery.

In recent years the pressures on existing stocks of fish traditionally caught has kindled interest in other species, the most obvious example being the rise and decline of mackerel fishing. Mackerel had always been viewed rather disdainfully as a poor substitute for herring but, when stocks of the latter began to disappear in the early 1960s, mackerel suddenly started to attract international attention. From being of only local significance at the beginning of the decade, more than 1 million tonnes were landed from the North Sea alone in

1967. Catches of this size simply could not be sustained and by 1972 the stock had collapsed and with it the fishery. A similar pattern, though rather less wildly fluctuating, was seen to the west of the UK, though even here mackerel stocks have only been stabilized by instituting catch quotas and fishing seasons.

Fish processing for both human and animal consumption has also widened the range of species of interest to the fishing industry. Whiting, coley and Norway pout have all begun to be taken in large quantities in the last twenty years as a result of the growth in these less discriminating demands for fish.

All the species described above predominantly inhabit shallow coastal waters and, with the exception of mackerel, are largely non-migratory. This means that their distribution is both limited and predictable and also that access to stocks is very unequally divided. The UK, the Faeroes, Iceland and Norway, surrounded as they are by shallow seas, have been particularly favoured, but the fleets from other countries have been anxious to share in their good fortune. The effects of the extensions of territorial waters have already been discussed, but the truth is that no amount of administrative manipulation can disguise the fact that there are too many fishermen chasing too few fish in the waters of northern Europe, no matter what species is being talked about. With access to coastal waters becoming more difficult everywhere in the world, European fishing fleets have begun to turn to deep-sea varieties as a substitute. In the North Atlantic, the most important of these so far has been the blue-whiting, catches of which are now believed to be well in excess of 1 million tonnes a year, but potentially a similar interest could be focused on several other species as well. The nature of the fishing industry is, therefore, changing rapidly as a result of the extended territorial waters and the increased pressure on the most easily harvested stocks. Regulation is becoming the norm in shallow coastal waters and, faced with increasing restrictions, fishermen are turning to more capital-intensive operations in the deep oceans where political restrictions are, at the moment, still virtually non-existent.

Management and regulation inside territorial waters

Western European governments have been faced with a novel and rather bewildering challenge by the sudden and mushrooming interest in the resources of their territorial waters. With little in the way of precedent to guide them from the annals of international or national law, they have each been faced with the problem of devising enforceable management policies, stringent enough to provide a sound return for their national exchequers and to guarantee the health and safety of those working in the hazardous environment of the sea, while at the same time not being so draconian as to dissuade investors from venturing risk capital.

Oil and natural gas

The most contentious and difficult problems have of course been caused by the burgeoning discoveries and exploitation of hydrocarbons in the North Sea and other coastal waters, following the massive discovery of gas in what is now known as the Groningen field on the northern coast of the Netherlands in 1959, and the first commercial offshore find in 1965 close to the shores of eastern England at West Sole. Since 1981, some 25 per cent of Western Europe's oil and virtually all its natural gas has been produced within the region, the bulk of it from offshore production sites. As invariably happens with such natural resources, however, the physical occurrence of both oil and natural gas in no way matches either the political map or the distribution of potential demand. The UK enjoys the luxury of 60 per cent of known offshore reserves of oil and 32 per cent of the gas reserves; the figures for Norway are 35 and 53 per cent respectively. These two countries dominate the picture; for the rest, the Netherlands controls 1.8 per cent of the offshore oil production and 9 per cent of the natural gas, while the comparable figures for Denmark are 3 and 6 per cent. Belgium, the Republic of Ireland, France and Germany have almost nothing to show for their exploration efforts, except the small gas field off the south coast of Ireland at Kinsale Head and the pipeline linking Emden in Germany with the Norwegian Cod gasfield (Offshore Promotion Services 1980).

These variations, coupled with the very different energy demands of the countries in Western Europe, have inevitably led to somewhat different policies and priorities for the development of oil and natural gas within the region. The UK sector has been developed fastest, because of the availability of supplies combined with a large and eager domestic market. On the other hand, in the Norwegian sector the pace has been markedly slower, reflecting a determination on the part of government that the benefits of oil and gas revenues be available to Norway's 4 million citizens for as long as possible. In the Danish sector, where the finds so far have been disappointing, the circumstances have been quite different and it is probable that none of the gasfields would have been brought into production at all had it not been for the fact that in doing so the production consortium guaranteed itself further exploration rights from the government.

The most direct control over development has been by governments deciding who shall prospect and where. Two basic systems have been used to allocate prospecting rights: either the sector has been divided into blocks which have then been auctioned for specific periods, or the whole sector has been leased to a consortium. As time has progressed, the block system has come to be almost universally favoured and now only Denmark persists in treating its sector as a single entity.

There are many variations on the block principle and the size varies from sector to sector. In general, the oil companies prefer to have blocks that are as large as possible, giving them maximum freedom of action, while governments try to restrict the size so that they gain as much control as possible. Nevertheless, in seeking to maximize their influence, governments must guard against the danger of deterring the oil companies from prospecting at all.

Off Western Europe the blocks are smallest in the UK sector and the government has carefully controlled both exploration and development, exercising an ever-closer rein as time has gone on. There have so far been seven rounds of offers for prospecting rights since 1964 and it is interesting to observe how the UK strategy has changed over time. In 1964 and 1965, in the first two rounds of bidding, over 2000 blocks were put on offer and 475 taken up, reflecting an eagerness on the part of both the government and the prospecting companies to push ahead with development as fast as possible, even though neither had any precise idea about the locations that were likely to prove fruitful. Subsequently, both the numbers of blocks put on offer and the uptake have declined sharply as knowledge about the physical extent of the deposits has become more accurate and the oil and gas companies have tailored their activity to better informed assessments of market demands and costs of extraction (Anon 1982a).

The relationship between the public and the private sectors of the economy in the development and exploitation of oil- and gasfields in the seas off the coasts of Western Europe has been both complex and dynamic. In general, governments were loath to commit public resources to this initial search but, once exploitable reserves were proven, they became increasingly interested in securing a direct stake in their development. In 1975, the Labour government in the UK established the British National Oil Corporation, a name changed later to Britoil, with a view to undertaking exploration and development on its own behalf and the company now has interests in five major oilfields. The then government also made it a condition of all new licences granted for prospecting blocks that Britoil had at least a 51 per cent share, thus ensuring

direct national control over development policy. After
the 1979 election and the change of political control to
a Conservative administration, however, much less
importance was attached to public ownership. The
requirement that Britoil have a majority shareholding
in new developments was abolished and in 1982 the
company was sold to private investors, although with a
number of safeguards to ensure that it remained predominantly in British hands. The Norwegian government has also taken steps to guarantee the state a
strong stake in the development of offshore hydrocar-
bon deposits, but in the Netherlands and Germany,
where the sectors have been markedly less productive,
companies have maintained a much freer hand.

Fisheries

Revision of the Common Fisheries Policy, worked out
by the original six members of the EC in 1970 and
supplanting the temporary, transitional arrangements
agreed to ease the enlargement of the Community in
1973, has proved predictably difficult and contentious.
A draft policy was published by the Commission in
mid-1982 and, after six months of hard negotiation, it
was accepted by nine of the ten members but
Denmark, the country whose fishing industry had
grown faster than any other since joining the
Community, refused to accept the limitations that the
policy would impose on its access to what it considered Community waters. Without agreement by the
end of 1982, fleets from all the member countries
would have had the right to fish anywhere at all in
these waters, and this situation arose briefly in early
1983. By the end of January in that year, however, the
Danes accepted the terms of the Common Fisheries
Policy and it finally became fully operational.

The accession of Spain and Portugal to the EC at the
beginning of 1986 imposed further strains on the policy. Spain has a larger fishing fleet than any other
member state and was eager to exploit the
Community's seas. Although limitations on access by
its fishing fleet have been agreed for a transitional
period up to 1995, there is still a threat of the Common
Fisheries Policy being disrupted if permanent arrangements are not decided for Spain by that date. Portugal,
which has a much smaller fleet of mainly inshore fishing vessels, poses less of a problem (Wise 1987).

The new policy established a 9.6 km wide exclusive
zone around all coasts, which are treated as part of the
national preserve and within which other member
states have absolutely no fishing rights. There is then a
zone between 9.6 and 19.2 km to which there is limited access. For instance, the French are still allowed
to fish off south-west England, the Irish and the British

have a reciprocal arrangement in the Irish Sea, and the
Danes can fish in the Baltic waters off Germany. The
major problem has arisen off northern Britain where
this limited-access zone has been considerably extended to protect the fishing industry in both Scotland and
the Shetland Islands. Since the Danes have traditionally
fished these waters, among the most productive in
Western Europe, and initially felt somewhat unfairly
treated, they were determined either to see the limited-
access zone reduced in size or the catches they are
allowed to take from it increased (Anon 1982b).
However, the whole question of access to specific
areas has subsequently become overshadowed by arguments about how to limit the total catch in Community
waters in the interests of conserving fish stocks.

In addition to stipulating where fishermen from each
member country may fish, the policy also lays down
overall catch quotas. The purpose of these quotas is to
allow some measure of Community control over the
level of fishing so as to ensure the permanent conservation of fish stocks, but it is an enormously complex
operation to try and police, not only the total quantity
of different sorts of fish being caught, but also from
where in Community waters they are being taken and
by fishermen from which country (Table 11.2). There
are some 300 000 fishermen in EC countries fishing
the fourteen major subdivisions of Community waters
and each of these major areas is further subdivided
into two or three smaller ones (Commission of the
European Communities 1991b). Another feature of the
policy is a three-year programme of financial aid from
Community funds to the tune of £150 million to help
the fishing industries in the individual member countries in adjusting to the revised provisions. Further proposals for conservation also figure prominently,
including minimum limits on the mesh size of nets so
that young fish are given some measure of protection.

Finally, there is a series of regulations governing the
relationships between the Community and other
Western European states as far as fisheries are concerned. These deal with matters such as catch quotas,
both for Community fishermen in the waters of third
states and vice versa, and the allowable level of fish
imports from these countries. The latter are included
not only because unrestricted imports could undermine
the whole basis of the Community fishing industry, but
also because there was a need to normalize relations
between individual member states and those outside
the Community. This applied particularly to the UK
and Iceland. The lucrative British market was all but
closed to Iceland after the Cod Wars, and the introduction of the new Common Fisheries Policy offered an
obvious basis for a more harmonious new start
(Commission of the European Communities 1982a).

Table 11.2 Total allowable catch (YAC) quotas for the main varieties of edible fish in the European Community waters (in tonnes of cod equivalent) under the terms of the Common Fisheries Policy 1991

Fish type	YAC
Herring	7 653 000
Sprat	149 000
Anchovy	42 000
Salmon	720
Haddock	81 000
Saithe	141 000
Pollack	19 100
Norway pout	200 000
Blue whiting	482 500
Whiting	199 140
Hake	135 000
Horse mackerel	378 000
Mackerel	536 070
Plaice	211 600
Sole	251 300
Megrins	37 240
Angler fish	63 690

Source: Bulletin of the European Communities 24 (12), 1991, pp 77–9.

The protracted negotiations and the conflicts of interest have underlined how difficult it is to agree on any joint policy among twelve independent states, each of which fears it may be compromising its national interest. Nor is the question of access to EC waters by any means totally settled. In addition to the difficulties associated with absorbing Spain and Portugal into the Community, discussed above, access to the waters in the Mediterranean has yet to be negotiated. This will almost certainly raise formidable barriers as so many countries that are not members of the EC are involved (Kliot 1987a).

Marine pollution

The growing use that is now made of the seas and the seabed in the coastal waters of Western Europe has focused attention on their long-established function as a free and unregulated rubbish dump. In the last two decades it has become increasingly evident that the growing demands on this means of waste disposal, together with the threat that massive marine pollution in the form of oil spillages could pose for the fisheries and the other economic activities in West European coastal waters, mean that some form of control is imperative.

Pollution of the sea is of course not a single problem, but rather a whole range of manifestations of the undesirable consequences of twentieth century industrial society. Nevertheless, Johnson (1979) has identified four major facets: the discharge of effluents from land; the deliberate dumping of waste at sea; the by-products of exploiting marine and submarine resources, especially the exploitation of the seabed; and waste spillages from sea transport and navigation.

The discharge of effluents into rivers or directly into the sea takes two forms: sewage and industrial waste. Throughout the nineteenth and early twentieth centuries, as urbanization and industrialization progressed, it was the accepted way of disposing of waste in all the countries of Western Europe, and the detrimental effects on both public health and freshwater river life were only gradually appreciated. Subsequently, great efforts have been made, not least by the EC, to clean up rivers and coastal waters by treating effluent and setting minimum standards for pollution (Commission of the European Communities 1992d). It has, however, been an uphill task, since it has traditionally been assumed that any effluent reaching the sea will quickly become so diluted and dissipated as to be rendered harmless and not even aesthetically unpleasant. It is now abundantly clear that such an attitude is completely unjustified: there is growing evidence of marine pollution causing a public health hazard and increasingly well-founded suspicions that it poses a long-term threat to wildlife. The dangers are particularly apparent in the enclosed waters of the Mediterranean and the Baltic and it was no accident that it was here that the coastal states first began to cooperate to reduce and, hopefully, eventually eliminate discharges of both domestic and industrial effluent. There is no doubting the enormity of the challenge, but the problems involved are more organizational than technical, as evidenced by the recent history of the UK. Some 3 billion m^3 of effluent is discharged annually around the British coasts, of which two-thirds is now treated, compared with virtually none a century ago. The difficulty is that the residual amount is produced by a large number of small sources scattered along the coast, and the cost of further extending treatment is therefore very high.

On its own, treatment is not always a complete answer to the use of the open seas as an option for waste disposal, as in most West European countries dumping is not only allowed, but officially sanctioned. In the UK, some 7 million tonnes a year of sewage sludge are dumped in coastal waters, most of it in the North Sea, and a similar amount of industrial waste. The largest component of the industrial waste is colliery spoil deposited off the coast of north-east England. The scale of industrial dumping by coastal states in Western Europe varies considerably and

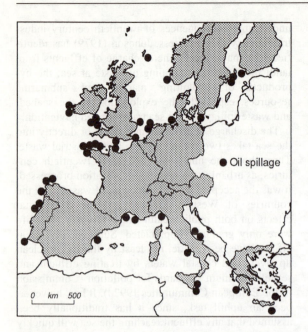

● Oil spillage

Fig. 11.2
Tanker oil spills (of over 5000 barrels) known to have
occurred in Western European waters 1970–90.

depends less on the amounts to be disposed of, than on
the policies of the individual governments. None of the
other countries bordering the North Sea currently contributes more than 1 million tonnes a year, with
Germany and Belgium generating only about half that
amount and Norway less than 200 000 tonnes. It needs
to be emphasized, however, that none of this officially
licensed dumping includes recognized toxic wastes,
which internationally agreed conventions required to
be containerized and disposed of more than 240 km
(150 miles) from land in waters more than 2000 m
(1.24 miles) deep.

Obvious, though happily so far potential rather than
actual, sources of marine pollution in coastal waters are
the burgeoning numbers of permanent installations for
extracting oil and natural gas from the seabed. There
have been several serious accidents in European waters.
A blow-out of gas and oil occurred at the Bravo drilling
rig on the Ekofisk oilfield in the Norwegian sector of
the North Sea in April 1977. It took a week to bring
under control and only resulted in the escape of 20 000
tonnes of crude oil, but it underlined the potential dangers associated with this new area of mineral exploitation and the difficulty and high cost of containing any
accident. The dangers were underlined further in July
1988 when the Piper Alpha oil rig caught fire and
exploded, killing over 150 workers. The fire took three
weeks to bring under control.

Finally there is the pollution generated regularly
from oil tankers, transporting crude oil from all parts
of the world to the unique concentration of industrial
cities in Western Europe. Figure 11.2 shows that in the
period 1970–90 some thirty-seven spillages of more
than 5000 barrels of oil are known to have occurred,
with a particularly heavy concentration in the English
Channel and the southern part of the North Sea. Since
it is thought that these large spillages account for less
than 5 per cent of the total number (International
Tanker Owners Federation 1980), the scale and seri-
ousness of the threat from this source of pollution are
self-evident, especially in light of the limited sanctions
that national governments can exercise over shipping
on the high seas.

As with all other matters relating to international
waters, controlling marine pollution, as opposed to
recognizing the actual and potential dangers, is a difficult task, involving several governments, all with different perceptions of the problems involved. However,
the three major semi-enclosed seas abutting the coasts
of Western Europe – the North Sea, the Baltic and the
Mediterranean – are now all covered by multilateral
agreements on the control of oil spillages. The Bonn
agreement, signed by Belgium, Denmark, France,
Germany, the Netherlands, Norway, Sweden and the
UK in 1969 covers the North Sea; a similar arrangement for the Baltic was signed by Denmark, Finland,
Germany (both East and West), Poland, Sweden and
the USSR in 1974; and in 1976, Cyprus, Egypt,
France, Greece, Israel, Italy, Lebanon, Libya, Malta,
Monaco, Morocco, Spain, Tunisia, Turkey and
Yugoslavia signed a convention covering the
Mediterranean. There is also an agreement signed by
most of the coastal states of Western Europe aimed at
controlling marine pollution from land-based sources
that came into force in 1978. Significantly though Italy,
whose industries contribute more than those of any
other country to the pollution in the Mediterranean,
has so far refused to be a party to it, thereby underlining the fundamental weakness of any international
action that depends on the goodwill of the countries
involved. Indeed, while international agreement on
procedures and policies may be highly desirable, effective preventative procedures are still very much the
preserve of national governments. In some parts of
Western Europe, bilateral arrangements, as for example the Manche Plan concluded between the UK and
France in 1978 to deal with oil spillages in the English
Channel, work well, but for the most part any corporate action is *ad hoc* and usually brought about by the
need to respond to a particular calamity.

The recent past has seen an unprecedented proliferation in formal arrangements for bringing offshore eco-

nomic activities under some form of political control. In the main the initiative has been taken by, and remains with, national governments, but they have had to act against a groundswell of support for international action. In Western Europe, the EC is now the key authority with respect to fisheries management and pollution control, through the Common Fisheries and Environmental Policies, but more general control is still blocked by the failure of world governments to ratify the Law of the Sea Treaty that has been under negotiation since 1973. One of the main objections is that the proposed International Seabed Mining Authority, which would be under the control of the United Nations, might adversely interfere with national programmes of exploration and recovery of minerals.

Note

The progression towards a Common Fisheries Policy was chronicled in numerous communications of the Commission of the European Communities during 1982: notably CPM (82) 338/final/5 (18.8.82); COM (82) 339/final (11.6.82); COM (82) 340/final (18.6.82); COM (82) 368/final (I 1.6.82); COM (82) 539/final (9.9.82).

Further reading

Allen R 1980 Fishing for a common policy. *Journal of Common Market Studies* **19**: 123–39

Blacksell M 1979 Frontiers at sea. *Geographical Magazine* **51**: 521–4

Churchill R 1980 Revision of the EEC's Common Fisheries Policy. *European Law Review* **5**: 3–37 and 95–111

Hodgson R D, Smith R W 1979 Boundary issues created by extended national maritime claims. *Geographical Review* **69**: 423–3

Ministry of Agriculture, Fisheries and Food 1981 *Atlas of the Sea around the British Isles*. MAFF, London

Mitchell B 1976 Politics, fish and international resource management: the British–Icelandic Cod War. *Geographical Review* **66**: 127–39

Wise M 1984 *The Common Fisheries Policy of the European Community*. Methuen, London

12

Trends in regional development

The diversity of Western Europe

An intricate patchwork

The states of Western Europe have shared many changes since the Second World War. Each experienced economic recovery and prolonged dynamism but then moved into a period of depression after 1973–74. Each underwent demographic revival and, although this decelerated after the mid-1960s, rural exodus and varying manifestations of urbanization served to alter the distribution of population profoundly. Each viewed the wider world through the apparatus of mass media and increasingly through tourists' eyes so that West Europeans tended to become more 'international' with regard to dress, popular music, sport and even diet and speech. Yet Western Europe remains an intricate patchwork of national and regional identities, with eighteen countries occupying a combined area only two-fifths that of the USA and each still retaining its own currency, language (or languages) and national 'system' for getting things done. However, greater cohesion of the member states of the EC may reduce their financial and administrative individuality in the years ahead.

Western Europe has been called a peninsula of Asia; it is also a realm of capes and bays and islands, of subtle variations in climate, soil quality and landscape that is quite different from the wide expanses of the greater continents. Its people display marked differences of physique, skin tone and colour of hair and eyes, but inter-regional and international migration has encouraged greater mixing than ever before. Local customs linger on but show a tendency to fade with every generation that passes. Fundamental uniqueness of place – the very substance of geography – survives but almost every state in Western Europe has, in its own way and in its own time, introduced schemes for attempting to reduce spatial manifestations of socio-economic

inequality. Some have been simple, others sophisticated; some have long pedigrees, others have appeared suddenly; some cover specific sectors, while others are much more integrated (Clout 1987a; Pinder 1983). Many are what may be thought of as 'open' or explicit regional policies, while other policies are implicit, with sectoral issues such as job creation as their focus but also having very important spatial expressions. Twenty-five years ago it seemed possible to believe that some of these brands of regional policy were quite successful, while others left something to be desired. Now all have been overtaken by world events and are tragically outpaced by recession, inflation and unemployment (Yuill, Allen and Hull 1980; Ashcroft 1982). The 'golden age' of regional policy in the countries of Western Europe spanned the quarter century preceding the mid-1970s; now that age is emphatically over.

Regional differences but human problems

Regional disparities are spatial generalizations, abstracted from the realities of life and work experienced by ordinary people at specific places. Harsh contrasts in well-being and class structure are softened by the calculation of regional indicators which collapse extreme conditions into the tidier and somewhat misleading form of spatial averages. Strictly speaking, there are no such things as 'regional problems' or indeed 'problem regions' but only varying permutations of the human condition cast in different spatial settings. Although the term 'regional development' is used widely – and will certainly be employed in this chapter – it should be considered as something of a misnomer since it is ultimately human life that needs to be enhanced; regions are merely convenient packages. This point is not just a matter of pedantry; to grasp it is fundamental to any meaningful understanding of the concept of spatial differentiation.

Regional differences in economic conditions

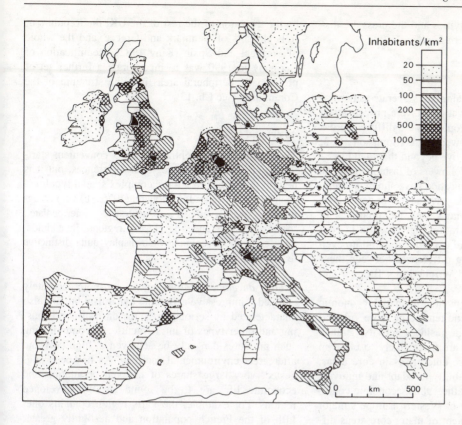

Fig. 12.1
Population density.

Inhabitants/km²

20	
50	
100	
200	
500	
1000	

embrace a multitude of contributory factors, ranging from density of population, location and array of natural resources in each territorial package, to the political power of its rulers, the entrepreneurial skills and inventiveness of its inhabitants, and the volume of finance available for productive investment (Fig. 12.1). Information- and capital-rich regions contrast with those that are less well endowed. Consideration must also be given to the relative demand for each region's products or expertise on the local market, and the efficiency of networks of communication which allow fuel, raw materials and other commodities to be brought in and manufactured goods and other outputs to be dispatched elsewhere (Brown and Burrows 1977).

Inevitably the significance of these interrelationships at each particular place undergoes change with the passage of time. New resources and alternative forms of technology are discovered, innovative modes of communication developed and administrative boundaries remodelled, while political and commercial empires rise and fall. Resources of capital and labour are, of course, reasonably mobile and may be directed in support of areas that lag in terms of economic dynamism or fail to enjoy qualities that endowed them with great-

ness in the past. But adjustments of this kind are far from automatic and Western Europe is littered with localities that once flourished but have now fallen into decline. The most striking examples are mining areas, textile towns and metallurgical districts that pioneered industrialization in the nineteenth century but subsequently lost their advantages because of competition from new sources of energy, raw materials and cheaper manufactured goods.

By contrast, capital cities in administratively centralized countries have flourished as dominant concentrations of labour, expertise and political and economic power. They form the largest single components of home markets, control links with the outside world and have succeeded in attracting investment capital on an impressive scale. The situation is less extreme in recently unified nations, such as Italy, or in federal states, like Switzerland or Germany, where economic and political powers are distributed more widely. For example, the former federal capital (Bonn) is literally a small town in Germany, coming behind a score of more populous urban centres in western Germany. Berlin, Hamburg, Frankfurt and Munich are the largest agglomerations in the country; Düsseldorf and Frankfurt are the major commercial cities; and manufacturing, service activities and the press are distributed widely in the western *Länder*.

Cores and peripheries

Capitals and many other major cities operate as normative centres in which innovations have a propensity to take place and from which opinions are diffused across wider territory (Keeble, Owens and Thompson 1982; Keeble 1989). People who live beyond them tend to be late receivers of ideas and followers of fashion, although modern systems of communication have shrunk space in ways that were inconceivable just a couple of generations ago. More peripheral locations lack the processes of agglomeration and concentration that function in core areas and enable capital and expertise to build up even further, thereby reinforcing differences between these types of region (Mény and Wright 1985; Brugger and Stuckey 1987).

Peripheral areas are fragmented in spatial, economic and organizational terms and tend to be more vulnerable to downward swings in business activity. Branch factories in the periphery are more likely to close in times of recession than are enterprises in core areas, which tend to display an ability to adapt and innovate more readily. However, in the 1980s, economic conditions deteriorated in many of Western Europe's major cities so that the environment of many core areas differs greatly from the stereotype. After enjoying the advantages of agglomeration for decades, firms in these localities have been forced to count the cost of congestion, pollution and expensive accommodation and labour. Planned decentralization has directed jobs away from many metropolitan areas and has contributed to the economic and social problems currently experienced in inner-city districts. the employment base of metropolitan areas will be weakened even more as new technology is further applied to routine office work.

The contrast between the prosperous industrialized 'centre' and the declining, largely rural 'periphery' of Western Europe was spelled out in a United Nations survey as early as 1954, and the model has been refined many times since then (Despicht 1980). Various scales of core and periphery have been identified, ranging from individual cities and their environs, through the contrasting regional structures of individual countries, to the great 'Lotharingian axis' (comprising the Rhinelands, northern Italy and Paris) which differed so markedly from the Mezzogiorno and western France in the Europe of the Six (Rokkan 1980). Successive enlargements of the EC in 1973, 1981 and 1986 added south-east England to the EC's developed axis and enlarged its complicated periphery

by the inclusion of the rest of the UK, the Republic of Ireland, parts of Denmark and Greece, and the whole of Portugal and Spain (King 1982). Reunification of Germany in 1990 was to incorporate a further set of problematic peripheral areas into the structure of the Community (see Ch. 13).

A regional typology

The core/periphery contrast offers a convenient starting point for discussing regional differences, but it is helpful to proceed to a more complex spatial typology, such as that advanced by Holland (1976a, b) who recognized 'over-developed', 'neutral', 'intermediate', 'depressed' and 'under-developed' regions. In addition, there are frontier regions which display quite distinctive features and problems.

1. *Over-developed regions* enjoy – or more accurately enjoyed – impressive economic dynamism but are also characterized by serious pressures on transport, housing and other types of infrastructure. Accommodating such pressures can only be achieved at high cost and after great environmental sacrifice. As has been suggested, very large shares of national population and economic life are to be found in over-developed regions. For example, the Paris region contains one-fifth of the French population and a slightly greater share of jobs on 2 per cent of national territory, while giving rise to 30 per cent of GDP. The London metropolitan region houses 23 per cent, greater Copenhagen 35 per cent, the Randstad 46 per cent, and the Rhine–Ruhr 15 per cent of their national populations (Guttesen and Nielsen 1975). Several other areas of western Germany, the industrial districts of northern Italy and arguably some cities in southern Europe also typify over-developed areas, albeit of differing magnitude and intensity of dynamism. These congested areas give rise to the most sophisticated life styles in Western Europe but also have serious social deprivation and environmental decay. In the recent past, decentralization has been seen as a remedy both for metropolitan congestion and for under-development elsewhere, but there was precious little dynamism anywhere in the 1980s and some argued that what did exist was desperately needed in the inner cities (Nicholson 1981).

2. *Neutral regions* are located in the readily accessible environs of over-developed areas and may be typified by outer parts of the Paris Basin and south-east England. They function as reception zones for urban deconcentration and tend to be characterized by important population growth, expanding employment and

rising incomes. They contain many light industries and office activities which have been moved to the pleasant living and working environments that market towns and greenfield sites can offer.

3. *Depressed regions* were formerly leaders in their respective fields but have been overtaken subsequently by other areas. Rates of unemployment and outmigration are typically high and the built environment of abandoned mines, spoil heaps, workers' housing and old factories displays many features that are sadly outmoded. Classic examples are provided by coal- and iron-mining districts (such as the Ruhr, Saar, northern France, Lorraine, Wallonia, Dutch Limburg and several parts of the UK), textile areas (as in Flanders, Lancashire and west Yorkshire), and shipbuilding towns (such as those in north-east England or on the Clyde in Scotland). New examples involve areas where flourishing activities have been turned into 'sunset' industries in the course of the present recession. The West Midlands, with its collapsing engineering and vehicle industries, is a particularly tragic case. The problems of depressed regions have been further aggravated by the fact that once-healthy core areas can no longer supply employment opportunities to bale them out.

4. *Under-developed regions* are those peripheral parts of Western Europe where modern industrial capitalism failed to take root to any great extent. Jobs and incomes are still quite strongly geared to farming but many farms are too small and under-capitalized to be viable. Younger people have tended to leave the land, abandoning farming to an ageing population that is unlikely to innovate. From mid-century to the onset of economic crisis in the 1970s, large numbers of labour migrants left the low wages, underemployment and high levels of unemployment in under-developed regions to move to labour-hungry urban areas both within and beyond their national boundaries. But since 1973 this palliative has come to an end and many unemployed migrants quit the industrial cities to return to their home areas where their presence aggravated employment problems even further. Rather than go back to the land, they often sought jobs in local market towns, but many traditional service and workshop activities were already overstaffed and their employees effectively underemployed.

The so-called black economy came to flourish under these circumstances in the Mezzogiorno, Iberia and many other regions where men, women and school-children earn a pittance in craft industries such as textiles, clothing, footwear and furniture. They have no social security, minimum wage, insurance or protection under health and safety regulations. In some towns

of southern Italy entire neighbourhoods function as clandestine workshops which disappear rapidly when word gets out that labour inspectors are about to visit (Hamilton and Just 1981). The challenge of under-developed regions is clearly not just to modernize agriculture but to create reasonable job opportunities in every other sector as well. The task is compounded by poor provision of roads, schools, hospitals and water supplies, with the latter problem being particularly serious in many Mediterranean areas, where poor soils and steep slopes hinder agricultural improvement. Remoteness from the economic core areas of Western Europe, lack of viable industrial expertise and absence of modern industrial training and local credit facilities make it all the more difficult to attract investment to peripheral regions.

5. *Intermediate regions* display employment, income and migration characteristics which reflect a mixture of conditions encountered in other types of region. For example some under-developed regions contain large, decaying urban areas which lack both the dynamism of over-developed areas and the residue of industrialization encountered in depressed areas; Naples is a case in point.

6. *Frontier regions* may be handicapped by inadequate cross-border infrastructure or by differences in income, currency and law in adjacent states. Some share general problems to which common solutions might prove advantageous, for example, many parts of the steel-making triangle of the Saar, Lorraine and southern Luxembourg face very similar difficulties that stem from the decline of their staple industry (Burtenshaw 1976). By contrast, a unique set of difficulties was experienced for forty-five years in the eastern fringe of West Germany and Austria which abutted on to the closed frontier of the Iron Curtain. The well-being of many long-recognized frontier regions may change considerably as the 'new Europe' emerges (Maillat 1990).

Systems of aid

General aims

Under-developed, depressed and, to a lesser extent, frontier regions have been perceived by individual countries as meriting special treatment to promote redistribution of capital and employment. Virtually every national government defined assisted areas and devised mechanisms of regional aid in response to its unique perception of political, economic and cultural forces at work within its borders. In many cases, policies and principles proved considerably more ambiguous than the vote-catching gestures involved in making financial awards (Vanhove and Klaasen 1987). In addition, assisted areas, mechanisms and rates of support were subject to considerable revision over the years, as the balance of political power shifted and economic circumstances changed. It would be impossible to cope with such intricacy in the present chapter, which will simply offer a basic outline.

The general aim of regional policy is to reduce spatial imbalances in economic activity and prosperity, once they have been perceived as political problems. Such perceptions derive from both moral and economic principles. It is arguable that unequal access to income and opportunity is socially unacceptable and morally unjust, and hence regional policy should be implemented as part of a broader drive for equity. In addition, it may be argued that maladaption in regional activity prevents national economies making the most efficient use of resources and thereby restricts the generation of wealth (Molle and Paelinck 1979). As a consequence of these and other arguments, the brands of regional policy implemented in Western Europe contain various strands, which include:

Managing economic development to promote not only the nation as a whole but also component regions, with respect to their distinct needs, problems and potentialities;

reducing disparities between regions in terms of economic activity, welfare and prosperity;

supporting the economic and social foundations of life in the regions; and undertaking physical planning of housing, roads and other features in a coherent programme in line with national and regional aims.

Constitutional contrasts

With so many countries involved it is not surprising that Western Europe's systems of regional aid were devised under very different constitutional conditions. *Aménagement du territoire* in France exemplifies an integrated approach which was elaborated during the 1960s but had its roots in the nation's highly centralized administration and its system of national economic planning which commenced in 1947. Regional policies expanded from attempting to reduce the disparity between Paris and the provinces to providing industrial jobs in the rural west and renovating old manufacturing and mining areas in the north and north-east. Every part of the country, and hence every type of region, had its place in the grand design and was allocated a specific policy. Needless to say, the

system proved more elegant and objective in theory than in practice (Frémont 1978). After the Socialist administration came to power in 1981, a share of the responsibility for regional development was transferred to the regions themselves. *Aménagement du territoire* has become more of a dialogue between Paris and the regions.

The British approach was also centralized but was far more pragmatic and sectoral in character, with a variety of policies being devised to meet a range of social and economic ends. By contrast, Germany operates a far more decentralized system since the *Länder* are responsible for regional policy and many other economic issues. However, links were strengthened between them and the central government in the 1960s and guidelines for regional development were drawn up. Broad matters of policy and sources of finance are worked out by both federal and state authorities, but implementation of policy is left to ministers in the *Länder* (Jung 1982; Kunzmann 1981). Internal bargaining is obviously central to all brands of regional policy but is particularly critical in Belgium, where claims for assistance from the Flemish north and the Walloon south have required extremely delicate handling and compromise (de Clercq and Naert 1985).

Individual countries devised policies at various times in the past and adopted quite different scales for defining assisted areas. For example, extensive parts of Italy, France and, for a long time, the UK qualified to receive regional aid, in contrast with much smaller units in West Germany, Belgium and the Netherlands (Fig. 12.2). Changes in political and economic circumstances gave rise to modifications in their territorial extent and to the type and value of assistance to which they were entitled. British schemes of regional aid began in 1934 as welfare measures in districts with high levels of unemployment. Objectives gradually broadened and policies proliferated, elaborating the map of regional assistance from the 'special areas' of the 1930s to the wide system of graded financial aid that operated in the 1970s. In the summer of 1979 the Conservative government announced fundamental changes. Further revisions led some to write of the demise of regional planning in the UK (Wood 1987; Damesick and Wood 1987).

Regional policies in Italy were prompted by political, economic and social problems in the Mezzogiorno at the end of the Second World War and the 'Fund for the South' (Cassa per il Mezzogiorno) was established in 1950 to complement ministerial resources. The first so-called pre-industrial phase of activity extended from 1950 to 1957, when strong emphasis was placed on land reform and improvement of infrastructures. In the following ten years, attention was directed towards

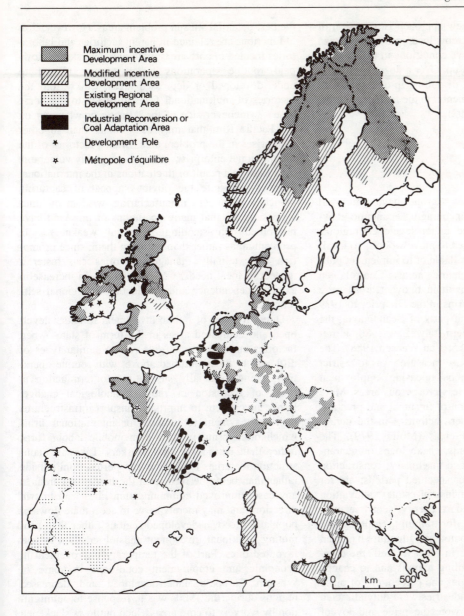

Fig. 12.2
National patterns of regional assistance, *c*. 1970.

Legend:
- Maximum incentive Development Area
- Modified incentive Development Area
- Existing Regional Development Area
- Industrial Reconversion or Coal Adaptation Area
- ★ Development Pole
- ☆ Métropole d'équilibre

0   km   500

encouraging modern industries by steering new branches of state-owned enterprises to southern locations which had been selected in partial deference to growth-pole theory. After 1967, policy became more concentrated spatially within the south and was aligned more closely with national planning objectives. The Fund continued to function until the end of 1980 with extensive backing from foreign sources, the EIB and the World Bank, as well as the Italian government. In 1984, liquidation was announced (Orlando and Antonelli 1982; Pacione 1982; King 1987 a, b).

Spanish regional policy underwent profound

changes after the death of Franco in 1975 and subsequent collapse of his highly centralized regime. The new constitution of 1978 granted a measure of self-government not only to the historically distinctive nationalities of Catalonia, Galicia and Euskadi (the Basque country) but also to other areas which had never displayed such a degree of separate identity (Hebbert 1982). A new kind of regional planning is taking shape that is quite different from the functional regionalism with highly centralized incentives and subsidies that existed under Franco but proved largely unsuccessful in steering investment and new jobs to the rural periphery. The new approach is a version of political regionalism whereby local groups organize themselves to try to improve conditions in their particular areas from the bottom up. Over a dozen regionally inspired plans run the risk of pulling in quite different directions and being detrimental to the national interest (Hebbert 1987; Naylon 1992).

Mechanisms

Sticks and carrots

The countries of Western Europe have employed numerous techniques for encouraging economic activity in their depressed and under-developed regions. But very few have attempted to move workers to localities where jobs may be available or to implement policies of restraint in dynamic areas. Controls on industrial and tertiary growth in buoyant areas were implemented in varying forms in the UK prior to 1979 and were also attempted in parts of Scandinavia, the Dutch Randstad and northern Italy (Law 1980; Jarrett 1975; van Hoogstraten 1985; Van Weesep 1988). The ultimate objective of all these measures was to restrict metropolitan growth, thereby releasing employment potential for relocation in less prosperous areas. More usual has been the operation of financial and practical measures to promote modern activities in the under-developed and depressed areas (Molle 1979). The range includes capital grants, cheap loans or guarantees for equipment, plant and machinery; construction of premises; tax relief over specified periods; aid for installing and equipping industrial estates; alleviation of certain costs (e.g. transport, energy, social security contributions); and finance for training, retraining and housing the workforce. Promotional literature has been produced in vast quantities in order to 'sell' the environmental advantages of ailing regions and to change long-established psychologies and perceptions of place (Burgess 1982). Modern industries were duly attracted and, as prestige projects, were the pride and joy of local politicians. Such schemes certainly increased regional productivity and economic output but their presence did not bring true salvation to problem areas. Most modern industries are highly mechanized and hence generate few jobs for large pools of local labour which may or may not contain skilled workers, depending on the recent economic history of each reception area. Many factories implanted in peripheral areas are highly dependent on external capital, decision-making and other production factors, while also being strongly geared to meeting external demands.

Multi-region and multinational corporations have made extensive use of development funds to capture the best potential sites in problem areas. However, R & D functions are retained in parent factories and laboratories located in core areas, with only less skilled, low-paid routine operations usually being moved to under-developed or depressed areas. Links from the factories of multinational firms are back to the home base – whichever side of the Atlantic or wherever on the Pacific Rim that may be – rather than to other enterprises in the problem area. Branch factories of the multinational enterprises prove to be highly vulnerable to closure as a result of fluctuations in the international economy. Despite the obvious appeal of capturing branches of modern manufacturing, wisdom of hindsight suggests that many operations of this kind have reinforced long-standing structural weaknesses in problem areas rather than reducing them, since promotion of externally managed factories can foster a neglect of local needs, while leading to an increase in external dependence and a reduction in regional self-reliance.

Direct transfer of public investment to under-developed and depressed areas in the form of state-owned factories and offices can give rise to similar types of difficulty to those outlined above, with routine operations being especially susceptible to termination at times of recession or rapid technological change. Public investment to improve transport infrastructures involves payment to national or international firms which often bring in their own mobile labour force rather than taking on local workers. Likewise, construction materials tend to be transported from outside rather than being derived from the immediate area. In any case, improved communication lines lead in two directions and may merely serve to accentuate flows of population to over-developed regions rather than promoting additional investment and job creation in less favoured ones. Part of the temporary improvement in economic and employment conditions in some of Western Europe's under-developed and depressed regions during the 1960s was in response to outmigration by workers to core areas. Such outflows slackened the pressure on local job markets, while the remittances that migrants sent home allowed improvement in living conditions for family members who stayed behind.

Growth centres

In many countries, attempts were made to counter the difficulties likely to stem from implanting individual branch factories by translating economic theories of agglomeration into spatial terms and designating growth centres at selected localities. Governments and other agencies sought to establish complexes of activities that were concentrated areally and linked functionally. Ideally, this kind of arrangement should reduce 'backwash' from under-developed regions to core areas and also encourage economic growth and the emergence of new activities locally. In addition, carefully planned growth centres should stimulate diffusion of dynamism and new job opportunities into their hinterlands. Complexes of this type, admittedly of varying dimensions and with differing titles, were duly established in many countries of Western Europe during the past four decades. The best-known and perhaps most ambitious experiment was in the Mezzogiorno, but results have been disappointing. After having devolved some of its regional policy-making to the regions in the 1970s, Italy readjusted its incentives and placed more emphasis on programmes to promote agriculture, small industries, services and tourism, rather than large capital-intensive manufacturing plant (Secchi 1980).

Judgements on growth-centre policies in all parts of Western Europe must be mixed. Some intended growth centres simply failed to flourish, others emerged as focal points for regional services and new jobs but did not diffuse economic growth into their hinterlands. Such partial success served to increase the scale of local socio-economic differences. Under such circumstances, growth centres contributed to a shift of disparities from the inter-regional to the intra-regional level. Only in rare cases did they manage to induce an overall reduction of spatial differences in levels of living (Stöhr and Tödtling 1978).

Rise and fall

Despite their deficiencies and the fact that practice often lagged behind theory, the numerous regional policies introduced in Western Europe did achieve a qualified measure of success during the economically buoyant 1960s. New factories and offices provided new jobs in problem areas and steered a proportion of available growth away from over-developed cores. The economic recession that followed the initial oil crisis of 1973–74 conspired to set much, if not all, of that to naught. New factories were still being opened in assisted areas during the 1970s but the overwhelming trends were for plants to close or to reduce their labour force. By the early 1980s, unemployment rates in formerly prosperous regions exceeded those that had prevailed a decade earlier in problem areas, while conditions in under-developed and depressed regions continued to deteriorate tragically (Keeble, Owens and Thompson 1981). Capital for investment by governments and businesses dwindled and it became painfully clear that traditional approaches to regional development had been rendered largely powerless by world events. Under such circumstances many governments in Western Europe decided in the late 1970s or early 1980s to cut back their regional policies and their old definition of assisted areas. These were partly economy measures but also represented a shift towards general rather than area-specific attempts at job creation. Several countries modified what remained of their regional policies by lessening emphasis on potentially mobile manufacturing and placing greater stress on stimulating the service sector and indigenous firms which are often small in size. The UK and France pioneered this line of assistance for labour-intensive service-related schemes, and West Germany, Italy and the Republic of Ireland duly directed more attention to this sector. To summarize: before the late 1970s, schemes for reducing pressure on over-developed areas by relocating activities in under-developed and depressed regions could have been shown to be making a positive contribution to national economic growth in the countries of Western Europe. The harsh fact is that such days are over and, unless the current trend of recession can be reversed, they may never return.

The Community's quest for a regional policy

Background considerations

Over and above complicated national attempts to encourage regional development, the EC gradually evolved a common regional policy to apply to its expanding territory. This should not be taken to imply an attempt to reduce the highly diverse regions of the member states to some kind of European uniformity. Instead, it is aimed at eliminating kinds of economic inequality that might be judged to be harmful. The case for a supranational approach may be argued on several grounds, including those of viewpoint and responsibility. First, poverty or under-development are relative concepts and what may appear to be a backward region in, say, the context of western Germany would certainly look prosperous when set against conditions in the Mezzogiorno or western Ireland. A Community approach to regional disparities should offer a broad, comparative perspective which individual states would be unlikely to take for themselves. Second, the very existence of the Common Market, with increasingly free movement of capital, goods and labour and the operation of various common policies, tended to favour the more productive, central areas of the Community and has deprived the periphery even more. Common regional policy has never been intended to subsume national schemes but rather to provide a forum for coordinating the strategies of member states and allocating a limited amount of financial support (Klein 1981). Despite a vast amount of discussion stressing its desirability, 'regional policy' emerged as a latecomer among Community policies. None the less, the first 'European' attempt to assist several very specific types of depressed areas dates back to the creation of the ECSC in 1952 when loans were introduced to create new jobs in coal-mining and steel-making areas. General ideas on coping with regional disparities were floated at the Messina Conference which preceded the creation of the EEC but the Treaty of Rome did not mention a common regional policy, although its preamble and several articles touched on regional matters. Heads of state expressed their desire to achieve harmonious development of their six economies by reducing harmful differences between regions and tackling the problems of less favoured areas. Certainly it was intended that regional disparities and likely spatial impacts of competition, agricultural, social and transport policies should be scrutinized. Article 130 enabled the EIB to be set up to make loans to approved schemes to improve economic and social conditions in backward areas. It may well have been that the founders of the Community simply anticipated that the general process of economic growth, that duly operated in the 1950s and 1960s, would of itself reduce the gap between rich and poor regions.

The 1960s formed a time of discussion and more discussion on regional problems in the Six; in 1965 the possibility of a common regional policy was aired. The first economic policy programme (1966–70) emphasized the need to improve inter-regional infrastructures, especially transport, and in 1969 there were formal proposals for coordinating regional policies, drawing up programmes for problem areas, establishing a special fund and a committee for regional matters, and encouraging public and private sources to invest in regional development schemes. Two years later, national governments were urged to pay greater attention to the 'European' dimension of regional problems, and Community officials agreed to consider regional difficulties when operating the EIB, the Social Fund and the guidance component of the Agricultural Fund.

A new imperative

By this time the matter of regional disparity had acquired extra political significance since Denmark, the Republic of Ireland and the UK were about to enter the Community (Jensen-Butler 1982). The largest element of the EC budget, namely the CAP, was thought to have little to offer the UK, but a common regional policy would appeal to all the newcomers since each had serious regional problems. In the final months before enlargement, government representatives of the Six agreed to coordinate regional subsidies. As soon as enlargement took place in 1973 it proved necessary to analyse regional disparities throughout the Nine, and that responsibility was given to the new commissioner for regional policy, George Thomson. There was no possibility of undertaking a special standardized survey; instead, Thomson and his team had the difficult job of collating and interpreting highly diverse sets of data, submitted quite independently by each member state. Precise definitions of socio-economic categories, dates of data collection and the size of spatial unit by which they were ordered varied enormously. Some bodies of information were incomplete or highly dubious and hence the researchers were restricted in the number of indicators they could use. They also had to tackle the fundamental problem of defining problem regions at the Community scale rather than relying on the political perceptions of each member state. The final compromise recognized such regions as those with per capita GDP below the EC average and with one or more of the following features:

(1) a greater share of the workforce in agriculture than the EC average;
(2) at least 20 per cent of employment in declining industries (e.g. coal mining and/or textile manufacture); persistently high unemployment (defined as over 3.5 per cent and more than 20 per cent above the appropriate national average) or annual outmigration exceeding 1 per cent over a long period.

The *Report on the Regional Problems in the Enlarged Community* (known as the Thomson Report) represented a brave first attempt to quantify regional differences according to a procedure that was applied throughout the Nine. However, the apparently rigorous and precise statements in the Report were drawn from diverse and somewhat imprecise data. In any case, information that is generalized to the regional scale serves to conceal diverse local conditions, hence similar regional levels of income or unemployment could well derive from very different combinations of local values and underlying causes. In addition, the Report tended to treat socio-economic conditions as static features; it would have been more meaningful to review them in an evolutionary sense. Doubtless, deficiency of data was the reason for adopting this narrow view; however, the Report did insist that future changes be monitored and the definition of problem regions reappraised.

Facts and figures

Thomson showed that in 1970, Denmark displayed the highest average per capita GDP in the Nine, with a figure almost 30 per cent over the EC average. West Germany, France, Luxembourg and Belgium were each characterized by above-average affluence, while the Netherlands, the UK, Italy and the Republic of Ireland fell below the mean. Every region in the last three countries had a per capita GDP that was below average. Only tiny Luxembourg was entirely above the mean; each of the remaining five countries had a scatter of regions above and below the average. Greater Paris was unquestionably the richest area with an average per capita GDP that was almost double the EC average and five times that recorded in the poorest areas of the Mezzogiorno.

Agricultural employment had been retreating sharply during the 1960s and only 9.8 per cent of the Community's workforce remained in that sector in 1971. None the less, many regions in the Mezzogiorno and the Republic of Ireland still had more than 30 per cent of their workers on the land and in some areas the figure exceeded 40 per cent. Western France and many remaining parts of Italy had over 20 per cent. Less than 5 per cent of employment was in agriculture in virtually all of the UK, the Ile de France, parts of Belgium and the German Rhinelands. Jobs in manufacturing came above the average (43.9 per cent) in most of the UK, northern France, northern Italy, the southern Netherlands, and much of Belgium and West Germany; while jobs in services exceeded the mean (46.3 per cent) in heavily urbanized areas and those with well-established tourism functions.

Statistics on industrial decline brought out the familiar pattern of Western Europe's coalfields together with less well-known distributions of quarrying and textile manufacture. What were judged to be high unemployment levels for the time were encountered in both agricultural and manufacturing regions in the Republic of Ireland, southern and central Italy, southern Belgium, northern Jutland, Scotland, Wales and Ulster. Heavy rates of outmigration affected much the same broad regions and also small pockets of job loss in mining districts and old industrial areas. The official collation of these indicators presented the full pattern of the Community's problem regions, which housed no less than 40 per cent of its residents (Fig. 12.3). Thomson's selection of criteria had stressed problems associated with poverty rather than affluence, but the Report did make due mention of the congestion, pollution and high costs of living that were experienced in economically 'healthy' conurbations and capital cities.

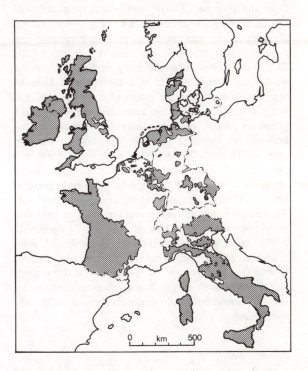

Fig. 12.3
Areas qualifying for aid, according to the Thomson Report.

Money matters

In addition to the technical operation of assembling indicators and the difficult political exercise of defining critical threshold values, it was necessary to embark on even more contentious matters. Existing national schemes of regional aid had to be coordinated, the cost of the proposed European Regional Development Fund (ERDF) agreed, the appropriate share to be borne by each member country approved, and a mechanism found for levying contributions. Real transfers of money from wealthy member states to the problem regions of poorer ones would be essential if the policy were to have any chance of success. Expectations that large contributions by some states would guarantee privileged treatment when awards came to be made needed to be dispelled. Not surprisingly, these pious hopes enshrined in the embryonic regional policy gave rise to massive political confrontation!

As well as defining problem regions, the Thomson Report indicated which countries were poor – by

European standards – and should be net beneficiaries from the proposed ERDF, and which were rich and would therefore be called upon to be net contributors. Basking in the success of its economic miracle and with relatively few problem areas recognized by the EC, West Germany was clearly destined to be the major net contributor. It was therefore in German interests to challenge the extent of assisted areas, the basic criteria analysed by Thomson, the size of the proposed ERDF, and the relationship between contributions and receipts. The West German government opposed Thomson's proposal of a fund of *c*. £900 million for 1974–76 and advocated *c*. £250 million, thereby requiring a rewrite of the regions that might hope to benefit. Moderately poor areas would be struck off Thomson's map and the absolute amount of help that very poor regions in Ireland and the Mezzogiorno could expect were trimmed down. The political situation degenerated, with West Germany refusing to contribute the large volume of resources that Thomson's recommendations required and France opposing a skeletal fund that would probably benefit only Italy and Ireland.

A little piggy bank

After eleven months of stalemate, the Nine agreed on 1 December 1974 to set up a relatively small ERDF of £540 million for the period 1975–78 (Talbot 1977). This amounted to little more than half the original proposal and was all the more pathetic when compared with the enormous sums being allocated at that time by individual countries to promote regional development and by multinational corporations to set up industrial facilities in depressed and under-developed regions. Even more disturbing was the fact that such a small ERDF was launched into a world of rising inflation, mounting unemployment and ultimately profound recession. The criteria of the boom years of the middle and late 1960s that had been used by Thomson had become quite inappropriate for judging the shape and scale of regional problems in the mid-1970s and were to become even more outmoded with the passage of time, as formerly buoyant cities and once healthy 'new' industrial regions fell into decline.

National definitions of problem areas were to be used, rather than Thomson's map, and the contents of the little piggy bank were to be shared among the Nine according to a quota system. Italy, the UK and the Republic of Ireland were entitled to 40, 28 and 6 per cent respectively, each being a net beneficiary. The remaining states were net contributors and could receive the following shares: France (15 per cent), West Germany (6.4), the Netherlands (1.7), Belgium (1.5), Denmark (1.3) and Luxembourg (0.1) (Kiljunen 1980).

Community regional policy in action

Quotas and quota-free allocations

Projects for which ERDF finance were sought had first to be vetted by national authorities and details then forwarded to Brussels for evaluation and decision (Kerr 1986). Financial support for favoured schemes was transmitted to national treasuries for payment to development agencies, local authorities or firms. During the first phase of the ERDF (1975–78), grants were directed to a wide range of projects, including factory estates in Belgium, West Germany, Italy, the Netherlands and the UK, new roads in France and improvements to the antiquated telephone system in the Republic of Ireland. As many had suspected, developed areas had enhanced their advantage to the detriment of poorer regions during the previous two decades; schemes supported by national systems of regional aid or by the infant ERDF had failed to counter the trend; and the volume of capital for all kinds of regional investment had slumped in the recession since 1973. Economic restructuring was operating at the world scale and worked to the detriment of high-cost European nations.

In the light of these trends, the EC's regional policy was modified for 1978–80. The regional implications of other Community policies were recognized, and fluctuations in regional well-being started to be monitored. The ERDF was increased to £1234 million for the three-year period, of which 5 per cent was outside the quota system. Modifications in 1979 raised the share for France and scaled down the quotas for six other countries by a fraction. Further adjustments were made to take account of Greece's entry to the EC in 1981. Grants from the quota section had to be used on investment projects, with 70 per cent being allocated to infrastructure schemes and the remainder to manufacturing, craft or service projects. Awards assisted the construction of a natural gas pipeline from North Africa to Sicily and the rest of the Mezzogiorno, irrigation and land-drainage schemes, improvements in telecommunications and railway lines, new roads and water-supply systems, as well as individual factories and hotels (Dean 1982).

The Commission had the right to suggest appropriate schemes for quota-free assistance from the ERDF. Member states wishing to seek support from the quota-free section were required to plan their programmes for several years ahead and to devise strategies for coordinating resources from national funds and private sources as well as the ERDF. For example, five programmes had been approved by early 1982, involving the Mezzogiorno, south-west France, the border region of Northern Ireland and the Republic, a number of steel-making areas (in Belgium, Luxembourg and the UK), and shipbuilding localities in the UK. In addition, experimental 'integrated operations' were launched in Naples and Belfast to assemble EC funds and national resources to promote economic development. These embraced public transport, roads and sewerage in the case of Naples, and water supply, training facilities and industrial and commercial development for Belfast (Spooner 1984; Noble 1985).

Modifications

Further guidelines for modifying regional policy were advanced in autumn 1981 and reflected the findings of the *Regions of Europe: First periodic report on the social and economic situation of regions of the Community* (Commission of the European Communities 1981c) which covered the period 1970–77 (Martins and Mawson 1980, 1982). It stressed that during those years, population growth had decelerated in the Nine, mainly due to falling birth rates. With rising rates of unemployment, migration flows tended to turn away from industrial cities and back to the periphery, while deconcentration of urban population became increasingly widespread. With the exception of some parts of Belgium, the highest rates of unemployment were found in the Republic of Ireland, the Mezzogiorno and south-west France. None the less, the most rapid rates of increase in unemployment involved regions that had been regarded as strong economically and had displayed very low levels of joblessness in the 1960s.

The regional patterns of employment that Thomson had identified for the late 1960s remained clearly in evidence a decade later (Fig. 12.4). The complex human conditions underlying regional disparities were crisply summarized by indicators of per capita GDP. In the late 1970s, the whole of Italy, the Republic of Ireland and the UK generated values below the EC average, as did western France and a scatter of areas in southern Belgium, the eastern Netherlands and northern Germany (Fig. 12.5a). The Mezzogiorno and Republic of Ireland were by far the poorest parts of the Nine, producing indices under half the mean. At the other extreme, prosperity was roughly six times higher in Hamburg, Bremen, the Ile de France and Greater Copenhagen than in Calabria and Donegal at the bottom of the regional league. The gap between the richest and the poorest areas was soon to widen to 1 : 10 when Greece entered the Community (and in a few

years' time was to widen even more to 1 : 12 when Portugal and Spain joined in 1986!). During the 1970s, richer regions had displayed above-average rates of growth in per capita GDP and all regions with below-average GDP had also experienced below-average rates of growth. Particularly low levels of investment worsened their already depressing positions. In absolute terms, every region raised its per capita GDP but in relative terms the rich grew richer and the poor poorer.

The 1981 revisions to EC regional policy paid attention to the implications of Greek accession and introduced the following modifications. Stress needed to be placed on job creation and raising productivity; financial support should be focused rather more on a smaller number of particularly poor regions; rather than concentrating on luring investment from outside, emphasis should be given to stimulating local potential for economic growth through manpower training and marketing schemes; and regional and local authorities should be given a greater role in handling ERDF allocations. The ERDF should function more explicitly as a development agency rather than just as a financing body. The 'sprinkler effect' of the old quota system had tended to dissipate resources and the situation had not been helped by ERDF money being used throughout the extensive assisted areas that had been designated by member countries. All this tended to make regional help from the EC somewhat haphazard. In addition, the Commission was concerned that ERDF finance had often been 'swallowed' by national treasuries for use by governments as a substitute for national funds instead of being a supplement to them.

In accordance with the 1981 revisions, the quota section of the Fund was reserved for areas with very low per capita GDP and very high long-term unemployment (Fig. 12.5b). This new 'poor man's' quota was allocated as follows: the Republic of Ireland (7.3 per cent); the Mezzogiorno (43.7 per cent); all of Greece, except Athens and Thessalonika (16.0 per cent); Northern Ireland and parts of Scotland, Wales, north and northwest England (29.3 per cent); French overseas *départements* (2.4 per cent); and Greenland (1.3 per cent). (Greenland automatically became a part of the EC when its colonial power (Denmark) joined in 1973. In 1979 it achieved home rule and following a plebiscite it negotiated withdrawal from the EC but without discontinuing relations with Denmark. Thus Greenland left the EC in February 1985, with a handsome 'golden handshake' and a generous arrangement over fish catches in North Atlantic waters. Conversely, Benelux, West Germany, metropolitan France and Denmark no longer qualified for assistance from the quota section. The non-quota section was increased from 5 to 20 per cent of the total ERDF and was for well-targeted

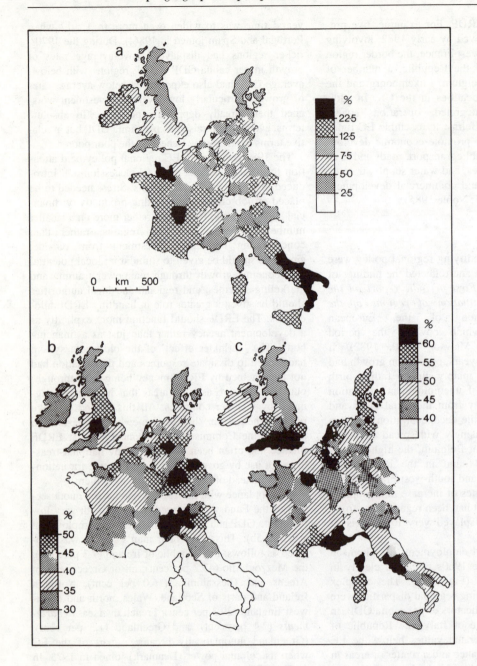

Fig. 12.4
Employment by sector,
1978 (a) primary sector;
(b) secondary sector;
(c) tertiary sector.

schemes, especially in areas harshly affected by recent industrial decline. Neither quota nor non-quota payments would be allocated to individual developments but only to well-coordinated packages of projects or to broad-based 'programmes' spanning several years, agreed between the Commission and the national governments concerned. Operations of the ERDF could now be applied more consistently and be dovetailed into overall development schemes for particular regions. Further proposals for revision were put forward in 1983 with, as we shall see, a new regulation for the ERDF being launched in 1985.

The other funds

In addition to the ERDF, four older funds continued to provide Community money for tackling social and economic problems.

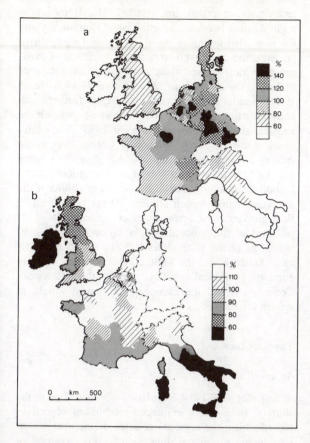

Fig. 12.5
Regional characteristics: (a) gross value added per inhabitant, 1978; (b) relative intensity of regional problems in the European Community, based on 1977 situation.

(a) The sectoral and spatial impact of assistance from the ECSC is restricted but large loans awarded to modernize production and to stimulate alternative job opportunities in areas concerned with coal- or iron-mining and steel manufacture (Collins 1975).

(b) The EIB made important loans for regional development purposes (Pinder 1978, 1986). Irrigation and other rural projects in the Mezzogiorno and numerous transport schemes received attention over many years. More recently money has been directed to industrial rehabilitation.

(c) The guidance section of the European Agricultural Fund makes awards to a wide range of modernization schemes, which include farm amalgamation, cooperatives, training workers in modern techniques, and retirement schemes for elderly farmers. Assistance is available to hill farmers and, in the

late 1970s, special programmes were started to improve farming in Mediterranean areas, to drain land in western Ireland, and to support rural renovation experiments in Italy, Ireland, the Western Isles of Scotland, the Massif Central and the Belgian Ardennes.

(d) The European Social Fund supports schemes for training and retraining unemployed workers, mainly through national training programmes, but local authorities and companies can also receive help, with higher rates of assistance available in the most seriously disadvantaged regions (Keating and Jones 1985; Croxford and Wise 1988).

In addition, a New Community Instrument was introduced in 1979 to give loans for modernizing infrastructure, developing energy resources, and boosting small and medium-sized businesses. Harsh experience has shown that EC assistance has greater beneficial effect if resources from various funds are closely coordinated. Greater attention has been duly paid to this kind of harmonization as part of the continuing reappraisal of all Community matters that impinge on regional development.

Evaluations

Even before the ERDF came into existence, distinguished economists were arguing that its operation should be merged with or closely allied to the established funds (Cairncross 1974). The point was to be repeated many times in subsequent years, with some insisting that the ERDF was inadequate and was no more than a 'cosmetic' treatment to disguise massively serious regional problems. (In 1975 the Fund had accounted for 4.8 per cent of the EC budget; by 1987 it had risen to 9.1 per cent of a much larger EC budget but was still completely dwarfed by the vast price-guarantee element of the European Agricultural Fund. To put that information in perspective, we must appreciate that the total EC budget was only about 1.2 per cent of the combined GDP of the twelve member states.)

Some academics dismissed the ERDF as a means of attempting to cope with national differences in contribution to and payment from the EC budget rather than a serious mechanism for tackling regional inequalities. Such a view of the ERDF as a system of budgetary redistribution has much to commend it. In a different vein, one eminent French politician likened the EC's regional policy to a farce that was being staged to amuse the gallery (i.e. the new members) and there is plenty of evidence that several member states were unwilling to take it seriously by simply subtracting the amount of money they received from the ERDF from the sums they would themselves have allocated to assist regional development.

The common regional policy, the ERDF and the older funds undoubtedly made some contributions to job-creation schemes and infrastructural projects at particular locations; however, their overall impact in narrowing regional disparities in the EC was less positive. Opponents argue that during the first decade of activity the modest finances of the ERDF were neither sufficiently targeted on the poorest regions nor adequately coordinated on schemes that had a reasonable chance of success. Exactly the same conclusion must be drawn from reading the Commission's own reports (Commission of the European Communities 1981c, 1984, 1987c) dealing with regional disparities and problems. Which employed successively more sophisticated indicators to describe regional conditions and demonstrated that the development gap was widening with the passage of time. This was in response to the worsening economic climate affecting Western Europe, and the inclusion of additional 'poor' regions in the new member states of the growing Community.

New neighbours: bigger problems

The three Mediterranean countries that joined the EC in 1981 and 1986 increased its total area and agricultural surface by roughly a half and its population by one-fifth.

Greece, Portugal and Spain retained sizeable proportions of their workforce in agriculture, but their farm structures were varied, with irrigated land contrasting with hill farming (Williams 1984). Average productivity was generally modest, and exportable commodities (e.g. fruit, vegetables, wine, olive oil) face severe competition from Italy and southern France. Industrial activities were equally problematic. Domestic firms were usually small, while a considerable number of the larger enterprises were owned by foreign capital and were therefore subject to decision-making which may have little to do with the welfare of people in Mediterranean areas. Many manufacturing activities were traditional, labour-intensive and of low productivity, while more modern firms tend to involve iron and steel, shipbuilding, textiles and clothing sectors which were already in dire straits in the Community. These industries face a difficult future and brought parts of Greece and Portugal and sections of the Spanish regions of Asturias, Catalonia and Aragon into the lengthening list of depressed industrial regions (Hudson and Lewis 1985).

Experience showed that structurally weak countries were much less able than rich ones to cope with dilem-

mas such as inflation, unemployment and outmoded techniques of production (Hall 1981). Removal of trade barriers between the established Community and the new Mediterranean members boosted competitive pressure and could well work to the detriment of existing activities in the 'south' and to the advantage of those in the 'north' of the Twelve (Jensen-Butler 1987). Spatial disparities increased even further, both between strong and weak countries, and between more and less developed regions (Secchi 1980). For example, one-third of the population of Greece lives in the Athens–Piraeus conurbation, which generated more than half the GDP. Over half of all Portuguese inhabit Lisbon–Setubal and four-fifths live on one-third of the national territory which produce 90 per cent of GDP. Four provinces of Spain together accommodated 30 per cent of the national population and gave rise to two-fifths of the GDP (Musto 1981). Faced with this vast challenge and the limited success of the existing common regional policy both the European Commission and the European Parliament sought to reform the ERDF.

The way forward?

An uphill struggle

A new Regional Fund Regulation was launched at the start of 1985 which embraced established objectives (coordinating the regional policies of member states; introducing a regional dimension to other policies of the Community; and providing financial aid to less favoured regions) and paid attention to the enormity of regional disparities and the modest funds available. Instead of adhering to quotas, grants would be distributed according to a system which identified the upper and lower limits of ERDF resources available to each member state. The lower limit was guaranteed on condition that enough eligible requests were sent to the Commission. Allocations above the minimum would depend on the significance to the Community of projects and programmes submitted by individual states. Three new principles were introduced for the post-1985 Regional Fund (Perrons 1992).

(a) More emphasis was placed on 'co-financing' programmes by the EC and member states, with 'programmes' being defined as coherent sets of projects within specified areas whose operations spread over a number of years. 'Community programmes' are initiated by the European Commission, while 'national programmes of Community interest' are proposed by member states but require the approval of the Commission.

The first Community programmes began in 1987 and included STAR (promoting advanced telecommunications) and VALOREN (developing the local energy potential of less favoured regions). Regulations for these programmes were adopted by the Council of Ministers late in 1986. The regions involved were Greece, Portugal, the Republic of Ireland, Northern Ireland, the Mezzogiorno, Spain, Corsica and the French overseas *départements*.

(b) Extra efforts were made to promote the indigenous development potential of problem regions rather than parachuting in manufacturing enterprises from outside. Instead of looking to external investors, the new approach gives greater support to local businesses to acquire modern technology, investigate market potential and obtain capital backing.

(c) Rather than viewing the ERDF in isolation, more effort was made to coordinate aid from the range of EC funds which could help regional development. Thus the ERDF participated in integrated schemes, particularly the Integrated Mediterranean Programmes (IMPs) which paid attention to socio-economic problems encountered in Greece, southern Italy and the southernmost parts of France, which had to face new competition from new Iberian members of the EC (Tommel 1987). The IMPs directed EC aid from various funds to job-creation schemes in agricultural modernization and off-farm jobs, in crafts, tourism, fishing and forestry. The first IMP was launched for Crete in autumn 1986 amid concern from ecologists that economically oriented IMPs might promote schemes that would cause irreparable damage to fragile Mediterranean ecosystems.

The 1985 Regulation also stressed the need to concentrate ERDF aid in the most problematical regions. Thus in the following year three-quarters of ERDF money went to the EC's 'priority regions', namely the Mezzogiorno, Greece (minus Athens), Northern Ireland, the Republic of Ireland, most of Portugal and Spain, and the French overseas *départements*. These changes represented a move in the right direction, but the challenge of narrowing the 'development gap' in the Twelve remained enormous.

The development gap

That point was hammered home in the *Third Periodic Report* from the Commission on the regions of the Community (May 1987) which incorporated information on Portugal and Spain. Analysis of socio-economic trends throughout the Twelve and calculation of a new index of regional well-being (combining GDP per capita and GDP per worker, with unemployment and changes in the size of the labour force) confirmed some disturbing facts. Regional disparities had widened during the mid-1980s; none of the regions of Iberia had an income level that approached the EC average; and conditions in 'old' industrial areas (especially in the UK and Belgium) had deteriorated. Contrasts in regional well-being could not be summarized as either north/south or core/periphery, since Ireland formed a substantial fragment of 'impoverished' Europe and old industrial areas and inner-city localities posed formidable problems within traditionally recognized 'core' regions (Fig. 12.6).

Under the Single European Act (signed 1986, operational 1987), which strengthened the Treaty of Rome, EC regional policy was incorporated in the framework of Community treaties and was assigned the task of helping convergence and cohesion. All member states undertook to conduct and coordinate their economic policies to reduce regional disparities as they worked towards these aims. The EC's drive to create a single market by the end of 1992 was deemed necessary if the Community was ever to challenge the commercial might of Japan and the USA. Such a single market would, of course, open the way for new competitive pressures throughout the Twelve and would affect the future relationship between 'rich' and 'poor' areas. Policies to assist 'disadvantaged' regions to modernize would be all the more important since the experience of recent years revealed a strong preference for high-technology, 'future' activities to cluster in localities characterized by excellent transport and communications facilities, supplies of highly skilled labour, and proximity to major research complexes and financial centres.

In July 1987 the Commission published three important recommendations, for financing Community budgets, reorganizing agricultural policies, and reforming the regional and social funds and the 'guidance' component of the farm fund (Doutriaux 1991). It proposed a regulation to come into effect at the start of 1989 to reorganize these 'structural funds' in order to concentrate on five objectives:

(1) promoting less developed regions;
(2) creating work in deindustrializing areas and frontier zones;
(3) establishing jobs in areas with high levels of long-term unemployment;
(4) promoting youth training; and
(5) stimulating agricultural and rural development.

Using the most recent statistical indicators, the 'problem regions' of the Community were redefined and labelled according to these 'objectives'. For

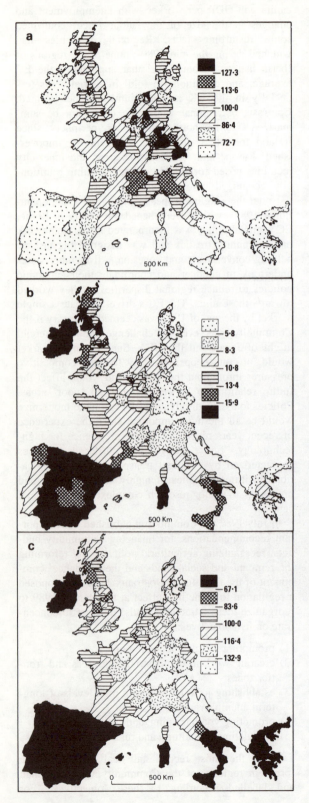

example, 'backward' (or 'objective 1') regions were characterized by incomes 25 per cent or more below the EC average. They embraced the Mediterranean and Atlantic peripheries and contained 21 per cent of the EC's total population (Fig. 12.7). Regions of industrial decline ('objective 2') displayed heavy and sustained job losses from old (and sometimes new) manufacturing, and typically had problems of dereliction and environmental pollution. Rural regions ('objective 5b') had relatively high proportions of their workforce in agriculture and needed alternative job opportunities (Perrons 1992).

The combined resources of the structural funds would be increased from 7.8 billion ECU (1988) to 13 billion ECU (1992) which would represent a quarter of the EC budget (still amounting to only 1.2 per cent of the combined GDP of the Twelve) (Murray 1992). In line with earlier thinking, the funds would be allocated to a series of multi-annual 'programmes' rather than dispersed among briefer projects. The thrust of the structural funds would be coordinated with that of the EC's other financial instruments, such as the EIB. Programmes would be prepared as a result of dialogue between local authorities, regions, member states and the Commission, rather than being devised in a 'top-down' manner. Use of allocated funds would be audited and attempts made to harmonize the regional policies of member states. In addition, more EC initiatives were launched to help tackle specific problems in areas dominated by steelmaking, shipbuilding and coalmining, to assist border areas, rural zones and the 'ultra-periphery' (French Caribbean territories and Spanish Atlantic islands), and to promote science, technology and advanced telecommunications (Hall and van der Wee 1992; Marques 1992).

The fourth periodic report, entitled *The Regions in the 1990s* (January 1991), provided a comprehensive review of conditions for 1988–90. Not surprisingly, it identified great spatial disparities in wealth, with the ten least developed regions (in Greece and Portugal especially) displaying mean incomes that were less than one-third of the average for the ten most advanced regions (in the West European 'mega-core'). Rates of unemployment in 1990 varied from approximately 2.5 per cent in the most favoured regions to roughly 22 per cent in the most problematic (especially in Spain and southern Italy). Although declining throughout the Community, birth rates remained relatively high in

Fig. 12.6
Regional characteristics in the Twelve, 1985–86; (a) GDP per head of population in purchasing power parities, 1985; (b) total unemployment rate, 1986; (c) synthetic index of regional disparity according to the *Third Periodic Report*.

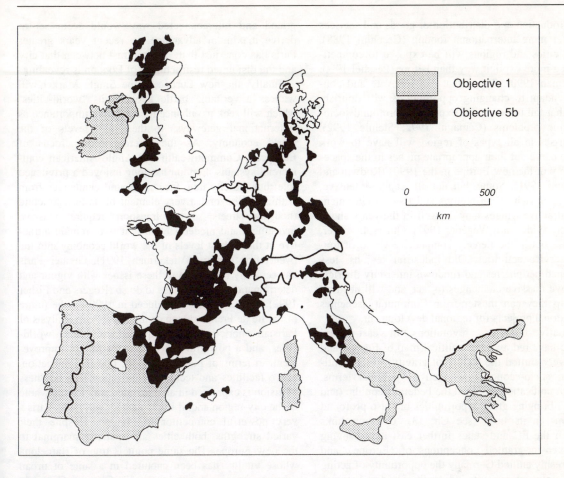

Fig. 12.7
Areas qualifying for assistance under the revised structural funds.

many regions characterized by high levels of unemployment. Of course, those areas still had to cope with earlier cohorts of young people on their problematic labour markets. With the exception of flows between the Republic of Ireland and the UK and between certain borderlands (such as the Rhinelands), there was not a great amount of migration between member states. However, family members of labour migrants continued to enter the EC and the number of asylum-seekers grew sharply (see Chs. 3 and 4). As well as movements from parts of the Third World, large numbers of people were moving from Eastern and east-central Europe to the EC. For example, out of an estimated inflow of 1 200 000 to the Community during 1989, roughly 1 million involved moves to West Germany from the German Democratic Republic and other (then) socialist countries. As well as these eminently quantifiable expressions of regional disparity, the Fourth Report identified spatial variations in telecommunications and transport infrastructure; in labour force qualifications and facilities for training and retraining; in availability of local credit facilities to support innovations; and in the existence of R & D activities, which were particularly concentrated in northern regions of the Community. Finally, attention was drawn to regional variations in the pace of sectoral shift to employment in service activities.

New challenges

The pattern of regional disparity in the EC will continue to evolve in the years ahead in response to at least four sets of processes. The first of these will be the completion of the Single European Market after 1992 (see Ch. 13) which will not only give rise to major changes in internal and external perceptions of what the (West) 'European' market involves but will set both production and distribution of goods and services on a far more international footing (Cecchini 1988). Firms, cities and regions will be exposed to competition in a more explicit way than before (Peschel 1990; Vickerman 1990). Those that can innovate and constantly adapt to changing opportunities will flourish; those that fail to do so will experience profound socio-economic problems (Camagni 1992; Steinle 1992). Enterprises in all types of region will have to work hard to carve out their appropriate niches in the space-economy of the new Europe in the 1990s (Rodwin and Sazanami 1991). Some, but not all, capital-intensive, information-rich urban regions and their environmentally attractive fringes may do well in the years ahead (Masser, Svidén and Wegener 1992). Others, however, will lose out in the fiercely competitive world of top-level service activities. Old industrial regions, less developed peripheries and rundown inner-city districts will have massive challenges to face and will need all the help they can muster from Community schemes and national projects for regional development.

Virtually bloodless revolutions in east-central Europe have redrawn the political map of Europe and effectively shifted the 'centre of gravity' in the continent in socio-economic as well as political terms. German unification enlarged the Federal Republic (and the EC) bringing great opportunities but also profound problems in its train (see Ch. 13). Growing links between the EC and states further east are changing our taken-for-granted conceptions of 'Europe', and offer freshly unified Germany the opportunity of acting as the point of contact and trading bridge between 'East' and 'West'. The spatial vacuum that existed for forty-five years in European affairs as a result of the Iron Curtain slicing through *Mitteleuropa* will be filled by the new Germany whose east/west transportation facilities will have to be improved. Realization of that scenario will strengthen the importance of Frankfurt, Berlin and, to some extent, Vienna, and will emphasize the marginality of the British Isles, Iberia and western France in the emerging European space-economy.

Transport innovations will play an important role in reorganizing the mosaic of regional inequality in Western Europe. New motorways, modernized ports and airports, high-speed train services, and innovative links between land masses will be of vital significance in shaping and reshaping the regional geography of unequal opportunity. The Channel Tunnel will offer great potential for the efficient movement of passengers and goods between the UK and the European mainland, but the spatial marginality of the British Isles and other parts of the Atlantic rim cannot be denied. Locations where provision of several new modes of transportation has been planned in an integrated and interconnected way would seem set to derive maximum advantage. In recent years greater Paris has done just that; the contrast between that city and the disjointed restructuring of London is revealing.

Finally, the new Europe of the Single Market will witness a veritable battle between its 'world-cities' which will seek to enhance their relative importance as 'international gateways' to the upper levels of the world economy. As the interconnecting focus of European, Commonwealth and Anglo-American commercial systems, London has long enjoyed a privileged role, but the city now faces profound challenges from within and beyond. Every element of its infrastructure (housing, streets, sewers, transport) requires massive investment and modernization if it is to remain attractive at the highest levels of the world economy into the twenty-first century (Vickerman 1991). Greater Paris has certainly responded to these issues with vigour and determination and continues to do so (Rogers and Fisher 1992). The *Livre Blanc* produced in 1990 for the future of the Ile de France region contained a sharp analysis of Parisian shortcomings, compared with other world-cities, and a powerful exhortation for further improvement, in terms of public transport, housing, education, office facilities and technical innovation.This boldness of vision is expressed in an integrated way for the whole of the city-region and, if implemented, will make Paris a very powerful competitor to London. Despite their varied strengths, both cities are spatially marginal to the new Europe. The same point is true of Barcelona whose vitality has been captured in a suite of urban schemes associated with hosting the Olympic Games in 1992. Frankfurt is already the rising star of international banking and financial functions, but it is Berlin that is the truly unknown quantity in the reshaping of urban Europe. Well placed in the heart of the new Europe, well supplied with surrounding land for development, and functioning as political capital of the strongest economy on the continent, its chance for success would seem very considerable when the world economy revives.

The enormity of recent changes in both 'Western Europe' and the 'new Europe' that is overtaking it ensures that we appreciate the ongoing nature of regional disparity. 'Regional problems' can never be solved – they keep on being created and redefined in differing ways! The best that can be hoped is that the starkest human inequalities as the heart of regional differences may be lessened and the development 'gap' reduced in however small a way. The experience of recent decades demonstrates with harsh clarity that – in both political and economic terms – every step along the road towards that goal will be a steeply uphill struggle (Molle and Cappelin 1988).

Further reading

Bachtler J, Clement K 1992. 1992 and regional development. *Regional Studies* **26**: 305–419

Clout H D (ed) 1987 *Regional Development in Western Europe* 3rd edn. Fulton, London

Hudson R, Lewis J R (eds) 1982 *Regional Planning in Europe*. Pion, London

Keating M, Jones B (eds) 1985 *Regions in the European Community*. Oxford University Press, Oxford

Knox P L 1984 *The Geography of Western Europe*. Croom Helm, London

Pinder D 1983 *Regional Economic Development Policy: theory and practice in the European Community*. George Allen and Unwin, London

Wise M, Gibb R 1993 *Single Market to Social Europe: the European Community in the 1990s*. Longman, London

13

Western Europe and the 'new Europe'

The coming of the 'new Europe'

The concluding sentences of the second edition of this book, which were written in 1988, noted growing cordiality between Moscow and Washington and looked forward to closer links between European neighbours across the Iron Curtain (Ante 1991; Ritter and Hajdu 1989). In the very next year a largely peaceful and virtually bloodless revolution swept through the states of east-central and Eastern Europe, dismissing administrations, rejecting Communism, and opening the way for economic reform and the espousal of capitalist principles (Murphy 1991). On 9 November 1989 the Berlin Wall was breached and following free elections in East Germany on 18 March 1990, the territory of the German Democratic Republic was formally incorporated into the Federal Republic on 3 October 1990. This brought a further 16 400 000 people into the EC and extended its territory by 108 000 km². The two Germanies had already recognized the eastern border with Poland and given up claims to 114 550 km² to the east of the rivers Oder and Neisse that had been held by Germany before the Second World War (Harris 1991). In July 1990 ministerial representatives of the four wartime allies (France, UK, USA, USSR), the two Germanies and Poland had confirmed that any suggestion that the German–Polish boundary was provisional would be removed from German laws. Four months later the foreign ministers of Germany and Poland signed a treaty guaranteeing that border.

Further to the east the openness of *glasnost* stimulated not only reform but also political fragmentation across the massive territory of the USSR, as first the three Baltic states and then the other republics declared independence. By the end of 1991 the USSR had ceased to exist. Bitter strife ripped apart the patchwork quilt of Yugoslavia and added more names to the list of the world's independent states. At a faster pace than almost anyone would have predicted, the political and economic geography of east-central and Eastern Europe was transformed. The taken-for-granted ideological and spatial distinction between Communist East and capitalist West had disappeared, along with the Iron Curtain and the Berlin Wall, to be replaced by a swathe of countries basically committed to individual rights, capitalism (tempered by state intervention), and peace with neighbouring states. (At the time of writing, the bloodbath of Yugoslavia was a tragic exception to this harmonious scenario.) In a less dramatic way, the member states of EFTA were seeking closer contacts with their twelve neighbours in the EC which, in turn, were working to complete the Single Market by the end of 1992. At the same time, the number of states seeking full membership of the EC continued to grow.

These momentous changes provide some of the key features of the 'new Europe' which is taking shape in the final decade of the twentieth century and whose characteristics must be acknowledged in the present volume, despite the fact that its general focus remains on the eighteen 'western' states of the continent. Processes of change and continuity with the past are not simply superimposed across Europe but are interwoven in a very complex way. Scratch the surface of the 'new Europe' and one will still find in place many of the old divisions of wealth, work habits, language, culture and nationalism (*Economist* 1992). German unification is involving painful and costly adjustments in both parts of the country, which displayed many important socio-economic contrasts. After their largely bloodless revolutions, the states of Eastern and east-central Europe are evolving their own distinctive routes towards capitalism. Not surprisingly, popular visions of the future of the EC are not entirely positive, hence '1992' is surrounded by fears as well as hopes. These important themes will be explored in more detail in the paragraphs that follow.

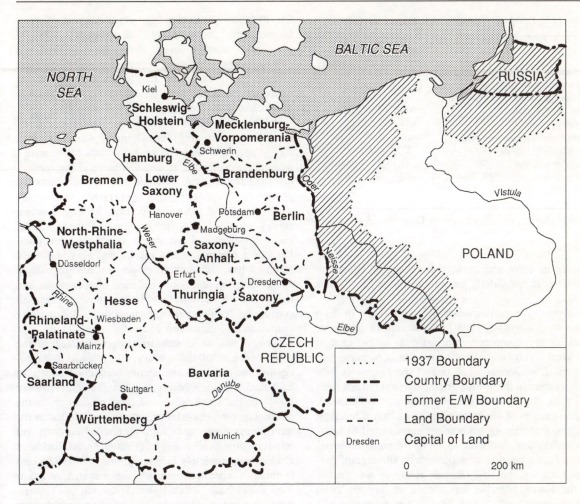

Fig. 13.1
The new Germany and surrounding states.

The legacy of eastern Germany to Western Europe

The former German Democratic Republic comprised the east-central section of inter-war Germany whose territory had extended much further eastward into what was to become part of Poland and a tiny section of the USSR (Fig. 13.1). With some 16.4 million inhabitants at the end of 1988 the population of East Germany was roughly a quarter of that of West Germany but its land surface was almost half that of its western neighbour (Table 13.1). Average population density (152 per km^2) was fractionally higher than that of the EC twelve (144 per km^2) but was substantially lower than that of West Germany (249 per km^2). With the exception of Berlin, industrial activities and main urban settlements were concentrated in the south of the

Democratic Republic, with half the state's population living in the cities of East Berlin, Chemnitz (Karl Marx Stadt between 1950 and 1990), Dresden, Halle and Leipzig (Commission of the European Communities 1992b). By contrast, a quarter lived in rural settlements of fewer than 2000 residents apiece. The age structure of East Germany was younger than that of the West, with 24 per cent of the eastern population under 18 years of age compared with 19 per cent in the West. All other age groups were somewhat smaller than in West Germany, especially those over 60 years which accounted for only 18 per cent of the eastern population, compared with 23 per cent in the West. However, younger, well-qualified people and their children formed the major component of the 600 000 migrants who were to move to the western

Table 13.1 Economic indicators for East and West Germany

	Former E. Germany	Former W.Germany
Population 1989 (million)	16.4	61.7
Population change 1989/1970(%)	−3.5	+1.1
Area ('000 km^2)	108	249
Density per km^2	154	248
Workforce in agriculture (%)	10.8	4.9
Workforce in industry (%)	47.1	40.1
Workforce in services (%)	42.1	55.0
Unemployed, Dec. 1990 ('000)	642	1 748
Short-time workers, Dec. 1990 ('000)	1 794	50

Source: Commission of the European Communities (1990).

Länder in 1989 and 1990, thereby accentuating the middle-aged and elderly proportions in the East (Jones 1990).

Labour-force participation was very high in East Germany, embracing almost 90 per cent of people of working age, compared with just over 60 per cent in the West. With 86 per cent of women of working age in jobs, East Germany had one of the highest female activity rates in the world. Unemployment did not exist officially but the underuse of labour was widespread. Three-quarters of the eastern labour force had received professional training but was not accustomed to western technology, attitudes to work or entrepreneurial skills. Partly because of a permanent shortage of foreign currency, the East German economy was geared to the lowest possible dependence on imports from Western countries. A low level of specialization was the result. Compared with Western industrialized countries, the structure of the East German economy had changed little in recent years. In 1988 47.1 per cent of the workforce of 9 million had jobs in manufacturing and construction, 10.8 per cent in agriculture, and only 42.1 per cent in services, giving East Germany a smaller share in service-sector employment than any other member of the EC (Wild 1992). Only a tiny fraction of the workforce was in the private sector, mainly in trade, handicrafts and repairing.

Three-quarters of East Germany's farmland was devoted to arable production and ranged from excellent soils to very poor sandy ones. Some 95 per cent of the agricultural land was worked by 4750 farms (including 465 owned by the state and 3855 operated by cooperatives). Most specialized in either crop production or animal husbandry, with arable farms averaging 4500 ha and dairy herds averaging 750 cows apiece. By virtue of their socialized organization they were vastly larger than their western counterparts (see

Ch. 9). Crop yields averaged three-quarters of West German ones, and output of milk and eggs reached 90 per cent of western levels. Fertilizers, pesticides and other agro-chemicals were important in permitting these results. Performance of the agricultural processing sector was weak and declined during the 1980s because of a lack of investment to modernize equipment.

Producing 310 000 000 tonnes of lignite (one-quarter of the world output) East Germany was the world's largest source of brown coal. Two-thirds of the state's production originated around Cottbus and the remainder from the environs of Halle and Leipzig. Some 85 per cent of electricity was generated from lignite which was used for most domestic heating. Nuclear energy provided one-tenth of East Germany's electricity requirements but safety standards were poor. Per capita consumption of energy was very high, as a result of low efficiency of power stations, unrealistic energy pricing, and poor insulation of apartments and houses. Power stations, factories and domestic fires, all consuming brown coal, were responsible for massive outputs of pollution to air and water.

Energy and basic industries were focused in the south and east of the country. Electronics, data processing and precision engineering were concentrated in East Berlin and southern cities, textiles and consumer goods in the south-west, and food processing in the north. The Siemens–Martin technique prevailed in the steel industry and made for high production costs and relatively low quality output. (That technique had been abandoned completely in West Germany in 1982.) East Germany's chemical industry concentrated on coal-fired works built before 1940. Production of synthetic materials and fertilizers would require considerable modernization to compete on the world market. Engineering and vehicle manufacture lagged behind West European standards but conditions were more competitive for precision engineering and optics. Investment in micro-electronics had been very high, with the objective of commanding that niche for all the COMECON countries. Textile production concentrated on the mass market where developing countries were strongly competitive on the world market. Food products lacked variety and were of relatively poor quality.

East German industrial productivity was low, with some commentators placing it no higher than one-third of the West German level, but conditions varied considerably between different sectors. Such low efficiency was due to the stifling weight of bureaucratic central planning, lack of incentives and consequent low motivation, and the use of outdated technology. The latter problem had become more intense during the 1980s as investment fell. East Germany's integration in COMECON and modest participation in world trade further contributed to the inefficiency of its economy. In 1989 East German GNP per capita was less than two-thirds that of the Federal Republic. According to some commentators, perhaps one-third of industrial production was competitive on the world market, a further third stood no chance of survival, while the remainder might be brought up to western standards. Other estimates were more pessimistic.

Not surprisingly, transport and other facilities were particularly concentrated in the southern part of East Germany and would require heavy investment for repair and modernization (Mellor 1992). By contrast, northern, predominantly rural, areas were poorly provided with infrastructure. Steam trains formed the most important means of transport throughout East Germany, having been given preferential treatment in order to save on oil consumption. The existing rail network needed repair and modernization, as did the road and telecommunications systems. Only half of households were connected to sewage treatment plants and the volume of sulphur dioxide emissions was more than twice that coming from West Germany. Many eastern rivers were polluted and forests were damaged by pollutants coming from Czechoslovakia and Polish Silesia as well as from East Germany itself (Elkins 1990). For example, the river Weser was polluted with salts from East German potash mines, and the Elbe reached the West with a worrying content of heavy metals. In some areas provision of pure drinking water posed serious problems. None the less, living standards were higher in East Germany than in other Eastern bloc countries but were well below those of West Germany. For example, in 1988 only 52 per cent of East German households had a car (compared with 97 per cent in West Germany), 52 per cent had a television (94 per cent) and a mere 9 per cent had a telephone (98 per cent). Living space was a quarter less than in West Germany and was often of poor quality. Per capita income in East Germany was probably higher than in Ireland, Greece or Portugal but was below that of Spain.

The German Democratic Republic was poorly integrated into the world trade system, with over two-thirds of transactions being with other COMECON countries, especially the USSR. Exports consisted largely of machinery and equipment, while imports included a high proportion of energy products and raw materials. Modest trade flows with the EC (see Table 13.3) showed East Germany to be a net exporter of energy- and labour-intensive products and a net importer of goods with high raw materials, R & D, and technology contents.

The process of unification

Such was the legacy of East Germany to the 'new Germany' that came into being on 3 October 1990. An earlier treaty (18 May 1990) had dealt with economic and financial issues. It was agreed that on 1 July 1990 West Germany would exchange Deutschmarks for less valuable East German marks and in so doing would allocate a substantial subsidy to the East. Thereafter the DM was the single currency in both parts. Basic rules governing market economies were duly accepted in eastern Germany, involving freedom of contract between economic agents, abolition of fixed price levels, and introduction of private property rights. West German rules affecting public finance, pensions, health, insurance and employment were adopted in the East.

The economy of the eastern part of Germany formally abandoned its protected existence, which had involved state ownership, centralized planning and commitment to COMECON, and started along the challenging route of exposure to competition from western Germany, the EC and the wider world (Peschel 1992). Unification on 3 October 1990 conferred certain advantages on the eastern *Länder* in their transition to a market economy. They had become part of the strongest economy in the EC and would benefit from inter-regional financial transfers as well as grants and loans; for example the German Unity Fund was set up by the West German Finance Ministry and the *Länder* to help finance the eastern budget from 1990 to 1994. In addition, the eastern part of Germany qualified for financial help from the EC. Transitional arrangements were made to ease East German integration into the EC but it was clear that many East German industries would have massive difficulties adjusting to the Community system, especially with regard to safety, quality standards, environmental legislation and structural policy.

The European Commission explored the implications of German unification for the EC and a score of transitional measures were identified involving agriculture, energy, structural policies, social affairs, environment and nuclear safety. Fundamental changes would be required to integrate East German farming into the CAP, including abolition of central planning and introduction of new pricing structures. For example, substantial reductions in milk production would be required during 1991. With respect to the environment, production and consumption of lignite would have to be cut and special technology introduced in power stations to reduce pollution by sulphur dioxide, nitrogen dioxide and dust. A 50 per cent reduction in the use of brown coal by 1989 was recommended, with

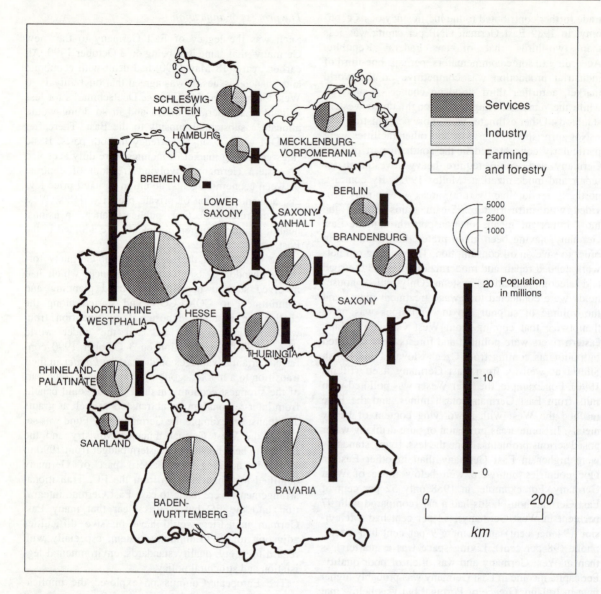

Fig. 13.2
Germany: employment structure, by *Land*.

substantially greater reliance to be placed on imported natural gas. Allocation of 3 billion ECU under the structural funds was proposed for 1991–93 to assist eastern Germany whose needs would then be incorporated into the revised distribution of Structural Funds. The EC legislation on the free movement of workers, equal treatment of men and women, job protection and vocational training took effect in eastern Germany in late 1990. Education and training procedures are being changed to the West German systems. Environmental

analyses in 1990 revealed serious pollution of soil, air and water, uncontrolled waste disposal, and dangerous nuclear installations. The latter problem required immediate action to improve protection against radiation or else nuclear plant would have to be closed. Against this background of economic and environmental problems the complex integration of the eastern *Länder* into federal Germany has been taking place (Fig. 13.2).

The uncompetitiveness of most aspects of the eastern economy became strikingly apparent during 1991. A massive increase in joblessness followed the sudden shift to the market economy. Large numbers of plant were recognized as unviable and were closed; surplus jobs in administration, education and, indeed, every aspect of activity were axed. The high level of female employment was reduced to a shadow of its former self. In just a matter of months the unemployment rate rose from zero to 30 per cent (by April 1991) and many workers were placed on short-time contracts. Large numbers (600 000 during 1990) moved to the western *Länder* in search of new opportunities, despite severe housing shortages in many reception areas; some 200 000 commuted on a daily basis from their homes in the east to jobs in the west (Roesler 1991). Anti-immigrant sentiments surfaced in a particularly virulent form in eastern and western cities, at this time when jobs and housing were at a premium (see Ch. 3).

Everything owned by the East German state is to be privatized, ranging from coal mines and factories to hotels, restaurants and shops. The Berlin-based Treuhandanstalt is supervising the disposal of 10 000 state enterprises, one-third of which had been sold by late 1991, mostly to West German investors. Unfortunately the majority of its offers are hardly sound propositions, with industrial plant characterized by antiquated equipment and environmental pollution. As early as December 1990 the federal government decided to close nuclear-power reactors on the Baltic coast (four at Greifswald and one at Rheinsberg) since they failed to meet minimum safety standards. Over a million claims have been lodged for land and buildings that had been confiscated during the Communist regime to be returned to their previous owners. The federal government has introduced generous tax incentives to encourage businesses to set up in the eastern *Länder* and has pumped in billions of DM to improve local administration, road and rail systems, and to deal with the most severe environmental problems. The Treuhandanstalt hopes that foreign firms, especially Japanese ones, will start to buy during 1992.

The eastern *Länder* have held varied appeal for investors. Brandenburg has proved very attractive to developers seeking sites for new housing, offices, factories and leisure activities in the hinterland of Berlin (see Ch. 8). Plenty of work is also under way in Thuringia and Saxony-Anhalt but the southernmost *Land* of Saxony is witnessing particularly intense activity. Attractions for West German investors include the region's strong industrial tradition, embracing products as diverse as vehicles, porcelain, toys and optical goods, and the presence of reasonable transport facilities, although Saxony's autobahn and rail networks, two airports, and telecommunications

systems need great investment to bring them up to modern standards. Technical institutes in the south are promoting R & D and new universities are being created. Major investors include Volkswagen (building car plants near Zwickau), Siemens and AEG (electrical), Quelle (mail order) and a range of large banks and insurance companies.

The revival of Saxony is benefiting from the rivalry between its three very different major cities. Chemnitz is rebuilding its manufacturing base; Dresden is developing service industries and restoring its war-torn baroque heritage; and Leipzig is investing heavily in its relocated trade fair. These three cities and the triangle of land between them form the main focus for investment. But, of course, there are difficulties. In autumn 1991 Saxony had 290 000 completely unemployed workers (11 per cent unemployment rate) and a further 490 000 on short time. Petty crime and attacks on foreigners sour social relations and environmental degradation includes air pollution blowing in from Czechoslovakia and the deadly legacy of the Wismart uranium-mining site, where waste disposal and radiation protection were feeble (Carter 1985, 1990). Despite these problems, Saxony is reckoned to have the strongest chance of economic revival outside Berlin and is particularly well positioned for developing trade links with east-central and Eastern Europe in the decades ahead.

In 1990 and 1991 unity produced a boom in the western *Länder*, as they rushed to supply goods to the former East Germany, but gave rise to a profound slump and mass unemployment in the eastern provinces. But during 1991 some industrial plant in the east improved the quality of their products and began to market them to western standards. The price of democracy and the start of economic revival in the eastern *Länder* has been a vast transfer of capital from west to east to cover a multitude of costs. One-quarter of the federal budget is earmarked to pay to the east, for social security, infrastructure, and merging the two armed forces. The German Unity Fund and the Treuhandanstalt require financing; the federal post office is investing heavily to modernize telecommunications in the eastern *Länder*; and dilapidated housing is being modernized by the Reconstruction Loan Corporation. Germany is committed to paying the USSR for withdrawing and resettling Soviet troops and for ensuring that trading obligations concluded by East Germany are honoured. Compensation must be paid to former property-owners in the east and environmental pollution must be cleaned up, even though costs are incalculable. Add to all this the cost of shifting the federal government from Bonn to Berlin and it is not surprising that many residents of the western *Länder* fear

Table 13.2 Basic indicators of the East European countries

	Albania	Bulgaria	Czecho-slovakia	East Germany	Hungary	Poland	Romania	Yugo-slavia	USSR
Population, millions (1988)	3.3	9.0	15.6	16.6	10.6	38.0	23.0	23.6	286.4
GDP, billion US$ (1988)	6	51	119	155	69	207	95	100	1 590
GDP, per capita, US$	2 000	5 633	7 605	9 361	6 491	5 453	4 117	4 240	5 552
Annual growth of GDP 1981–85 (%)	*	0.8	1.2	1.9	0.7	0.6	–0.1	*	1.7
Annual growth of GDP 1986–89 (%)	*	1.9	1.5	1.7	1.5	1.0	1.0	*	2.3
Cars per '000 population	2	127	182	206	153	74	11	122	50
Telephones per '000 population	*	248	246	233	152	122	111	179	124
Radios per '000 population	170	220	256	660	590	290	290	235	690
Televisions per '000 population	85	190	285	750	402	263	175	210	310
Workforce in agriculture (%)	49	20	12	10	18	28	29	35	22
Workforce in industry (%)	40	47	48	50	38	37	45	39	39
Share of private enterprise in GDP (%)	*	*	3	4	15	15	3	13	3
Workforce with secondary education (%)	*	*	29	*	34	29	*	63	27
Birth rate (‰)	25	13	13	13	12	17	16	14	18
Death rate(‰)	6	12	12	14	13	10	11	9	10
Area ('000 km²)	29	110	130	108	90	310	240	256	22 402
Population under 15 (%)	32	22	24	19	21	26	25	23	26
Density of population per km²	114	81	122	154	114	120	97	92	13

*no data available

that all this investment in the east will jeopardize their own employment prospects and economic viability in the future.

The peaceful revolution

Until 1989 the states to the east of the Iron Curtain were largely unknown territory to most West Europeans (Table 13.2). Scarce mention was made of their affairs in Western newspapers or radio and television broadcasts; few West Europeans visited them. For more than a quarter of a century most Westerners perceived them simply as satellites of the USSR, lacking in individuality, characterized by rigid political structures and low standards of living, suffering from a lack of hard currency, and producing poor quality manufactured goods (Fig. 13.3). This image was not entirely false. By the mid-1980s the Eastern economies were approaching a state of collapse; investment in factories and equipment had been starved in recent years; the infrastructure suffered neglect; and environmental pollution was running at dangerous levels. However, this image ignored fierce pride in nationality, language, culture and religion. Unlike most of Western Europe, the states of east-central and Eastern Europe are creations of the present century. Their boundaries result from settlements concluded at the end of each world war, bringing together distinct cultural groups in single

states and giving rise to sizeable ethnic minorities in many areas (Fig. 13.2). For example, there are 2 600 000 ethnic Hungarians in Romania; Greek minorities in Albania, Yugoslavia and Bulgaria; Albanians in Yugoslavia; Turks in Bulgaria; Romanians in the Ukraine and especially in Moldavia (where they make up 65 per cent of the population); Ukrainians and Belorussians in Poland; German communities in Czechoslovakia, Hungary and Romania; and important Russian minorities in the Baltic republics (Merritt 1991). Most complex of all is the diverse array of peoples within the borders of Yugoslavia (Fig. 13.4).

West Europeans are no longer so ignorant of the 'other' Europe. Rarely has a daily newspaper or a new programme been produced since 1989 without reports on the achievements and failures in the states of east-central and Eastern Europe (Sword 1990). Their largely peaceful revolution had its first major expression in the legalization of Solidarity in Poland in spring 1989 and was followed by its resounding success in partially free elections in June. During the summer months Hungarians were surprisingly critical of their government to Western visitors and by July sections of the 'Iron Curtain' between Hungary and Austria were being opened. East German tourists travelling through Czechoslovakia to Hungary made use of this new escape route to the west. The position of Solidarity continued to strengthen and by October powerful criti-

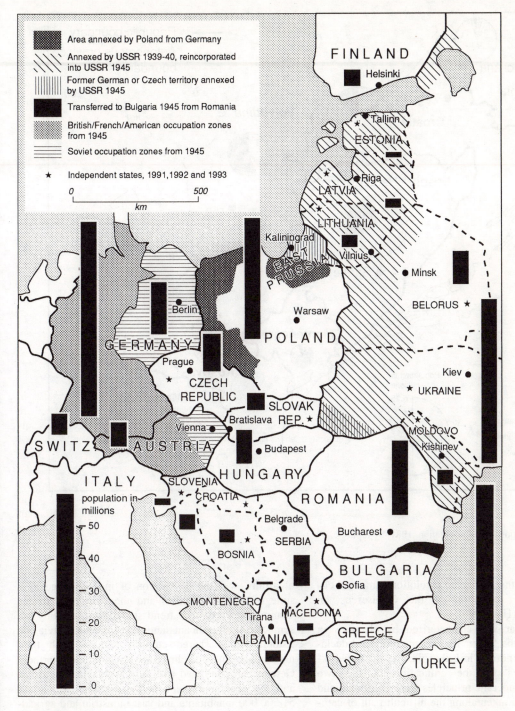

Fig. 13.3
East-central Europe.

cisms of the East German regime led to removal of its
leader. During November the Berlin Wall was
breached, the Bulgarian leader was deposed, and the
government of Czechoslovakia resigned. Next month a
new federal government was elected in
Czechoslovakia, the East German leadership

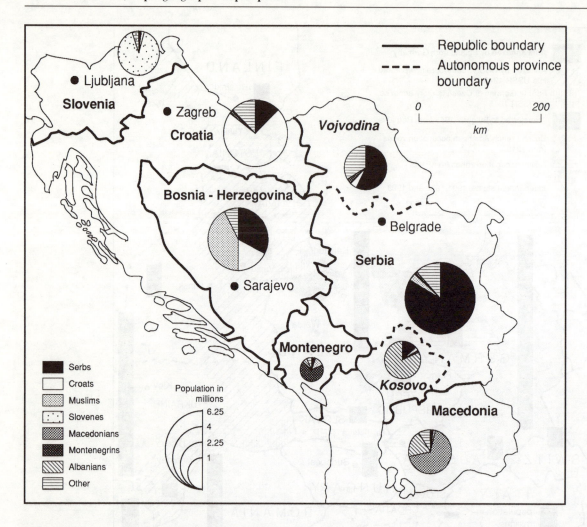

Fig. 13.4
Yugoslavia: composition of population, 1981, by republic.

foundered, and the Romanian revolution came to a climax with the execution of the Ceausescus on Christmas Day (Hall 1990).

The next year was characterized by elections in many states of the 'other' Europe, the unilateral declaration of independence in Lithuania, and the outworking of policies for the unification of the two Germanies. Democratic principles were reinforced and early steps were taken along the difficult path of capitalism. Following a year of consolidation, 1991 saw the fragmentation of Yugoslavia, with Slovenia, Croatia and Macedonia each declaring independence, and a tragic and highly destructive civil war being fought between Croatia and Serbia (Fig. 13.5). Far less predictable, but far more significant in the world order

of affairs, was the declaration of independence by a suite of Soviet republics and the overthrow of the Communist Party. The trend started with the tiny but symbolically significant states of Lithuania, Latvia and Estonia which shared cultural features with their neighbours in Scandinavia, Poland and northern Germany. It continued with tiny Moldavia, Belorussia and the economic giant of the Ukraine which has 51 500 000 inhabitants and vast industrial and agricultural reserves. Before 1991 had ended the USSR had ceased to exist; and in the spring of 1992 the former Yugoslav republic of Bosnia-Herzegovina was recognized as independent by the EC and the USA.

Commercial contacts between the EC and east-central Europe started some twenty years ago but the

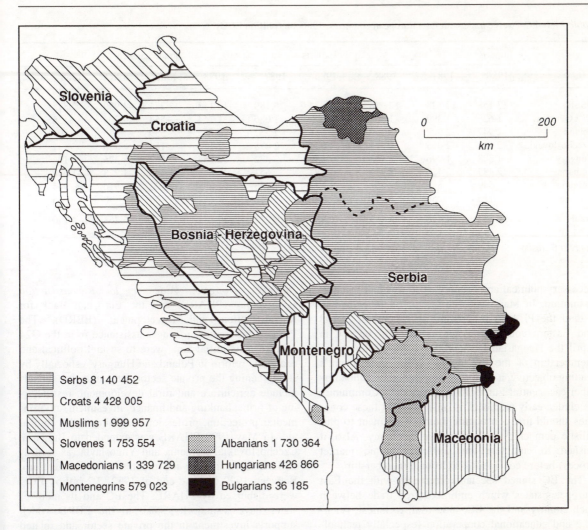

Fig. 13.5
Yugoslavia: majority populations, 1981.

Legend:
- Serbs 8 140 452
- Croats 4 428 005
- Muslims 1 999 957
- Slovenes 1 753 554
- Macedonians 1 339 729
- Montenegrins 579 023
- Albanians 1 730 364
- Hungarians 426 866
- Bulgarians 36 185

USSR insisted that these be channelled through COMECON. The EC offered to conclude trade agreements with individual countries and in 1980 a contract was signed with Romania. Accords restricted to steel and textiles were concluded with the other countries. Patterns of trade for 1986–88 between the EC twelve and East European countries are shown on Table 13.3. Not until June 1988 did the EC and COMECON agree to establish formal relations. This declaration opened the way for individual trade agreements concluded with Hungary and Czechoslovakia in 1988, Poland and the USSR in 1989, and East Germany and Bulgaria in 1990. The agreement with Czechoslovakia was revised in May 1990.

As the peaceful revolution was taking place in 1989 so the EC attempted to define its new working relationship with Eastern Europe. In July the G7 summit in Paris allocated the EC the task of coordinating Western aid to Poland and Hungary via the PHARE programme. During November EC foreign ministers agreed a common strategy for easing COMECON restrictions on high-technology exports to Eastern Europe. In December the EC agreed to create a European Bank for Reconstruction and Development to provide loans to Eastern Europe. In the same month the group of 24 met in Brussels to define their aid programme for Poland and Hungary and expressed willingness to help other East European countries once

Table 13.3   EC Twelve trade with East European countries, 1986–88 (million ECU)

	Imports			Exports			Trade balance		
	1986	1987	1988	1986	1987	1988	1986	1987	1988
USSR	13 158	13 128	12 988	9 874	9 189	10 113	–3 284	–3 939	–2 875
East Germany	1 626	1 390	1 400	1 072	1 086	1 264	–554	–304	–136
Poland	2 947	2 907	3 359	2 388	2 332	2 755	–559	575	–604
Czechoslovakia	2 108	2 055	2 211	1 944	2 078	2 170	–164	23	–41
Hungary	1 888	1 996	2 158	2 450	2 372	2 354	562	376	196
Romania	2 483	2 429	2 234	987	651	614	–1 496	–1 778	–1 620
Bulgaria	549	517	461	1 472	1 453	1 406	923	936	945
Albania	125	56	72	65	56	67	-60	0	–5
Total	24 884	24 478	24 883	20 252	19 217	20 743	–4 632	–5 261	–4 140

Source: Eurostat.

necessary political and economic reforms had been set in motion. In May 1990 the Group of 24 agreed to extend the PHARE programme to other east-central and Eastern countries. After its election victory in April the Hungarian Democratic Forum declared that membership of the EC would be the government's leading objective, a wish that would soon be echoed by other east-central European countries. The Community precluded early membership, insisting that these countries should first consolidate their commitment to pluralistic democracy and the market economy. Also, it wished to complete the single economic market process before considering additional membership.

The EC signed special agreements with the East European states which embraced free trade between each country and the EC; industrial, technical, scientific and educational cooperation (especially technology transfer and direct investment); a programme of financial assistance via subsidies and loans; and the operation of political dialogue. These agreements did not represent a transitional phase towards membership of the EC but they did not exclude the possibility of application. Thus in mid-December 1991 the 'troika countries' (Poland, Hungary and Czechoslovakia) each signed special 'Europe agreements' with the EC and emphasized their objective of eventually joining. These special agreements will lead to free trade over ten years, with the EC lowering its barriers to industrial imports within five or six years. Over the period 1988–90 Poland's exports to the EC had already risen by 53 per cent, Hungary's by 27 per cent, and Czechoslovakia's by 22 per cent. Over the same period Poland's imports from the EC had risen by 59 per cent, Hungary's 22 per cent, and Czechoslovakia's 17 per cent.

As agreed in 1989 the EC coordinates Western aid efforts to east-central and Eastern Europe through

Pologne, Hongrie, Aide à la Reconstruction Economique (PHARE) and the European Bank for Reconstruction and Development (EBRD). The PHARE programme manages assistance from the G24 countries. Its initial aims were to sustain political and economic reform in Poland and Hungary, especially by strengthening the private sector. Key areas for support include agriculture and rural development, restructuring of firms, banking and finance, investment, environmental protection, professional training and technical aid. In May 1990 PHARE was extended to Bulgaria, Czechoslovakia, Romania and Yugoslavia (as well as to East Germany for the months up to unification). Between 1990 and the end of 1992, 2.2 billion ECU were disbursed by PHARE. The EC and its member states have a majority stake in the EBRD which supports investment in the private sector and related infrastructure and advises governments on privatization. It is based in London and is presided over by Jacques Attali.

The countries of the 'other' Europe have each progressed along the path of economic liberalization at differing paces. Poland is the most dependent on Western financial support and was the first to obtain it via PHARE and an agreement with EFTA which covered economic, scientific, industrial and technical cooperation. As elsewhere in Eastern Europe, unemployment has risen dramatically in Poland since 1989, as inefficient plant and enterprises have been closed. Privatization of state enterprises has operated at a fairly restrained pace with assets being distributed over a wide section of the population. None the less, Poles have espoused the entrepreneurial spirit with over 100 000 small businesses coming into existence by mid-1991 (Dawson 1990; Merritt 1991).

Hungary is the most experienced economic

reformer, having attempted liberalization since the early 1970s, hence economic changes since 1989 have been rather less turbulent than elsewhere (Compton 1990). It has moved particularly fast to sell off those state enterprises for which buyers can be found. American and other foreign investors have gained access to the Hungarian market more readily than those elsewhere in east-central Europe. Western franchise partners have been welcomed to operate shops, restaurants and hotels in Budapest and other cities, thereby introducing Western marketing practices and controlling quality. Hungary also signed a cooperation agreement with EFTA in 1990 and during that year the EC ousted COMECON as Hungary's leading trade partner.

Czechoslovakia appeared to be the industrially most advanced state among the Eastern countries, with viable firms in the automobile, engineering, glass and footwear sectors, a well-educated workforce, and a favourable location in the heart of east-central Europe. It was cautious in implementing privatization and spread ownership quite widely. In reality it comprised two distinct economies experiencing differing degrees of health and which asserted their independence in 1993. With 10 million inhabitants, the Czech republic contains industries that stand a reasonable chance of modernization, unlike the 5-million strong Slovak republic, whose defence and textile industries are ailing and where a steep rise in unemployment threatens.

Unlike the other states of east-central Europe, the federation of republics until recently known as Yugoslavia has long been treated as a special case by the nations of Western Europe (Thomas 1990). Its distinctive form of economic self-management was seen as an alternative to both the rigid planning of Communism and the market economy. Refusing to function as part of the Soviet economic or military system, it held a special relationship with the EC since 1970 when a trade agreement was signed. Ten years later President Tito, charismatic leader of the federation, died and tensions between republics surfaced. In the same year Yugoslav exports were given preferential access to the EC market and the country qualified for loans from the EIB. The Dalmatian coast and Slovenia became the parts of east-central Europe that were most visited by Western tourists. But the unity of this patchwork of republics was fragile in the extreme and broke apart in horrifically bloody circumstances.

The European Community and its immediate neighbours

The objective of a free and unified market within the EC by 1992 was agreed at the Milan Summit of June 1985 and was the heart of the Single European Act of 1986 which incorporated the 1992 deadline for the completion of an internal market. This was fully in the spirit of the founding fathers of the Community who had sought to create a 'common market' in which goods, people, services and capital would move without obstacle. Indeed, 1992 is as much a state of mind as it is a date (Palmer 1989). The main drive for the '1992' programme had been presented in a European Commission White Paper of June 1985 which included 300 proposals (directives), later reduced to 279, for removing three types of barrier which prevented the internal market from functioning properly. 'Physical barriers' to commercial unity included border stoppages, customs controls and complex paperwork which plagued trading firms and lorry drivers alike. 'Technical barriers' comprised different product standards, technical regulations and business laws, and nationally protected procurement markets, whereby state agencies insisted on buying goods produced in their own country. In addition, different rates of value added tax and excise duty represented 'fiscal barriers' (Cecchini 1988).

Removal of physical barriers to trade and factor mobility involves the following features (Lintner and Mazey 1991). Administrative checks are to be simplified and removed from borders. All internal frontiers and controls for people and capital, as well as goods and services, are to be eliminated by the end of 1992. Monetary compensation amounts are to be phased out and mutually accepted health certificates introduced for agricultural trade. A common transport policy is to be developed, safety standards are to be agreed, and frontier checks removed so that hauliers may work more freely throughout the Community.

Removal of technical barriers involves attempting to bring in common standards and practices or else ensuring that national procedures are recognized mutually between countries. Thus, common standards for foodstuffs, pharmaceutical products, electrical goods, and health and safety conditions should be introduced. A common market for services will embrace deregulation of banking, insurance and other financial services; establishing an integrated securities market; deregulating modes of travel (especially air transport); developing common standards for transmitting and receiving television and radio programmes; and recognizing the equivalence of appropriate qualifications between member states to facilitate the free movement of professional labour.

Removal of fiscal barriers hinges on adjustments to value added tax among countries and revisions to excise duties. These topics have proved particularly difficult and some of the Commission's most controversial ideas have been modified. Other aspects of the '1992' package include a range of social policy measures to be implemented alongside the completion of the internal market. These embrace proposals to promote the free movement of labour, measures to encourage social and economic cohesion (for example, to curb poverty and create jobs), initiatives to improve training and living and working conditions, and the promotion of social dialogue between management and trade unions via worker participation in companies.

The Social Charter, produced in outline by the Commission, comprised limits on working hours and made provision for fair and reasonable remuneration, participation in trade unions, and receipt of minimum rates of social security. It also embraced free collective bargaining, access to training, sexual equality, worker participation, protection of children and adolescents, and support for the elderly. These features were acceptable to eleven of the heads of government at the Madrid Summit (June 1989), who voted to implement legislation to cover them, but not to the UK which employed economic and ideological arguments to defend its opposition.

In retaliation the European Commission insisted that a stronger social policy was essential for many reasons. Firstly, it was needed to cope with the spatially differentiated impact of the Single Market. Workers in backward regions, contracting industries and inefficient companies would need the help contained in its provisions. To fail to assist them would be politically unacceptable. In an increasingly competitive world it would be essential if free movement of labour was to be possible and the labour force of the Community be trained in the use of new technologies for the 1990s. It was judged necessary to prevent international companies locating in parts of the EC where workers' rights were poorly defended and wages were low. And, finally, the Commission argued that the Social Charter was simply the logical continuation of the EC treaties and the Single European Act (February 1986) which committed all signatories to promoting not only economic union but also social union (Lintner and Mazey 1991).

Advocates of a 'Western Europe without frontiers' envisage immense opportunities for economic growth, job creation, improved productivity, economies of scale, greater consumer choice, healthier competition, and easier professional and business mobility after 1992 (Cecchini 1988). Production and distribution will become increasingly 'Europeanized' and internationalized. Dynamic, viable and innovative firms, cities and regions will benefit: their less efficient counterparts will miss out. More than ever before Japanese and other overseas investors will perceive 'Europe' as the

'Europe of the Single Market'. Those parts of the continent that do not belong to the club may well be among the losers. That point was not lost on the new regimes in east-central Europe or on the member states of EFTA.

The combined populations of the seven EFTA countries amount to only one-tenth of that of the EC and their total GDP is only half that of Germany. However, EFTA is the largest trading partner of the EC, receiving 10 per cent of its exports, and the EC purchases 58 per cent of EFTA's exports. After more than two years of negotiation the members of the EC and of EFTA agreed in October 1991 to establish a European Economic Area (EEA) of 380 000 000 consumers, extending from the Arctic to the Mediterranean, and embracing 45 per cent of world trade. Each EFTA member already had its own free-trade agreement with the EC for most goods, but the new arrangement would go further, being planned to come into operation in 1993. Important points of disagreement had included EC demands for greater fishing quotas in EFTA waters, for guaranteed road transit rights for heavy lorries through Alpine passes in Austria and Switzerland, and for $2.4 billion 'cohesion' money to be paid by the EFTA countries to help the Community's poorer, southern regions to modernize and to compensate them for permitting the EFTA countries to have free access to the Single Market. These issues were resolved and it was pronounced that the EEA agreement would allow the EC breathing space before launching into further enlargement and would give valuable experience of dealing with the EC to those EFTA countries (Austria, Sweden) that had sought membership of the Community.

Key points of the agreement involved free movement of goods produced in the EEA (not imports from outside) from 1993; special arrangements to cover food, fish, energy, coal and steel; and EFTA to assume EC rules on consumer protection, education, the environment and social policy. From 1993 individuals should be able to live, work and offer services throughout the trading bloc. Professional qualifications would be recognized mutually. Switzerland, with its strict limits on immigration, was granted an extra five years to implement the rules. The EFTA countries will maintain domestic agricultural policies rather than join the CAP. Despite close cooperation, the EEA is to be a free-trade area, not a customs union with common external tariffs. Its members will maintain border controls with the EC.

The EEA Treaty needed to be approved by the European Parliament and ratified by the parliament of each EC and EFTA state. In return for better access to the EC's Single Market the EFTA countries would incorporate over 1500 EC rules (regarding consumer protection, the environment, competition, mergers, public procurement and state subsidies) into national law and would have to accept future EC legislation concerning the Single Market. They would have no formal say in EC legislation but would be consulted. The imbalance of this arrangement offered the EFTA countries economic obligations but not political rights. This mattered less for Austria and Sweden (which had applied to join the EC) than for the EFTA countries that would either apply to join or would oppose the deal. For example, immediately after the agreement had been signed the Swiss government declared its intention to become a full member of the EC. Finland planned to confirm its intention to apply in 1992 and Norway's decision would await another report. In December 1991 the European Court of Justice in Luxembourg declared the future EEA to be incompatible with the EC treaties. Disputes between EFTA, the EC and their members, as well as competition cases, were to have been settled by an EEA court consisting of five judges from the European Court and three from EFTA. The Luxembourg judges argued that the EEA court would infringe their autonomy by pronouncing on the powers of the EC and its members. However, in February 1992 officials from the EC and EFTA overcame the legal obstacles raised by the European Court of Justice thereby enabling the EEA to come into existence at the start of 1993. (In a referendum the Swiss voted not to join.)

And the future?

In the 1990s Europe is certainly not a taken-for-granted continent. Regimes have fallen, federations have crumbled, and new boundaries have been drawn. The USSR, COMECON and the Communist system have been shattered. The new Germany has appeared, commanding more economic power and housing more people than any other European state. Its geographic, political and economic centrality in the new Europe of the final years of the twentieth century is beyond question. The dimensions of the EC have expanded as a result of German unification and are surely set to change again as members of EFTA and the 'troika' nations of east-central Europe seek to obtain membership. Already the terms 'Western Europe' and 'Eastern Europe' have been shorn of the politico-economic meaning with which they had been charged ever since the Iron Curtain descended across the continent at the end of the Second World War. Now they survive simply as locational descriptions. With every year that passes Gorbachev's vision of a 'common European home' seems to be becoming more of a reality; but that home is not without its arguments, disputes and down-

right conflicts. De Gaulle's vision of Europe extending from the Atlantic ocean to the Ural Mountains seems increasingly valid, although the Russian Republic stretches far beyond. Perhaps we shall need to adopt a concept of 'Europe' that extends as far as the Pacific?

It would be quite improper to try to write formal conclusions to the essays that make up this book. The places and national and regional systems that compose 'Western Europe' have changed dramatically in recent decades; now they form part of the 'new Europe' that is emerging after 1989 and whose future is so uncertain. An optimistic scenario would embrace closer relations between states throughout the continent, with easier trading links being paralleled by the freer movement of people and a harmonious exchange of ideas on a truly European scale. A pessimistic vision would include fragmentation of the continent among commercial blocs, continued suspicion between states, and ongoing tension stemming from rival claims to territory and from hostility towards recent arrivals. In addition, a 'two-speed' Western Europe might be envisaged, with a distinction between economies linked to the strength of the Deutschmark and other Western economies operating on a weaker financial basis. At the time of writing the 'new Europe' displays aspects of each scenario (Masser, Svidén and Wegener 1992).

As the EC has moved towards political and economic union and many aspects of responsibility have shifted to the supranational level so, some commentators would have us believe, the significance of the dozen nation-states has declined (Ascherson 1992). In a counterbalancing fashion several Western countries have seen the emergence of regional power and regional governments, which are perceived in Brussels as favoured instruments of negotiation for the allocation of structural funds and other forms of area-specific assistance. As well as the well-established German *Länder*, new and quite politically meaningful regional structures have been introduced in Belgium, France, Italy and Spain, and Portugal is working towards such a system. Already trans-border, inter-regional networks have been created whereby economic, technical and cultural cooperation takes place without reference to national capitals (Labasse 1991). Certainly the map of Europe looks very different when conventional national borders are ignored and less familiar regional boundaries are highlighted (Fig. 13.6). Will the Europe of the third millennium be a continent of responsible regions rather than a jigsaw of powerful nation-states?

Europe continues to be an exciting, challenging and surprising continent; to think about its future is to peer into the unknown. Hence it is appropriate to pose

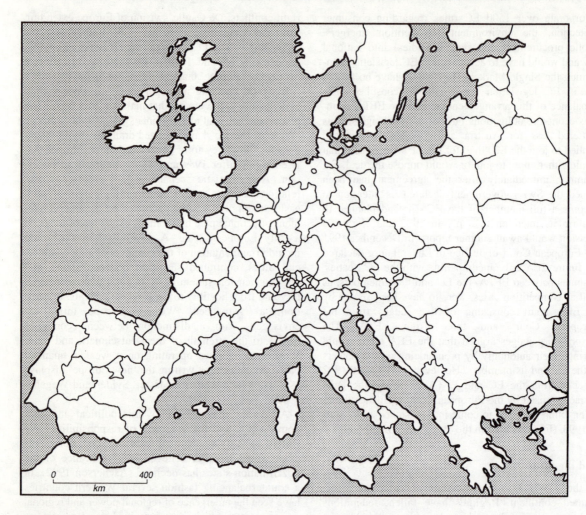

Fig. 13.6
Europe of the regions, 1990.

fundamental questions at the end of this book, rather than to attempt to make final statements. By the year 2000, how wide will the EC have become? Will national interests retreat as the '1992' package implies? Will regional disparities decline or intensify in the Single European market? Will regional responsibilities increase as national powers retreat? How will the international energy situation change during the 1990s and how will the nuclear debate evolve? How will Europeans cope with the future world of work, which may mean 'non-work' for many of them during greater parts of their lives? Will there be a 'rediscovery of the family' and a return to a more traditional view of human relations? Will the notion of 'fortress Europe' emerge more strongly, excluding new flows of

migrants from the Mediterranean Basin and beyond? What kinds of migration flow may develop between the states of east-central Europe and the 'elderly' societies further west? To what extent will environments continue to be ravaged by individual and collective thoughtlessness or will Europeans become more conscious of the fragility of their continent's resources? What kinds of social, economic and ecological principles will guide the rising generation of planners and managers? And to what extent will social justice be remembered in the desperate quest for economic efficiency?

We offer no conclusions; the 'new Europe' is only just beginning.

Further reading

Cecchini P 1988 *The European Challenge*. Wildwood, Aldershot

Cole J, Cole F 1993 *The Geography of the European Community*. Routledge, London

Dawson A H 1993 *The Geography of European Integration*. Belhaven, London

Economist 1992 *Atlas of the New Europe*. The Economist, London

Harris C D 1991 Unification of Germany in 1990. *Geographical Review* 81: 170–82

Labasse J 1991 *L'Europe des régions*. Flammarion, Paris

Lintner V, Mazey S 1991 *The European Community: economic and political aspects*. McGraw-Hill, Maidenhead

Masser I, Svidén O, Wegener M 1992 *The Geography of Europe's Futures*. Belhaven, London

Merritt G 1991 *Eastern Europe and the USSR*. Kogan Page, London

O'Loughlin J, Vander Wusten H (ed) 1993 *The New Political Geography of Eastern Europe*. Belhaven, London

Sword K (ed) 1990 *The Times Guide to Eastern Europe* 2nd edn. Times Books, London

Wise M, Gibb R 1993 *Single Market to Social Europe: the European Community in the 1990s*. Longman, London

References

Adamson D M 1985 European gas and European security. *Energy Policy* **13**: 13–26

Aldcroft D H 1978 *The European Economy, 1914–1970.* Croom Helm, London

Alden J D, Saha S K 1978 An analysis of second jobholding in the EEC. *Regional Studies* **12**: 639–50

Alexander L M 1963 *Offshore Geography of Northwestern Europe: the political and economic problems of delimitation and control.* Rand McNally, Chicago

Allen K (ed) 1970 *Balanced National Growth.* Lexington Books, Lexington

Allen R 1980 Fishing for a common policy. *Journal of Common Market Studies* **19**: 123–39

Anderiessen G 1981 Tanks in the streets: the growing conflict over housing in Amsterdam. *International Journal of Urban and Regional Research* **5**: 83–95

Anon 1982a On or off? Western Europe's oil and gas: a survey. *The Economist* 12 June 1982

Anon 1982b At last a common fish policy. *The Economist* 6 Nov. 1982, 72

Ante U 1991 Some developing and current problems of the eastern border landscape of the Federal Republic of Germany. In Rumley D, Minghi J V (eds) *The Geography of Border Landscapes.* London, Routledge, pp 63–85

Anwar M 1979 *The Myth of Return: Pakistanis in Britain.* Heinemann, London

Appleyard D (ed) 1979 *The Conservation of European Cities.* MIT Press, Cambridge, Mass.

Archer M S, Giner S (eds) 1971 *Contemporary Europe: class, status and power.* Routledge and Kegan Paul, London

Aron R 1964 Old nations, new Europe. *Daedalus* **93**: 43–66

Ascherson N 1992 The new Europe. *Independent on Sunday* 9.2.92: 31–4

Ashcroft B 1982 The measurement of the impact of regional policies in Europe. *Regional Studies* **16**: 287–305

Ashworth G J 1980 Language, society and the state in Belgium. *Journal of Area Studies* **1**: 28–33

Ashworth G J, Tunbridge J 1990 *The Tourist Historic City.* Belhaven, London

Bacchetta M 1978 The crisis of oil refining in the European Community. *Journal of Common Market Studies* **17**: 87–119

Bachtler J, Clement K (eds) 1992 1992 and regional development. *Regional Studies* **26**: 305–419

Bailey R 1977 *Energy: the rude awakening.* McGraw-Hill, New York and London

Baillet P 1975 Une population urbanisée; les rapatriés d'Algérie. *Bulletin de l'Association de Géographes Français* **52**: 269–78

Bailly A 1987a Les services et la production: pour un réexamen des secteurs économiques. *L'Espace Géographique* **16**: 5–13

Bailly A 1987b Service activities and regional development: some European examples. *Environment and Planning A* **19**: 653–68

Barberis C 1968 The agricultural exodus in Italy. *Sociologia Ruralis* **8**: 179–88

Barberis C 1977 Tre o sette milioni di attivi agricoli? Paradossi dell'occupazione a mezzo tempo. *Rivista di Economia Agraria* **32**: 571–8

Barbier B 1978 Ski et stations de sports d'hiver dans le monde. In Sinnhuber K, Jülg T (eds) *Studies in the Geography of Tourism and Recreation,* vol. 1. Ferdinand Hirt, Vienna, pp 131–46

Barker M L 1982 Traditional landscape and mass tourism in the Alps. *Geographical Review* **72**: 395–415

Barrère P, Cassou-Mounat M 1980 *Les villes françaises.* Masson, Paris

Barton A P, Young S 1975 The structure of the dairy industry in the EEC. *Oxford Agrarian Studies* **4**: 181–200

Bassett K 1984 Corporate structure and corporate change in a local economy: the case of Bristol. *Environment and Planning A* **16**: 879–900

Bassett K, Short J 1980 *Housing and Residential Structure.* Routledge and Kegan Paul, London

Bateman M 1985 *Office Development: a geographical analysis.* Croom Helm, London

Baučić I 1973 Yugoslavia as a country of emigration. *Options Méditerranéennes* **22**: 56–66

Beaujeu-Garnier J, Delobez A 1979 *Geography of Marketing.* Longman, London

Benko G 1991 *Géographie des technopôles.* Masson, Paris

Berger J, Mohr J 1975 *A Seventh Man: the story of a migrant worker in Europe.* Penguin, Harmondsworth

Berry B J L (ed) 1976 *Urbanization and Counterurbanization*. Sage, Beverly Hills and London

Berry B J L 1978 The counterurbanization process: how general? In Hansen N M (ed) *Human Settlement Systems*. Ballinger, Cambridge, Mass., pp 25–49

Bertelsmann Foundation 1991 *Challenges in the Mediterranean – the European Response*. Bertelsmann Foundation, Gutersloh

Bethemont J, Pelletier J 1983 *Italy: a geographical introduction*. Longman, London

Bird J H, Witherick M E 1986 Marks and Spencer: the geography of an image. *Geography* 71: 305–19

Blackaby F 1979 *De-industrialization*. Heinemann, London

Blackbourn A 1982 The impact of multinational corporations on the spatial organization of developed nations: a review. In Taylor M J. Thrift N J (eds) *The Geography of Multinationals*. Croom Helm, London and Canberra, pp 147–57

Blacksell M 1976 The role of *le parc naturel et régional*. *Town and Country Planning* 44: 165–70

Blacksell M 1979 Frontiers at sea. *Geographical Magazine* 51: 521–4

Blacksell M 1981 *Post-War Europe: a political geography* 2nd edn. Hutchinson, London

Blacksell M 1982 Reunification and the political geography of the Federal Republic of Germany. *Geography* 67: 310–19

Blacksell M, Brown M 1983 Ten years of Ostpolitik. *Geography* 68: 260–2

Blake G (ed) 1987 *Maritime Boundaries and Ocean Resources*. Croom Helm, London

Blanc M 1987 Family and employment in agriculture: recent changes in France. *Journal of Agricultural Economics* 38: 289–301

Blayo C 1989 L'avortement en Europe. *Espace, Populations, Sociétés* 1989(2): 225–38

Boal F W, Douglas J N H (eds) 1982 *Integration and Division: geographical perspectives on the Northern Ireland problem*. Academic Press, London

Böhning W R 1972 *The Migration of Workers in the United Kingdom and the European Community*. Oxford University Press, Oxford

Böhning W R 1979 International migration in Western Europe: reflections on the last five years. *International Labour Review* 118: 401–14

Boltho A (ed) 1982 *The European Economy: growth and crisis*. Oxford University Press, Oxford

Boucher J, Smeers Y 1985 Gas trade in the European Communities in the 1970s. *Energy Economics* 7: 102–16

Boucher J, Smeers Y 1987 Economic forces in the European gas market – a 1985 perspective. *Energy Economics* 9: 2–16

Bourgeois-Pichat J 1981 Recent demographic change in Western Europe: an assessment. *Population and Development Review* 7: 19–40

Bowler I R 1976 The CAP and the space-economy of agriculture in the EEC. In Lee R, Ogden P E (eds) *Economy and Society in the EEC: spatial perspectives*. Saxon House, Farnborough, pp 235–55

Bowler I R 1985 *Agriculture under the Common Agricultural*

Policy: a geography. Manchester University Press, Manchester

Bradshaw R 1972 Internal migration in Spain. *Iberian Studies* 1: 68–75

Bradshaw R 1985 Spanish migration : economic causes, grave social consequences. *Geography* 70: 175–8

Bransden S 1985 Land abandonment in Sicily. *Land Use Studies* 2: 244–6

Bretell C B 1979 Emigrar para voltar: a Portuguese ideology of return migration. *Papers in Anthropology* 20: 1–20

Brewin C 1987 The European Community – a union without unity of government. *Journal of Common Market Studies* 26: 1–23

British Petroleum 1992 *Statistical Review of World Energy*. BP, London

Brown A J, Burrows E M 1977 *Regional Economic Problems*. George Allen and Unwin, London

Brown E D 1978 Rockall and the limits of national jurisdiction of the UK. *Marine Policy* 2: 181

Brownill S 1990 *Developing London's Docklands*. Paul Chapman, London

Brugger E A, Stuckey B 1987 Regional economic structure and innovative behaviour in Switzerland. *Regional Studies* 21: 241–54

Brusco S 1982 The Emilian model: productive decentralization and social integration. *Cambridge Journal of Economics* 6: 167–84

Bruyelle P 1987 Le tunnel sous la Manche et l'aménagement régional dans la France du Nord. *Annales de Géographie* 96: 145–70

Bruyelle P 1988 Tunnel sous la Manche, systèmes de relations en Europe du nord-ouest et développement régional. *Hommes et Terres du Nord* 1988 (1 and 2): 1–121

Budd S A, Jones A R 1991 *The European Community: a guide to the maze*, 4th edn. Kogan Page, London

Burgess J A 1982 Selling places: environmental images for the executive. *Regional Studies* 16: 1–17

Burkart A J, Medlik S 1974 *Tourism*. Heinemann, London

Burtenshaw D 1976 *Saar-Lorraine*. Oxford University Press, Oxford

Burtenshaw D, Bateman M, Ashworth G J 1981 *The City in West Europe*. Wiley, Chichester

Burtenshaw D, Bateman, M, Ashworth G J 1991 *The European City: a western perspective*. Fulton, London

Cabouret M 1982 Traits permanents et tendances récentes de l'agriculture finlandaise. *Annales de Géographie* 91: 87–118

Cabral M V 1978 Agrarian structures and recent rural movements in Portugal. *Journal of Peasant Studies* 5: 411–45

Cairncross A (ed) 1974 *Economic Policy for the European Community: the way forward*. Macmillan, London

Camagni R P 1992 Development scenarios and policy guidelines for the lagging regions in the 1990s. *Regional Studies* 26: 361–74

Caralp M F 1989 Le réseau aérien européen. *Revue Géographique de l'Est* 29: 259–72

Carney J, Hudson R, Lewis J (eds) 1980 *Regions in Crisis: new perspectives in European regional theory*. Croom Helm, London

Carter F W 1985 Pollution in post-war Czechoslovakia. *Transactions, Institute of British Geographers* **10**: 17–44

Carter F W 1990 Czechoslovakia: geographical prospects for energy, environment and economy. *Geography* **75**: 253–5

Castells M 1978a Urban social movements and the struggle for democracy: the citizens' movement in Madrid. *International Journal of Urban and Regional Research* **2**: 133–46

Castells M 1978b *City, Class and Power*. Macmillan, London

Castells M 1983 *The City and the Grassroots*. Arnold, London

Castles S 1984 *Here for Good: Western Europe's new ethnic minorities*. Pluto, London

Castles S, Kosack G 1985 *Immigrant Workers and Class Structure in Western Europe* 2nd edn. Oxford University Press, Oxford

Catudal H M 1975 The plight of the Lilliputians: an analysis of five European microstates. *Geoforum* **6**: 187–204

Cavazzani A 1976 Social determinants of part-time farming in a marginal region of Italy. In Fuller A M, Mage J M (eds) *Part-time Farming: problem or resource in rural development*. Geo Abstracts, Norwich, pp 101–13

Cazès G 1988 Les grands parcs de loisirs en France. *Travaux de l'Institut de Géographie de Reims* **73–74**: 57–89

Cecchini P 1988 *The European Challenge*. Wildwood, Aldershot

Cerase F P 1974 Migration and social change: expectations and reality. A study of return migration from the United States to Italy. *International Migration Review* **8**: 245–62

Champion A (ed) 1989 *Counterurbanization*. Arnold, London

Champion A, Illeris S 1990 Population redistribution trends in Western Europe: mosaic of dynamics and crisis. In Hebbert M, Hansen J C (eds) *Unfamiliar Territory: the reshaping of European geography*. Avebury, Aldershot, pp 236–53

Chapman K 1992 Continuity and contingency in the spatial evolution of industries: the case of petrochemicals. *Transactions, Institute of British Geographers* **17**: 47–64

Cheshire P, Carbonaro G, Hay D 1986 Problems of urban decline and growth in EEC countries. *Urban Studies* **23**: 131–49

Cheshire P C, Hay D G 1989 *Urban Problems in Western Europe*. Unwin Hyman, London

Chisholm M 1986 The impact of the Channel Tunnel on the regions of Britain and Europe. *Geographical Journal* **152**: 314–34

Christaller W 1964 Some considerations of tourism locations in Europe. *Regional Science Association Papers* **12**: 95–105

Church A 1988 Urban regeneration in London Docklands. *Environment and Planning* **6**: 187–208

Church A, Hall J 1986 Discovery of Docklands. *Geographical Magazine* **58**: 632–9

Churchill R 1980 Revision of the EEC's Common Fisheries Policy. *European Law Review* **5**: 3–37 and 95–111

Clark G 1979 Farm amalgamations in Scotland. *Scottish Geographical Magazine* **95**: 93–107

Cloke P, McLaughlin B 1989 Policies of the alternative land use and rural economy (ALURE) proposals of the UK: crossroads or blind alley? *Land Use Policy* **6**: 235–48

Clout H D 1968 Planned and unplanned changes in French farm structure. *Geography* **53**: 311–15

Clout H D 1971 *Agriculture*. Macmillan Studies in Contemporary Europe, London

Clout H D 1974 Agricultural plot consolidation in the Auvergne region of central France. *Norsk Geografisk Tidsskrift* **28**: 181–94

Clout H D 1976 Rural–urban migration in Western Europe. In Salt J, Clout H D (eds) *Migration in Post-War Europe: geographical essays*. Oxford University Press, Oxford, pp 30–51

Clout H D 1984 *A Rural Policy for the EEC?* Methuen, London

Clout H D (ed) 1987a *Regional Development in Western Europe* 3rd edn. Fulton, London

Clout H D 1987b Rural space. In Clout H D (ed) *Regional Development in Western Europe*. Fulton, London, pp 87–108

Clout H D 1987c The neighbourhood project for urban rehabilitation in France. *Planning Outlook* **29**: 70–8

Clout H 1988 The chronicle of La Défense. *Erdkunde* **42**: 273–84

Clout H 1991 The recomposition of rural Europe. *Annales de Géographie* **100**: 714–29

Cohen R 1987 *The New Helots: migrants in the international division of labour*. Gower, Aldershot

Coleman D, Salt J 1992 *The British Population: patterns, trends and processes*. Oxford University Press, Oxford

Collingridge D 1984a Lessons of nuclear power: US and UK history. *Energy Policy* **12**: 46–67

Collingridge D 1984b Lessons of nuclear power: French 'success' and the breeder. *Energy Policy* **12**: 189–200

Collins D 1975 *The European Communities: the social policy of the first phase* 2 vols. Martin Robertson, London

Commission of the European Communities 1968 *First Guidelines for a Common Energy Policy*. Brussels

Commission of the European Communities 1973 *Report on the Regional Problems in the Enlarged Community*. Brussels

Commission of the European Communities 1976 Environment Programme, 1977–81. *Bulletin of the European Communities, Supplement, 6/76*

Commission of the European Communities 1980 *Outlook for the Long-term Coal Supply and Demand Trend in the Community*. COM(80) 117/final, Brussels

Commission of the European Communities 1981a The development of an energy strategy for the Community. *Bulletin of the European Communities, Supplement* **4/81**: 7–20

Commission of the European Communities 1981b The European automobile industry. *Bulletin of the European Communities, Supplement* **2/81**

Commission of the European Communities 1981c *The Regions of Europe: first periodic report on the social and economic situation in the regions of the Community*. Brussels

Commission of the European Communities 1981d *Basic Statistics*. Brussels

Commission of the European Communities 1982a *Commission Communication to Council on a Mediterranean policy for the enlarged Community.* COM(82) 353/final

Commission of the European Communities 1982b *Energy Statistics Yearbook 1980.* Statistical Office of the European Communities, Luxembourg

Commission of the European Communities 1984 *The Regions of Europe: second periodic report on the social and economic situation and development of the regions of the Community.* Luxembourg

Commission of the European Communities 1985a Advanced manufacturing equipment in the Community. *Bulletin of the European Communities, Supplement* **6/85**

Commission of the European Communities 1985b Perspectives for the Common Agricultural Policy. *Green Europe* 33

Commission of the European Communities 1987a The European energy policy. *European File* **2/87**

Commission of the European Communities 1987b The European Community and environmental protection. *European File* **5/87**, 3–11

Commission of the European Communities 1987c *Third Periodic Report from the Commission on the Social and Economic Situation and Development of the Regions of the Community.* Brussels

Commission of the European Communities 1987d *Basic Statistics of the Community.* Eurostat, Luxembourg

Commission of the European Communities 1988a The future of rural society. *Bulletin of the European Communities Supplement* **4/88**

Commission of the European Communities 1988b *Basic Statistics of the Community.* Eurostat, Luxembourg

Commission of the European Communities 1990a *Environment Statistics.* Commission of the EC, Luxembourg

Commission of the European Communities 1990b *The European Community and its eastern neighbours.* European Commission, Luxembourg

Commission of the European Communities 1991a *Employment in Europe 1991.* European Commission, Luxembourg

Commission of the European Communities 1991b *The Regions in the 1990s: fourth periodic report on the social and economic situation and development of the regions of the Community.* European Commission, Brussels

Commission of the European Communities 1991c *The Common Fisheries Policy.* European File **3/91**, Commission of the European Communities, Brussels

Commission of the European Communities 1992a *Promotion of Energy Technology for Europe: Thermie.* Directorate-General for Energy, Brussels

Commission of the European Communities 1992b *Community Support Framework 1991–3 for Eastern Areas of the Federal Republic of Germany.* European Commission, Luxembourg

Commission of the European Communities 1992c *Panorama of EC Industry 1991–1992.* European Commission, Brussels

Commission of the European Communities 1992d *The State of the Environment in the European Community.* COM(92) 23/final vol. III, Commission of the European Communities, Brussels

Commission of the European Communities 1992e *Basic Statistics of the European Community.* Eurostat, Luxembourg

Commission on Energy and the Environment 1981 *Coal and the Environment.* HMSO, London

Compton P A 1990 The Republic of Hungary bids farewell to Marxism–Leninism. *Geography* **75**: 255–7

Condon S, Ogden P E 1991 Afro-Caribbean migrants in France. *Transactions, Institute of British Geographers* **16**: 440–57

Constantinides P 1977 The Greek Cypriots: factors in the maintenance of ethnic identity. In Watson J L (ed) *Between Two Cultures: migrants and minorities in Britain.* Basil Blackwell, Oxford, pp 269–300

Cooke P, da Rosa Pires A 1985 Productive decentralization in three European regions. *Environment and Planning A* **17**: 527–54

Coull J R 1972 *The Fisheries of Europe.* Bell, London

Council of Europe 1987 *Management of Europe's Natural Heritage. Twenty-five years of activity.* Council of Europe, Strasbourg

Council of Europe 1990 *European Campaign for the Countryside: conclusions and declarations.* Council of Europe, Strasbourg

Couper A D 1972 *The Geography of Sea Transport.* Hutchinson, London

Courgeau D 1978 Les migrations internes en France de 1954–1975: vue d'ensemble. *Population* **33**: 525–45

Cowan P 1969 *The Office: a facet of urban growth.* Heinemann, London

Cranfield J 1992 Italian exploration moves into high gear. *Petroleum Review,* April, 174–6

Croxford G, Wise M 1988 The European Social Fund. *Regional Studies* **22**: 65–8

Croxford G J, Wise M, Chalkley B S 1987 The reform of the European Regional Development Fund; a preliminary assessment. *Journal of Common Market Studies* **26**: 25–38

Damesick P J 1979 Offices and inner urban regeneration. *Area* **11**: 41–7

Damesick P J, Wood P A (eds) 1987 *Regional Problems, Problem Regions and Public Policy in the United Kingdom.* Oxford University Press, Oxford

Daniels P W 1975 *Office Location: an urban and regional study.* Bell, London

Daniels P W (ed) 1979 *Spatial Patterns of Office Growth and Location.* Wiley, Chichester

Daniels P W 1982 *Service Industries: growth and location.* Cambridge University Press, Cambridge

Daniels P W 1983 Service industries: supporting role or centre stage? *Area* **15**: 301–10

Daniels P W 1985 *Service Industries: a geographical appraisal.* Methuen, London

Daniels P W (ed) 1991 *Services and Metropolitan Development: international perspectives.* Routledge, London

Dawson A H 1990 Tides of influence and the economic geography of Poland. *Geography* **75**: 258–60

Dawson J A (ed) 1979 *The Marketing Environment.* Croom Helm, London

Dawson J A (ed) 1980 *Retail Geography.* Croom Helm, London

Dean C J 1982 The trans-Mediterranean gas pipeline. *Geography* **67**: 258–60

Dean K G 1986 Counterurbanization continues in Brittany. *Geography* **71**: 151–4

De Barros A 1980 Portuguese agrarian reform and economic and social development. *Sociologia Ruralis* **20**: 82–96

De Bauw R 1992 Government promotion of markets: the European Energy Charter. *Energy Policy* **20**: 430–2

De Clercq M, Naert F 1985 Regional economic disparities in Belgium and regional, national and supra-national policies: an analysis of collective failure. *Planning Outlook* **28**: 13–20

Decroly J M 1992 Les naissances hors mariage en Europe. *Espace, Populations, Sociétés* 1992 (2): 259–64

De Lannoy W 1975 Residential segregation of foreigners in Brussels. *Bulletin de la Société Belge d'Etudes Géographiques* **44**: 215–38

De Reparaz A 1990 La culture en terrasses, expression de la petite paysannerie méditerranéenne traditionelle. *Méditerranée* **71**, 3–4: 23–9

Dematteis G 1982 Repeuplement et revalorisation des espaces périphériques: le cas de l'Italie. *Revue Géographique des Pyrénées et du Sud-Ouest* **53**: 129–43

Department of Energy 1974 *United Kingdom Offshore Oil and Gas Policy:* Cmnd. 5696. HMSO, London

Department of Energy 1976 *Development of the Oil and Gas Resources of the United Kingdom, 1976.* HMSO, London

Department of Energy 1982 *Development of the Oil and Gas Resources of the United Kingdom, 1982.* HMSO, London

Department of Energy 1987 *Development of the Oil and Gas Resources of the United Kingdom, 1987.* HMSO, London

Department of the Environment 1975 *The Flixborough Disaster: report of the Court of Inquiry.* HMSO, London

Department of the Environment 1981a *Coal and the Environment.* HMSO, London

Department of the Environment 1981b *The Vale of Belvoir Coalfield Inquiry.* HMSO, London

Department of Trade and Industry 1992 *Development of the Oil and Gas Resources of the UK.* HMSO, Alton

De Smidt M 1985 Relocation of government services in the Netherlands. *Tijdschrift voor Economische en Sociale Geografie* **76**: 232–6

De Smidt M 1992 International investments and the European challenge. *Environment and Planning A* **24**: 83–95

De Smidt M, Wever E (eds) 1984 *A Profile of Dutch Economic Geography.* Van Gorcum, Assen and Maastricht

Despicht N 1980 Centre and periphery in Europe. In de Bandt J (ed) *European Studies in Development.* Macmillan, London, pp 38–41

Deubner C 1980 The southern enlargement of the European Community: opportunities and dilemmas from a West German point of view. *Journal of Common Market Studies* **18**: 229–45

De Vries B, Dijk D 1985 Electric power generation options for the Netherlands to 2000: an evaluation of government's and environmentalists' scenarios. *Energy Policy* **13**: 230–42

Dicken P 1986 Multinational enterprises and the local economy: some further observations. *Area* **18**: 215–21

Dicken P 1990 European industry and global competition. In D A Pinder (ed) *Western Europe: challenge and change.* Belhaven, London, pp 37–55

Dicken 1992a *Global Shift: the internationalization of economic activity,* 2nd edn. Paul Chapman, London

Dicken 1992b Europe 1992 and strategic change in the international automobile industry. *Environment and Planning A* **24**: 11–31

Dicken P, Lloyd P E 1981 *Modern Western Society: a geographical perspective on work, home and well-being.* Harper and Row, New York

Dokopolou E 1986a Foreign manufacturing investment in Greece, competition and market structure. In Hamilton F E I (ed) *Industrialization in Developing and Peripheral Regions.* Croom Helm, London, pp 188–204

Dokopolou E 1986b Multinationals and manufacturing exports from the enlarged European community: the case of Greece. In Hamilton F E I (ed) *Industrialization in Developing and Peripheral Regions.* Croom Helm, London, pp 205–31

Donnay J P 1985 Méthodologie de la localisation des bureaux. *Annales de Géographie* **94**: 152–73

Donnison D, Middleton A (eds) 1987 *Regenerating the Inner City: Glasgow's experience.* Routledge and Kegan Paul, London

Douglass W A 1971 Rural exodus in two Spanish Basque villages. *American Anthropologist* **73**: 1100–14

Doutriaux Y 1991 *La politique régionale de la CEE.* Presses Universitaires de France, Paris

Dovring F 1965 *Land and Labour in Europe in the Twentieth Century.* Nijhoff, The Hague

Dower M 1965 Fourth wave: the challenge of leisure. *Architects' Journal* 20 Jan. 1965, 122–90

Dower M et al 1981 *Leisure Provision and People's Needs.* HMSO, London

Drewett R 1979 A European perspective on urban change. *Town and Country Planning* **48**: 224–6

Drudy P J 1978 Depopulation in a prosperous agricultural sub-region. *Regional Studies* **12**: 49–60

Duclaud-Williams R 1978 *The Politics of Housing in Britain and France.* Heinemann, London

Economist 1992 *Atlas of the New Europe.* The Economist, London

Edwards R 1991 *Fit for the Future. Report of the National Parks Review Panel.* Countryside Commission, Cheltenham

Elkins T H 1988 *Berlin: the spatial structure of a city.* Methuen, London

Elkins T H 1990 Developments in the German Democratic Republic. *Geography* **75**: 246–49

Ellger C 1992 Berlin: legacies of division and problems of unification. *Geographical Journal* **158**: 40–6

Ellington A, Burke T 1981 *Europe: environment.* Ecobooks, London

Environmental Resources Ltd 1983 *Acid Rain: a review of the phenomenon in the EEC and Europe*. Graham and Trotman, London

Erz W 1979 Katalog der Naturschutzgebiete in der Bundesrepublik Deutschland. *Naturschutz Aktuell* **3**

Esher V 1969 *York: a study in conservation*. HMSO, London

Estrada J, Bergesen H O, Moe A, Sydnes A K 1988 *Natural Gas in Europe: markets, organisation and politics*. Pinter, London

European Coal and Steel Community 1964 *Study on the Long-term Energy Outlook for the European Community*. High Authority of the ECSC, Luxembourg

European Free Trade Association 1981 *EFTA Bulletin* 1981–1, 2–12

European Investment Bank 1980 EIB support for energy investment. *Information* **23**: 1–11

European Investment Bank 1980 *Annual Report, 1979*. EIB, Luxembourg

European Investment Bank 1982 *Annual Report, 1981*. EIB, Luxembourg

European Investment Bank 1987 *Annual Report, 1986*. EIB, Luxembourg

European Investment Bank 1991 *Annual Report, 1990*. EIB, Luxembourg

European Investment Bank 1992 *Annual Report, 1991*. EIB, Luxembourg

Eurostat 1991a *A Social Portrait of Europe*. European Commission, Luxembourg

Eurostat 1991b *Agricultural Statistics Yearbook 1990*. Eurostat, Luxembourg

Evenson N 1981 The city as an artifact: building control in modern Paris. In Kain R J P (ed) *Planning for Conservation*. Mansell, London, pp 177–98

Everling U 1980 Possibilities and limits of European integration. *Journal of Common Market Studies* **18**: 217–28

Fabbri 1992 *Ocean Management and Global Change*. Elsevier, London and New York

FAO 1990 *FAO Production Yearbook 1989*. Rome

Federici N, Golini A 1972 Les migrations entre les grandes régions des six pays du Marché Commun. *Genus* **28**: 27–68

Fennell R 1982 Farm succession in the European Community. *Sociologia Ruralis* **21**: 19–42

Fennell R 1985 A reconsideration of the objectives of the Common Agricultural Policy. *Journal of Common Market Studies* **23**: 257–77

Fielding A J 1975 Internal migration in Western Europe. In Kosiński L A, Prothero R M (eds) *People on the Move: studies on internal migration*. Methuen, London, pp 237–54

Fielding A J 1982 Counterurbanization in Western Europe. *Progress in Planning* **17**: 1–52

Fielding A J (1986) Counterurbanization in Western Europe. In Findlay A, White P E (eds) *West European Population Change*. Croom Helm, London, pp 35–49

Flockton C H 1982 Strategic planning in the Paris region and French urban policy. *Geoforum* **13**: 193–208

Fofi G 1970 Immigrants in Turin. In Jansen C J (ed) *Readings in the Sociology of Migration*. Pergamon, Oxford, pp 269–84

Fontaine A 1971 The real divisions of Europe. *Foreign Affairs* **49**: 302–14

Ford L 1978 Continuity and change in historic cities: Bath, Chester and Norwich. *Geographical Review* **68**: 253–73

Franklin S H 1969 *The European Peasantry: the final phase*. Methuen, London

Franklin S H 1971 *Rural Societies*. Macmillan Studies in Contemporary Europe, London

Freeman C (ed) 1984 *Long Waves in the World Economy*. Frances Pinter, London

Frémont A 1978 L'aménagement régional en France: la pratique et les idées. *L'Espace Géographique* **7**: 73–84

Fröbel F, Heinrichs J, Kreye O 1977 The tendency towards a new international division of labour. *Review* **1**: 73–88

Fröbel F, Heinrichs J, Kreye O 1980 *The New International Division of Labour*. Cambridge University Press, Cambridge and Maison des Sciences de l'Homme, Paris

Fuller A M 1975 Linkages between social and spatial systems: the case of farmer mobility in northern Italy. *Sociologia Ruralis* **15**: 119–24

Galt A H 1979 Exploring the cultural ecology of field fragmentation and scattering in the island of Pantellaria. *Journal of Anthropological Research* **35**: 95–108

Gasson R 1974 *Mobility of Farm Workers*. Department of Land Economy Occasional Paper no. 2. University of Cambridge

Georgakopoulos T A 1980 Greece and the EEC. *Three Banks Review* **128**: 38–50

George P 1972 Questions de géographie de la population en République fédérale allemande. *Annales de Géographie* **81**: 523–37

Gerholm T, Lithman Y G (eds) 1988 *The New Islamic Presence in Western Europe*. Mansell, London

Ghosh B 1991 Trends in world migration: the European perspective. *The Courier,* **129**: 46–51

Gibb R A 1986 The impact of the Channel Tunnel rail link on south-east England. *Geographical Journal.* **152**: 335–50

Gillespie A E, Goddard J B 1986 Advanced telecommunications and regional economic development. *Geographical Journal* **152**: 383–97

Glassner, M 1990 *Neptune's Domain*. Unwin Hyman, London

Glyptis S A 1981 Leisure life styles. *Regional Studies* **15**: 311–26

Goddard J B 1975 *Office Location in Urban and Regional Development*. Oxford University Press, Oxford

Goddard J B, Champion A G (eds) 1983 *The Urban and Regional Transformation of Britain*. Methuen, London

Goosens M, de Rudder J 1976 Management centres and location of top management in Belgium. *Bulletin de la Société Belge d'Etudes Géographiques* **45**: 121–33

Gorbachev M 1987 *Perestroika. New Thinking for our Country and the World*. Collins, London

Gordon I 1991 *The Impact of Economic Change on Minorities and Migrants in Western Europe*. Discussion Paper 2, Department of Geography, University of Reading

Gordon R L 1970 *The Evolution of Energy Policy in Western Europe: the reluctant retreat from coal*. Praeger, New York

Gordon R L 1987 *World Coal: economics, policies and prospects*. Cambridge University Press, Cambridge

Gottmann J 1950 *A Geography of Europe*. Holt, Rinehart and Winston, New York

Green A E, Howells J 1987 Spatial prospects for service growth in Britain. *Area* **19**: 111–22

Greene R P, Gallagher J M 1980 *Future Coal Prospects: country and regional assessment*. Ballinger, Cambridge, Mass.

Greenwood D 1976 The decline of agriculture in Fuenterrabia. In Aceves J B, Douglass W A (eds) *The Changing Faces of Rural Spain*. Schenkman, New York, pp 29–44

Guedes M 1981 Recent agricultural land policy in Spain. *Oxford Agrarian Studies* **10**: 26–43

Guilmot J, McGlue D, Valette P, Waeterloos C (eds) 1986 *Energy 2000: a reference projection and alternative energy outlooks for the European Community and the World*. Cambridge University Press, Cambridge

Guruswamy L D, Papps I, Storey D J 1983 The development and impact of an EEC Directive: the control of discharges of mercury into the aquatic environment. *Journal of Common Market Studies* **22**: 71–100

Guttesen R, Nielsen F H 1975 Regional development in Denmark. *Geografisk Tidsskrift* **75**: 74–87

Haigh N 1986 *European Community Environmental Policy in Practice*, Volume 1, *Water and Waste in Four Countries*. Graham and Trotman, London

Haines R C 1988 *A Study on the Safety Aspects Relating to the Handling and Monitoring of Hazardous Wastes*. Commission on the European Communities, Brussels

Hajdu J G 1978 The German city today. *Geography* **63**: 23–30

Hajdu J G 1979 Phases in the post-war German urban experience. *Town Planning Review* **50**: 267–86

Hajdu J G 1981 Pedestrian precincts in Germany. *Town and Country Planning* **50**: 173–4

Hall D R 1990 The changing face of Eastern Europe and the Soviet Union: introduction. *Geography* **75**: 239–44

Hall P G 1967 Plannning for urban growth: metropolitan area plans. *Regional Studies* **1**: 101–34

Hall P G 1977 *The World Cities* 2nd edn. Weidenfeld and Nicolson, London

Hall P G 1981 Regional development in the EEC: a look back and a look forward. *Built Environment* **7**: 229–32

Hall P G 1984 *World Cities* 3rd edn. Weidenfeld and Nicolson, London

Hall P G, Cheshire P 1987 The key to success for cities. *Town and Country Planning* **56**: 50–3

Hall P G, Hay D 1980 *Growth Centres in the European Urban System*. Heinemann, London

Hall P G, Markusen A 1985 *Silicon Landscapes*. Allen and Unwin, London

Hall R 1976 Demographic and social change in the EEC. In Lee R, Ogden P E (eds) *Economy and Society in the EEC*. Saxon House, Farnborough, pp 85–107

Hall R 1988 Recent patterns and trends in western European households. *Espace, Populations, Sociétés* **1988**(1): 13–32

Hall R, van der Wee D 1992 Community regional policies for the 1990s. *Regional Studies* **26**: 399–404

Hamilton L, Just W D 1981 The evolution of regional dispar-

ities in the European Community: an ethical perspective. *Ecumenical Research Exchange (Rotterdam), Discussion Papers* **288**: 1–38

Hammar T (ed) 1985 *European Immigration Policy: a comparative study*. Cambridge University Press, Cambridge

Hamnett C 1983 The condition in England's inner cities on the eve of the 1981 riots. *Area* **15**: 7–13

Hamnett C, Williams P R 1980 Social change in London: a study in gentrification. *Urban Affairs Quarterly* **15**: 469–87

Harris C D 1991 Unification of Germany in 1990. *Geographical Review* **81**: 170–82

Harris R C 1977 The simplification of Europe overseas. *Annals of the Association of American Geographers* **67**: 469–83

Harrison A 1982 Land policies in the member states of the European Community. *Agricultural Administration* **11**: 159–74

Hartner P, Jones G 1986 *Multinationals: theory and history*. Gower, Aldershot

Harvey D R, Thomson K J 1985 Costs, benefits and the future of the Common Agricultural Policy. *Journal of Common Market Studies* **24**: 1–20

Haskins G, Gibb A, Hubert A (eds) 1986 *A Guide to Small Firm Assistance in Europe*. Gower, Aldershot

Hass-Klau C H M 1982 The housing shortage in Germany's major cities. *Built Environment* **8**: 60–70

Hayter R 1982 Truncation, the international firm and regional policy. *Area* **14**: 277–82

Health and Safety Executive 1979 *Advisory Committee on Major Hazards, Second Report*. HMSO, London

Health and Safety Executive 1981 *Canvey: a review of potential hazards in the Canvey Island/Thurrock area*. HMSO, London

Heater D 1992 *The Idea of European Unity*. Leicester University Press, Leicester

Hebbert M 1982 Regional policy in Spain. *Geoforum* **13**: 107–20

Hebbert M 1987 Regionalism: a reform concept and its application to Spain. *Environment and Planning C* **5**: 239–50

Hechter M, Brustein W 1980 Regional modes of production and patterns of state formation in Western Europe. *American Journal of Sociology* **85**: 1061–94

Heineberg H 1977 Service centres in East and West Berlin. In French R A, Hamilton F E I (eds) *The Socialist City*. Wiley, Chichester, pp 305–34

Heinritz G, Lichtenberger E (eds) 1986 *The Take-off of Suburbia and the Crisis of the Central City*. Steiner, Stuttgart

Hengchen B, Melis C 1980 Contribution to the study of the Brussels urban social movement. *International Journal of Urban and Regional Research* **4**: 116–26

Her Majesty's Stationery Office 1969 *Bath: a study in conservation*. HMSO, London

Herbert D T, Smith D M (eds) 1979 *Social Problems and the City: geographical perspectives*. Oxford University Press, Oxford

Herbert D, Smith D (eds) *1989 Social Problems and the City: new perspectives*. Oxford University Press, Oxford

Herzfeld M 1980 Social tension and inheritance by lot in

three Greek villages. *Anthropological Quarterly* **53**: 91–100

Hill K 1974 Belgium: political change in a segmented society. In Rose R (ed) *Electoral Behavior*. Free Press, New York, pp 29–208

Hirsch G P, Maunder A M 1978 *Farm Amalgamation in Western Europe*. Saxon House, Farnborough

Hodgson R D, Smith R W 1979 Boundary issues created by extended national maritime claims. *Geographical Review* **69**: 423–33

Hoffmann G W (ed) 1953 *A Geography of Europe*. Ronald, New York

Hoffmann G (ed) 1989 *Europe in the 1990s*, 6th edn. Wiley, New York

Hoffmann-Nowotny H J 1978 European migration after World War II. In McNeill W H, Adams R S (eds) *Human Migration: patterns and policies*. Indiana University Press, Bloomington, pp 85–105

Holland S H 1976a *Capital versus the Regions*. Macmillan, London

Holland S H 1976b *The Regional Problem*. Macmillan, London

Holland S H 1980 *The UnCommon Market*. Macmillan, London

Holliday I, Vickerman R 1990 The Channel Tunnel and regional development. *Regional Studies* **24**: 455–64

Horner A A, Parker A J (eds) 1987 *Geographical Perspectives on the Dublin Region*. Geographical Society of Ireland, Dublin

House J W 1978 *France: an applied geography*. Methuen, London

Houston J M 1964 *The Western Mediterranean World: an introduction to its regional landscapes*. Longman, London

Howells J 1988 *Economic, Technological and Locational Trends in European Services*. Avebury, Aldershot

Hoyle B S, Pinder D A, Husain M S (eds) 1988 *Revitalising the Waterfront*. Belhaven, London

Hudman L E 1978 Origin regions of international tourism. In Sinnhuber K, Jülg T (eds) *Studies in the Geography of Tourism*, vol. 2. Ferdinand Hirt, Vienna, pp 43–9

Hudson R, Lewis J (eds) 1985 *Uneven Development in Southern Europe*. Methuen, London

Hume I M 1973 Migrant workers in Europe. *Finance and Development* **10**: 2–6

Ilbery B W 1984 Core–periphery contrasts in European social well-being. *Geography* **69**: 289–302

Ilbery B W 1986 *Western Europe: a systematic human geography* 2nd edn. Oxford University Press, Oxford

Ilbery B W 1990 Adoption of the arable set-aside scheme in England. *Geography* **75**: 69–73

Inglehart R 1971 Changing value priorities and European integration. *Journal of Common Market Studies* **10**: 1–36

Illeris S 1989 *Services and Regions in Europe*. Avebury, Aldershot

Insall D and Associates 1969 *Chester: a study in conservation*. HMSO, London

International Energy Agency 1980a *Energy Policies and Programmes of IEA Countries*. OECD, Paris

International Energy Agency 1980b *Energy Research*

Development and Demonstration in the IEA Countries. OECD, Paris

International Energy Agency 1984 *Energy Policies and Programmes of IEA Countries*. OECD, Paris

International Energy Agency 1991 *Energy Statistics of OECD Countries, 1980–1989*. OECD, Paris

International Labour Office 1986 *Yearbook of Labour Statistics*. ILO, Geneva

International Labour Office 1989 *Yearbook of Labour Statistics*. ILO, Geneva

International Tanker Owners' Federation 1980 *Measures to Combat Oil Pollution*. Graham and Trotman, London

Ion D C 1977 The North Sea countries. In Mangone J G (ed) *Energy Policies of the World*, vol. 2. Elsevier, New York, Oxford and Amsterdam, pp 121–216

IUCN 1974 *United Nations List of National Parks and Equivalent Reserves*. Morges

IUCN 1986 *United Nations List of National Parks and Equivalent Reserves 1985:* Gland

Jacoby E H 1959 *Land Consolidation in Europe*. International Institute for Land Reclamation and Improvement, Wageningen

Jarrett R J 1975 Disincentives: the other side of regional development policy. *Journal of Common Market Studies* **13**: 379–90

Jenkins J R G 1986 *Jura Separatism in Switzerland*. Oxford University Press, Oxford

Jensen-Butler C 1982 Capital accumulation and regional development: the case of Denmark. *Environment and Planning A* **14**: 1307–40

Jensen-Butler C 1987 The regional economic effects of European integration. *Geoforum* **18**: 213–27

Johnson S P 1979 *The Pollution Control Policy of the European Communities*. Graham and Trotman, London

Johnston D 1976 *Marine Policy and the Coastal Community: the impact of the Law of the Sea*. Croom Helm, London

Johnston R J, Gardiner V (eds) 1991 *The Changing Geography of the United Kingdom 2nd edn*. Routledge, London

Jolliffe W 1977 Farming in the rural–urban fringe in Britain. In Gasson R (ed) *The Place of Part-time Farming in Rural and Regional Development*. Centre for European Agricultural Studies, Wye, pp 35–41

Jones A R 1986 The role of institutions in technical despecialization under CAP: an example from Languedoc-Roussillon. In Bowler I R (ed) *Agriculture and the Common Agricultural Policy*. Leicester University Geography Department, Occasional Paper 15, pp 66–79

Jones A R 1989a The role of the SAFER in agricultural restructuring. *Land Use Policy* **6**: 249–61

Jones A R 1989b Reform of the European Community's table wine sector: agricultural despecialization in the Languedoc. *Geography* **74**: 29–37

Jones A R 1991 The impact of the EC's set-aside programme: the response of farm businesses in Rendsburg–Eckernforde, Germany. *Land Use Policy* **8**: 108–24

Jones A R, Munton R 1990 Set-aside story. *Geographical Magazine (Analysis Supplement)* **62**(12): 1–3

Jones P 1984 Retail warehouse developments in Britain. *Area* **16**: 41–6

Jones P N 1990 Recent ethnic German migration from Eastern Europe to the Federal Republic of Germany. *Geography* **75**: 249–52

Jones P N 1991 The French census 1990: the southward drift continues. *Geography* **76**: 358–61

Jordan T G 1973 *The European Culture Area*. Harper and Row, New York

Jougla E 1990 Disparités géographiques de la mortalité par SIDA en France. *Espace, Populations, Sociétés* **1990**(3): 533–40

Jung H-U 1982 Regional policy in the Federal Republic of Germany. *Geoforum* **13**: 83–96

Kain R J P 1981 Conservation planning in France: policy and practice in the Marais, Paris. In Kain R J P (ed) *Planning for Conservation*. Mansell, London, pp 199–233

Kain R J P 1982 Europe's model and exemplar still? The French approach to urban conservation. *Town Planning Review* **53**: 403–22

Kariel H G, Kariel P E 1982 Socio-cultural impacts of tourism: an example from the Austrian Alps. *Geografiska Annaler (Series B)* **64**: 1–16

Karpat K H 1976 *The Gecekondu: rural migration and urbanization*. Cambridge University Press, Cambridge

Kearns K C 1979 Intraurban squatting in London. *Annals of the Association of American Geographers* **69**: 589–98

Kearns K C 1982 Preservation and transformation of Georgian Dublin. *Geographical Review* **72**: 270–90

Keating M, Jones B (eds) 1985 *Regions in the European Community*. Oxford University Press, Oxford

Keeble D 1989 Core–periphery disparities, recession and new regional dynamisms in the European Community. *Geography* **74**: 1–11

Keeble D, Owens P L, Thompson C 1981 EEC regional disparities and trends in the 1970s. *Built Environment* **7**: 154–61

Keeble D, Owens P L, Thompson C 1982 Regional accessibility and economic potential in the European Community. *Regional Studies* **16**: 419–31

Keeble D, Owens P L, Thompson C 1983 The urban–rural manufacturing shift in the European Community. *Urban Studies* **20**: 405–18

Kemp A G, Rose D, Dandie R 1992 Development and production prospects for UK oil and gas post-Gulf crisis. *Energy Policy* **20**: 20–9

Kerr A J C 1986 *The Common Market and how it works* 3rd edn. Pergamon, Oxford

Kiljunen M L 1980 Regional disparities and policy in the EEC. In Seers D, Vaitsos C (eds) *Integration and Unequal Development: the experience of the EEC*. St. Martin's Press, New York, pp 199–222

King R L 1970 Structural and geographical problems of south Italian agriculture. *Norsk Geografisk Tidsskrift* **24**: 83–95

King R L 1973 *Land Reform: the Italian experience*. Butterworths, London

King R L 1976 The evolution of international labour migration movements concerning the EEC. *Tijdschrift voor Economische en Sociale Geografie* **67**: 66–82

King R L 1978 Shifting progress in Portugal's land reform. *Geography* **63**: 118–19

King R L 1979 Return migration: a review of some case studies from southern Europe. *Mediterranean Studies* **1**: 3–30

King R L 1982 Southern Europe: dependency or development? *Geography* **67**: 221–34

King R L 1987a *Italy*. Harper and Row, London

King R L 1987b Italy. In Clout H D (ed) *Regional Development in Western Europe*. Fulton, London, pp 129–63

King R L 1991 Europe's metamorphic demographic map. *Town and Country Planning* **60**: 111–13

King R L 1992 Italy: from sick man to rich man in Europe. *Geography* **77**: 153–69

King R L, Killingbeck J 1990 Agricultural land-use change in central Basilicata: from Carlo Levi to the Comunità Montana. *Land Use Policy* **7**: 7–26

King R L, Mortimer J, Strachan A J 1984 Return migration and tertiary development: a Calabrian case-study. *Anthropological Quarterly* **57**: 112–24

King R L, Took L J 1983 Land tenure and rural social change: the Italian case. *Erdkunde* **37**: 186–98

Klatte E 1986 The past and future of European Environmental Policy. *European Environmental Review*. **1**: 32–4

Klein L 1981 The European Community's regional policy. *Built Environment* **7**: 182–9

Kliot N 1987a Maritime boundaries in the Mediterranean: aspects of co-operation and dispute. In Blake G (ed) *Maritime Boundaries and Ocean Resources*. Croom Helm, London, pp 200–26

Kliot N 1987b The era of homeless Man. *Geography* **72**: 109–21

Kofman E 1982 Differential modernization, social conflicts and ethno-regionalism in Corsica. *Ethnic and Racial Studies* **5**: 300–12

Komninos N 1992 Science parks in Europe. In Dunford M, Kafkala G (eds) *Cities and Regions in the New Europe*. Belhaven, London, pp 86–101

Kosiński L 1970 *The Population of Europe*. Longman, London

Kostrubiec B 1988 La géographie du SIDA. *Travaux et Documents du Laboratoire de Géographie Humaine, Université de Lille*, **3**: 1–100

Krätke S 1992 Berlin: the rise of a new metropolis in a post-Fordist landscape. In Dunford M, Kafkala G (eds) *Cities and Regions in the New Europe*. Belhaven, London, pp 213–38

Kromarek P 1986 The Single European Act and the environment. *European Environmental Review* **1**: 10–12

Kunzmann K R 1981 Development trends in the regional and settlement structure of the Federal Republic of Germany. *Built Environment* **7**: 243–54

Labasse J 1984 Les congrès, activité tertiaire de villes privilégiées. *Annales de Géographie* **93**: 687–703

Labasse J 1991 *L'Europe des régions*. Flammarion, Paris

Laborde P 1989 *Les espaces urbains dans le monde*. Nathan, Paris

Lambert A M 1963 Farm consolidation in Western Europe. *Geography* **48**: 31–48

Landes D S 1969 *The Unbound Prometheus: technological change and industrial development in Western Europe from 1750 to the present*. Cambridge University Press, Cambridge

Lane D 1980 Mini-farming in the Italian south. *Geographical Magazine* **53**: 177–9

Law C M 1980 *British Regional Development since World War I*. Methuen, London

Law C M 1986 The geography of exhibition centres in Britain. *Geography* **71**: 359–62

Lawless P 1981 *Britain's Inner Cities*. Harper and Row, London

Lebon A, Falchi G 1980 New developments in intra-European migration since 1974. *International Migration Review* **14**: 539–79

Lee R, Ogden P E (eds) 1976 *Economy and Society in the EEC*. Saxon House, Farnborough

Leigh M, van Praag N 1978 *The Mediterranean challenge: 1*. Sussex European Papers no. 2. Sussex European Research Centre, Brighton

Lenon B 1987 The geography of the big bang: London's office building boom. *Geography* **72**: 56–9

Leontidou L 1990 *The Mediterranean City*. Cambridge University Press, Cambridge

Lewis J, Williams A 1985 Portugal: the decade of return. *Geography* **70**: 178–82

Lichtenberger E 1970 The nature of European urbanism. *Geoforum* **4**: 45–62

Lichtenberger E 1976 The changing nature of European urbanization. In Berry B J L (ed) *Urbanization and Counter-urbanization*. Sage, Beverly Hills and London, pp 91–107

Limouzin P 1980 Les facteurs de dynamisme des communes rurales françaises. *Annales de Géographie* **89**: 549–87

Ling J J 1973 Early state building and late peripheral nationalisms against the state: the case of Spain. In Eisenstadt S N, Rokkan S (eds) *Building States and Nations* vol. 2. Sage, Beverly Hills, pp 32–116

Lintner V, Mazey S 1991 *The European Community: economic and political aspects*. McGraw-Hill, Maidenhead

Lodge J 1986 The Single European Act: towards a new Euro-dynamism? *Journal of Common Market Studies* **24**: 203–23

Lopreato J 1967 *Peasants no more*. Chandler, San Francisco

Lopriano N 1986 Radiation knows no frontiers: Chernobyl's challenge to science and government. *European Environmental Review* **1**: 2–9

Lowenthal D 1968 The American scene. *Geographical Review* **58**: 61–88

Lucas N J D 1979 *Energy in France: planning, politics and policy*. Europa Publications, London

Lucas N J D 1985 *Western European Energy Policies: a comparative study of the influence of institutional structure on technical change*. Clarendon Press, Oxford

McDermott P J 1979 Multinational manufacturing firms and regional development: external control in the Scottish electronics industry. *Scottish Journal of Political Economy* **23**: 279–94

McGrath F 1991 *Emigration and Landscape: the case of Achill Island*. Trinity Papers in Geography 4, Department of Geography, Trinity College Dublin

Maffenini W, Rallu J L 1991 Les accidents de la circulation en Italie et en France. *Population* **46**: 913–40

Maillat D 1990 Transborder regions between members of the EC and non-member countries. *Built Environment* **16**: 38–51

Mallet S 1975 *The New Working Class*. Spokesman Book, Nottingham

Manganara J 1977 Some social aspects of the return movement of Greek migrant workers from West Germany to rural Greece. *Greek Review of Social Research* **29**: 65–75

Manne A S, Roland K, Stephan G 1986 Security of supply in the Western European market for natural gas. *Energy Policy* **14**: 52–64

Manners G 1981 *Coal in Britain: an uncertain future*. George Allen and Unwin, London

Marceau J 1977 *Class and Status in France: economic change and social immobility 1945–75*. Oxford University Press, Oxford

Marie C V 1991 Immigration – an awkward issue. *The Courier* **129**: 41–5

Marmot A F, Worthington J 1986 Great Fire to Big Bang: private and public designs on the City of London. *Built Environment* **12**: 216–33

Marques A 1992 Community competition policy and economic and social cohesion. *Regional Studies* **26**: 404–7

Marsh J S 1979 Agriculture. In Tsoukalis L (ed) *Greece and the European Community*. Saxon House, Farnborough, pp 68–83

Marshall J N 1985 Services and regional policy in Great Britain. *Area* **17**: 303–8

Marstrand P (ed) 1984 *New Technology and the Future of Work and Skills*. Frances Pinter, London

Martin R, Rowthorn B 1986 *The Geography of Deindustrialization*. Macmillan, London

Martins M R, Mawson J 1980 Regional trends in the European Community and revision of the common regional policy. *Built Environment* **6**: 145–52

Martins M R, Mawson J 1982 The programming of regional development in the EC. *Journal of Common Market Studies* **20**: 229–44

Mason C M, Harrison R T 1990 Small firms: phoenix from the ashes? In D A Pinder (ed) *Western Europe: challenge and change*, Belhaven, London, pp 72–90

Masser I, Svidén O, Wegener M 1992 *The Geography of Europe's Futures*. Belhaven, London

Massey D 1979 In what sense a regional problem? *Regional Studies* **13**: 233–43

Massey D, Meegan R 1982 *The Anatomy of Job Loss*. Methuen, London

Maull H 1980 *Europe and World Energy*. Butterworths, London

Mayhew A 1970 Structural reform and the future of West German agriculture. *Geographical Review* **60**: 54–68

Mead W R 1982 The discovery of Europe. *Geography* **67**: 193–202

Mellor R 1992 Railways and German unification. *Geography* **77**: 261–64

Mellor R E H, Smith E A 1979 *Europe: a geographical survey of the continent*. Macmillan, London

Mendosa E L 1982 Benefits of migration as a personal strategy in Nazare, Portugal. *International Migration Review* **16**: 625–45

Mendras H 1970 *The Vanishing Peasant: innovation and change in French agriculture*. MIT Press, Cambridge, Mass

Mennell S 1976 *Cultural Policy in Towns*. Council of Europe, Strasbourg

Mény Y, Wright V (eds) 1985 *Centre–Periphery Relations in Western Europe*. Allen and Unwin, London

Merritt G 1991 *Eastern Europe and the USSR*. Kogan Page, London

Metton A 1982 L'expansion du commerce périphérique en France. *Annales de Géographie* **91**: 463–79

Michael R 1979 Metropolitan development concepts and planning policies in West Germany. *Town Planning Review* **50**: 287–312

Minet G, Siotis J, Tsakaloyannis P 1981 *Spain, Greece and Community Politics*. Sussex European Papers no. 11. Sussex European Research Centre, Brighton

Ministry of Agriculture, Fisheries and Food 1981 *Atlas of the Seas around the British Isles*. MAFF, London

Minshull G N 1978 *The New Europe: an economic geography of the EEC*. Hodder and Stoughton, London

Mitchell B 1976 Politics, fish and international resource management: the British–Icelandic Cod War. *Geographical Review* **66**: 127–38

Molle W 1979 *Regional Economic Development in the European Community*. Saxon House, Farnborough

Molle W, Cappelin R 1988 *Regional Impact of Community Policies in Europe*. Avebury, Aldershot

Molle W, Paelinck J 1979 Regional policy. In Coffey P (eds) *Economic Policies in the Common Market*. Macmillan, London, pp 146–71

Molle W, Wever E 1984a *Oil Refineries and Petrochemical Industries in Western Europe: buoyant past, uncertain future*. Gower, Aldershot

Molle W, Wever E 1984b Oil refineries and petrochemical industries in Europe. *Geo Journal* **9**: 421–30

Monke J 1986 Portugal on the brink of Europe: the CAP and Portuguese agriculture. *Journal of Agricultural Economics* **37**: 317–31

Monkhouse F J 1959 *A Regional Geography of Western Europe*. Longman, London

Montanari A 1991 The southern regions of the EEC on the threshold of 1992: environment, agriculture and economic development. In Montanari A (ed) *Growth and Perspectives of the Agrarian Sector in Portugal, Italy, Greece and Turkey*. Edizioni Scientiche Italiane, Naples, pp 9–32

Morris A 1985 Tourism and town planning in Catalonia. *Planning Outlook* **28**: 77–82

Morris A, Dickinson G 1987 Tourist development in Spain. *Geography* **72**: 16–25

Moseley M J 1980 Strategic planning and the Paris agglom-

eration in the 1960s and 1970s: the quest for balance and structure. *Geoforum* **8**: 179–223

Murphy A B 1991 The emerging Europe of the 1990s. *Geographical Review* **81**: 1–17

Murray R 1992 Europe and the new regionalism. In Dunford M, Kafkalas G (eds) *Cities and Regions in the New Europe*. Belhaven, London, pp 299–308

Musto S A 1981 Regional consequences of the enlargement of the European Community. *Built Environment* **7**: 172–81

Mutton A F A 1961 *Central Europe: a regional and human geography*. Longman, London

Nake-Mann B, Nake R 1979 *Schutzwürdige Gebiete von europaische Bedeutung*. Bonn

Nature Conservancy Council 1982 *National Nature Reserves*. NCC, London

Naylon J 1967 Tourism: Spain's most important industry. *Geography* **52**: 23–40

Naylon J 1981 Barcelona. In Pacione M (ed) *Urban Problems and Planning in the Developed World*. Croom Helm, London, pp 223–57

Naylon J 1992 Ascent and decline in the Spanish regional system. *Geography* **77**: 46–62

Netting R McC 1981 *Balancing on an Alp: ecological change and continuity in a Swiss mountain community*. Cambridge University Press, Cambridge

Newby H 1979 *Green and Pleasant Land? Social change in rural England*. Penguin, Harmondsworth

Nicholson B 1975 Return migration to a marginal area: an example from north Norway. *Sociologia Ruralis* **15**: 227–45

Nicholson B M 1981 The role of the inner city in the development of manufacturing industry. *Urban Studies* **18**: 57–72

Noble M K 1985 Reconstructing Naples. *Geography* **70**: 246–9

Noin D, Warnes A 1987 Personnes agées et vieillissement. *Espace, Populations, Sociétés* **1987**(1): 1–300

Noreng Ø 1980 *The Oil Industry and Government Strategy in the North Sea*. Croom Helm, London, and the International Research Center for Energy and Economic Development, Boulder, Colorado

North J, Spooner D 1977 The great UK coal rush. *Area* **9**: 15–27

North J, Spooner D 1978 The geography of the coal industry in the UK in the 1970s. *Geo Journal* **2**: 25–72

Nuclear Energy Agency 1986 *Severe Accidents in Nuclear Power Plants*. OECD, Paris

Öberg S, Oscarsson G 1987 Northern Europe. In Clout H D (ed) *Regional Development in Western Europe*. Fulton, London, pp 319–51

Ochel W, Wegner M 1987 *Service Economies in Europe*. Pinter, London

Odell P R 1981 The energy economy of Western Europe: a return to the use of indigenous resources. *Geography* **66**: 1–14

Odell P R 1986 *Oil and World Power* 8th edn. Penguin, Harmondsworth

Odell P R 1990 Energy: resources and choices. In Pinder D A (ed) *Western Europe: challenge and change*. Belhaven, London, pp 19–36

Offshore Promotion Services 1980 *European Continental Shelf Guide and Atlas*. OPS, Maidenhead

O'Flanagan T P 1982 Land reform and rural modernization in Spain: a Galician perspective. *Erdkunde* **36**: 48–53

Ogden P E 1980 Migration, marriage and the collapse of traditional peasant society in France. In White P, Woods R (eds) *The Geographical Impact of Migration*. Longman, London, pp 152–79

Ogden P E 1985a France: recession, politics and migration policy. *Geography* **70**: 158–62

Ogden P E 1985b Counterurbanization in France: the results of the 1982 population census. *Geography* **70**: 24–35

O'Loughlin J 1980 Distribution and migration of foreigners in German cities. *Geographical Review* **70**: 253–75

Openshaw S 1982 The geography of reactor siting policies in the UK. *Transactions of the Institute of British Geographers* **7**: 150–62

Organization for Economic Cooperation and Development 1964 *Oil Today*. OECD, Paris

Organization for Economic Cooperation and Development 1977 *The OECD Programme on Long-range Transport of Air Pollutants: measurements and findings*. OECD, Paris

Organization for Economic Cooperation and Development 1979a *The Siting of Major Energy Facilities*. OECD, Paris

Organization for Economic Cooperation and Development 1979b *The State of the Environment in OECD Member Countries*. OECD, Paris

Organization for Economic Cooperation and Development 1980a *OECD Economic Surveys: the Netherlands*. OECD, Paris

Organization for Economic Cooperation and Development 1980b *The Impact of Tourism on the Environment*. OECD, Paris

Organization for Economic Cooperation and Development 1983 *Environmental Effects of Energy Systems*. OECD, Paris

Organization for Economic Cooperation and Development 1984 *Emission Standards for Major Air Pollutants from Energy Facilities in OECD Member Countries*. OECD, Paris

Organization for Economic Cooperation and Development 1985a *Transfrontier Movements of Hazardous Wastes: legal and institutional aspects*. OECD, Paris

Organization for Economic Cooperation and Development 1985b *Petroleum Industry: energy aspects of structural change*. OECD, Paris

Organization for Economic Cooperation and Development 1987a *OECD Economic Surveys: Norway*. OECD, Paris

Organization for Economic Cooperation and Development 1987b *Energy Statistics 1970–85*. OECD, Paris

Organization for Economic Cooperation and Development 1988 *Tourism Policy and International Tourism in OECD Member Countries*. OECD, Paris

Organization for Economic Cooperation and Development 1992 *OECD Economic Surveys: Norway*. OECD, Paris

Orlando G, Antonelli G 1982 Regional policy in the European Community and Community regional policy. *European Review of Agricultural Economics* **8**: 213–46

Orme E 1990 *Nature Conservation and the European Community: the present and the future*. Centre for Rural Studies, Royal Agricultural College, Cirencester

Pacione M 1974 The Venetian problem: an overview. *Geography* **59**: 339–43

Pacione M 1977 Tourism: its effects on a traditional landscape in Ibiza and Formentera. *Geography* **62**: 43–7

Pacione M 1979 Housing policies in Glasgow since 1880. *Geographical Review* **69**: 395–412

Pacione M 1982 Economic development in the Mezzogiorno. *Geography* **67**: 340–3

Pacione M 1985 Inner-city regeneration: perspectives on the GEAR project. *Planning Outlook* **28**: 65–9

Pacione M 1987 Socio-spatial development of the south Italian city: the case of Naples. *Transactions, Institute of British Geographers* **12**: 433–50

Padgett S 1992 The Single European Energy Market: the politics of realisation. *Journal of Common Market Studies* **30**: 53–76

Palmer J 1989 *1992 and Beyond*. European Commission, Luxembourg

Papageorgiou-Venetas A 1981 Conservation of the architectural heritage of Greece: means, methods and policies. In Kain R J P (ed) *Planning for Conservation*. Mansell, London, pp 235–58

Parker S 1976 *The Sociology of Leisure*. George Allen and Unwin, London

Pathe D C 1986 Simulator key to successful plant start-up. *Oil and Gas Journal* **84**: 49–53

Peach C 1992 The new Islamic presence in Europe. *Geography Review* **5**(3): 2–6

Pearson L F 1981 *The Organization of the Energy Industry*. Macmillan, London

Peet R 1980 Capital accumulation and regional crisis in Western Europe. *Environment and Planning A* **12**: 1317–24

Pepelasis A A et al 1980 *The Tenth Member: economic aspects*. Sussex European Papers no. 7. Sussex European Research Centre, Brighton

Perrons D 1992 The regions and the Single market. In Dunford M, Kafkalas G (eds) *Cities and Regions in the New Europe*. Belhaven, London, pp 170–89

Perroux F 1955 Note sur la notion de pôle de croissance. *Economie Appliquée*. Appears in English translation in McKee D L, Dean R D, Leahy W H (eds) 1970 *Regional Economics*. Collier Macmillan, London and New York

Peschel K 1990 Spatial effects of the completion of the European Single Market. *Built Environment* **26**: 387–97

Peschel K 1992 European integration and regional development in northern Europe. *Regional Studies* **16**: 11–129

Petroleum Economist. A monthly survey of developments in the oil, gas and petrochemical industries

Pickvance C G (ed) 1976 *Urban Sociology: critical essays*. Tavistock, London

Pieroni O 1982 Positive aspects of part-time farming in the development of a professional agriculture: remarks on the Italian situation. *Geo Journal* **6**: 331–6

Pilat J F 1980 *Ecological Politics: the rise of the green movement*. Sage, Beverly Hills and London

Pinder D A 1976 *The Netherlands*. Dawson, Folkestone,

Hutchinson, London, and Westview Press, Boulder, Colorado

Pinder D A 1978 Guiding economic development in the EEC: the approach of the European Investment Bank. *Geography* **63**: 88–97

Pinder D A 1981 Community attitude as a limiting factor in port growth: the case of Rotterdam. In Hoyle B S, Pinder D A (eds) *Cityport Industrialization and Regional Development: spatial analysis and planning strategies.* Pergamon, Oxford, pp 181–99

Pinder D A 1983 *Regional Economic Development and Policy: theory and practice in the European Community.* George Allen and Unwin, London

Pinder D A 1986 Small firms, regional development and the European Investment Bank. *Journal of Common Market Studies* **24**: 171–86

Pinder D (ed) 1990 *Western Europe: challenge and change.* Belhaven, London

Pinder D A 1992 Seaports and the European energy system. In Hoyle B S, Pinder D A (eds) *European Port Cities in Transition.* Belhaven, London, pp 20–39

Pinder D A, Hoyle B S 1981 Cityports, technologies and development strategies. In Hoyle B S, Pinder D A (eds) *Cityport Industrialization and Regional Development: spatial analysis and planning strategies.* Pergamon, Oxford, pp 323–38

Pinder D A, Husain M S 1987a Innovation, adaptation and survival in the West European oil refining industry. In Chapman K, Humphrys G (eds) *Technical Change and Industrial Policy.* Blackwell, Oxford, pp 100–20

Pinder D A, Husain M S 1987b Oil industry restructuring in the Netherlands and its European context. *Geography* **72**: 300–08

Poinard M 1979 Le million des immigrés. *Revue Géographique des Pyrénées et du Sud-Ouest* **50**: 511–39

Postan M M 1977 The European economy since 1945: a retrospect. In Griffiths R T (ed) *Government, Business and Labour in European Capitalism.* Europotentials Press, London, pp 23–39

Pounds N J G 1953 *Europe and the Mediterranean.* McGraw-Hill, New York

Pourcher G 1970 The growing population of Paris. In Jansen C J (ed) *Readings in the Sociology of Migration.* Pergamon, Oxford, pp 179–202

Prioux F 1989 Fécondité et dimension des familles en Europe occidentale. *Espace, Populations, Sociétés* **1989**(2): 161–76

Puchala D J 1971 Patterns of West European integration. *Journal of Common Market Studies* **9**: 117–42

Pugliese E 1985 Farmworkers in Italy: agricultural working class, landless peasants or clients of the welfare state? In Hudson R, Lewis J (eds) *Uneven Development in Southern Europe.* Methuen, London, pp 123–39

Rallu J L 1990 Conduite automobile et accidents de la route. *Population* **45**: 27–62

Ray G F 1976 Impact of the oil crisis on the energy situation in Western Europe. In Rybczynski T M (ed) *The Economics of the Oil Crisis.* Macmillan, London, pp 94–130

Ray G F, Uhlmann L 1979 *The Innovation Process in the Energy Industries.* Cambridge University Press for the National Institute of Economic and Social Research, Cambridge, London and New York

Reid G L, Allen K, Harris D J 1973 *The Nationalized Fuel Industries.* Heinemann, London

Reitel F 1976 Les bourses de valeurs. *Mosella* **6**: 1–72

Richmond A H 1969 Sociology of migration in industrial and post-industrial societies. In Jackson J A (ed) *Migration.* Cambridge University Press, Cambridge, pp 238–81

Ritson C 1982 Impact on agriculture. In Seers D, Vaitsos C (eds) *The Second Enlargement of the EEC: the integration of unequal partners.* Macmillan, London, pp 92–108

Ritter G, Hajdu J G 1989 The east–west German boundary. *Geographical Review* **79**: 326–44

Robert S, Randolph W G 1983 Beyond decentralization: the evolution of population distribution in England and Wales 1961–1981. *Geoforum* **14**: 75–102

Robinson G M 1990 *Conflict and Change in the Countryside.* Belhaven, London

Robinson G M 1991 EC agricultural policy and the environment. *Land Use Policy* **8**: 95–107

Rodwin L, Sazanami H (eds) 1991 *Industrial Change and Regional Economic Transformation: the experience of Western Europe.* Harper Collins, London

Roesler J 1991 Mass unemployment in Eastern Germany. *Journal of European Social Policy* **1**: 129–50

Rogers R, Fisher M 1992 *A New London.* Penguin, Harmondsworth

Rokkan S 1980 Territories, centres and peripheries. In Gottmann J (ed) *Centre and Periphery: spatial variations in politics.* Sage, Beverly Hills and London, pp 163–204

Romanos M C 1979 Forsaken farms: the village-to-city movement in Western Europe. In Romanos, M C (ed) *Western European Cities in Crisis.* Lexington Books, Lexington, pp 3–19

Roussel L 1987 Deux décennies de mutations démographiques (1965–85) dans les pays industrialisés. *Population* **42**: 429–48

Routledge I 1977 Land reform and the Portuguese revolution. *Journal of Peasant Studies* **5**: 79–98

Rustow D, Penrose T 1981 *Turkey and the Community.* Sussex European Papers no. 10. Sussex European Research Centre, Brighton

Sallnow J, John A 1982 *An Electoral Atlas of Europe 1968–81.* Butterworths, London

Salt J 1976 International labour migration: the geographical pattern of demand. In Salt J, Clout H D (eds) *Migration in Post-War Europe: geographical essays.* Oxford University Press, Oxford, pp 80–125

Salt J 1981 International labor migration in Western Europe: a geographical review. In Kritz M M, Keely C B, Tomasi S M (eds) *Global Trends in Migration.* Center for Migration Studies, New York, pp 133–57

Salt J 1984 High level manpower movements in north-western Europe and the role of careers. *International Migration Review* **17**:633–51

Salt J 1985a West German dilemma: little Turks or young Germans? *Geography* **70**: 162–8

Salt J 1985b Europe's foreign labour migrants in transition. *Geography* 70: 151–8

Salt J, Kitching R 1991 Movimenti migratori: analisi delle tendenze storiche. *Politica Internazionale* 19(5): 5–19

Sampson A 1968 *The New Europeans.* Hodder and Stoughton, London

Saville J 1957 *Rural Depopulation in England and Wales, 1851–1951.* Routledge and Kegan Paul, London

Scargill I 1991 Regional inequalities in France. *Geography* 76: 343–57

Schaller C, Motamen H 1985 Structural changes in the UK energy market. *Energy Policy* 13: 559–63

Schipper L 1983 Residential energy use and conservation in Denmark, 1965–80. *Energy Policy* 11: 313–23

Schipper L 1987 Energy conservation policies in the OECD – did they make a difference? *Energy Policy* 15: 538–48

Schmid J 1982 The family today: sociological highlights on an embattled institution. *European Demographic Information Bulletin* 13: 49–72

Schmoll F 1990 Metropolis Berlin? Prospects and problems of post-November 1989 urban developments. *International Journal of Urban and Regional Research* 14: 676–86

Seabrook J 1985 *Landscapes of Poverty.* Blackwell, Oxford

Secchi C 1980 The effects of enlargement on the EEC periphery. In de Bandt J (ed) *European Studies in Development.* Macmillan, London, pp 42–8

Simmons I G 1975 *Rural Recreation in the Industrial World.* Edward Arnold, London

Sivini G 1976 Some remarks on the development of capitalism and specific forms of part-time farming in Europe. In Fuller A M, Mage J A (eds) *Part-time Farming in Europe: problem or resource in rural development.* Geo Abstracts. Norwich, pp 274–83

Slater M 1979 Migrant employment, recessions and return migration: some consequences for migration policy and development. *Studies in Comparative International Development* 14: 3–22

Smidelius B, Wiklund C 1979 The Nordic Community: the ugly duckling of regional co-operation. *Journal of Common Market Studies* 18: 59–75

Smith N, Williams P (eds) 1986 *Gentrification of the City.* Allen and Unwin, Winchester, Mass.

Smith R W 1980 Trends in national maritime claims. *Professional Geographer* 32: 216–23

Spooner D 1981 The geography of coal's second coming. *Geography* 66: 29–41

Spooner D J 1984 The southern problem, the Neapolitan problem and Italian regional policy. *Geographical Journal* 150: 11–26

Stamp L D 1965 *Land-use Statistics for the Countries of Europe.* World Land-use Survey Occasional Paper no. 3. Bude

Steed G F P 1978a Global industrial systems: the case of the clothing industry. *Geoforum* 9: 35–47

Steed G F P 1978b Product differentiation, location protection and economic integration: Western Europe's clothing industries. *Geoforum* 9: 307–18

Steinle W J 1992 Regional competitiveness and the Single Market. *Regional Studies* 26: 307–18

Stephens M 1978 *Linguistic Minorities in Western Europe.* Gower, Llandysal

Stephenson G 1972 Cultural regionalism and the unitary state idea in Belgium. *Geographical Review* 62: 501–23

Stern J 1986 After Sleipner: a policy for UK gas supplies. *Energy Policy* 14: 9–14

Stöhr W, Tödtling F 1978 An evaluation of regional policies: experiences in market and mixed economies. In Hansen N M (ed) *Human Settlement Systems.* Ballinger, Cambridge, Mass., pp 85–119

Stungo A 1972 The Malraux Act. *Journal of the Royal Town Planning Institute* 58: 357–62

Sword K (ed) 1990 *The Times Guide to Eastern Europe,* 2nd edn. Times Books, London

Symes D 1992 Agriculture, the state and rural society in Europe. *Sociologia Ruralis* 32: 193–208

Talbot R B 1977 The European Community's regional fund: a study in the politics of redistribution. *Progress in Planning* 8: 183–281

Taylor D, Diprose G, Duffy M 1986 EC environmental policy and the control of water pollution: the implementation of Directive 76/464 in perspective. *Journal of Common Market Studies* 24: 225–46

Teumis H B 1978 The early state in France. In Claessen H J M, Skalnik P (eds) *The Early State.* Mouton, The Hague, pp 235–56

Thieme G 1983 Agricultural change and its impact on rural areas. In Wild M T (ed) *Urban and Rural Change in West Germany.* Croom, Helm, London, pp 220–47

Thomas C 1990 Yugoslavia: the enduring dilemmas. *Geography* 75: 265–8

Thompson I B 1970 *Modern France: a social and economic geography.* Butterworths, London

Thompson K 1963 *Farm Fragmentation in Greece.* Centre for Economic Research, Athens

Thornley A (ed) 1992 *The Crisis of London.* Routledge, London

Tommel I 1987 Regional policy in the European Community: its impact on regional policies and public administration in the Mediterranean member states. *Environment and Planning C* 5: 369–81

Torre J de la, Bacchetta M 1980 The uncommon market: European policies towards the clothing industry in the 1970s. *Journal of Common Market Studies* 19: 95–122

Townsend A 1980 The role of returned migrants in England's poorest region. *Geoforum* 11: 353–69

Townsend P 1987 *Poverty and Labour in London.* Low Pay Unit, London

Treasure C 1991 Germans search for a homeland. *Geographical Magazine* 64: 24–7

Trébous M 1980 Moslem workers in France and the practice of their faith. In de Bandt J (ed) *European Studies in Development.* Macmillan, London, pp 142–8

Tugendhat C 1987 *Making Sense of Europe.* Penguin, Harmondsworth

Tuppen J 1977 Redevelopment of the city centre: the case of Lyon–Part Dieu. *Scottish Geographical Magazine* 93: 151–8

Tuppen J, Bateman M 1987 The relaxation of office

development controls in Paris. *Regional Studies* **21**: 69–74

Tuppen J N 1988 *France in Recession, 1981–86*. Macmillan, London

Tuppen J N, Mingret P 1986 Suburban malaise in French cities. *Town Planning Review* **57**: 187–201

Turner B, Nordquist G 1982 *The Other European Community: integration and co-operation in Nordic Europe*. Weidenfeld and Nicolson, London

Uhrich R 1987 *La France inverse? Les régions en mutation*. Economica, Paris

Unger K 1986 Return migration and regional characteristics: the case of Greece. In King R L (ed) *Return Migration and Regional Economic Problems*. Croom Helm, London, pp 129–51

United Nations 1976 *World Energy Supplies, 1950–1974*. United Nations, New York

United Nations 1977 *Statistical Yearbook 1976*. United Nations, New York

United Nations 1981a *Statistical Yearbook 1979/80*. United Nations, New York

United Nations 1981b *Yearbook of Energy Statistics 1980*. United Nations, New York

United Nations Economic Commission for Europe 1983 *An Efficient Energy Future: prospects for Europe and North America*. Butterworths, London

Unwin T 1989 *North Portuguese Agriculture in the European Context: progress and prospect*. Royal Holloway and Bedford New College, Papers in Geography **6**

Urry J 1990 *The Tourist Gaze*. Sage, London

Van Hoogstraten P 1985 Stages in Dutch regional policy, 1945–1984. *Planning Outlook* **28**: 20–8

Vanhove N, Klaasen L H 1987 *Regional Policy: a European Approach* 2nd edn. Avebury, Aldershot

Van Valkenburg S 1959 Land use within the European Common Market. *Economic Geography* **35**: 1–24

Van Valkenburg S 1960 An evaluation of the standard of land use in Western Europe. *Economic Geography* **36**: 283–95

Van Valkenburg S, Held C C 1952 *Europe*. Wiley, New York

Van Voorden F W 1981 The preservation of monuments and historic townscapes in the Netherlands. *Town Planning Review* **52**: 433–53

Van Weesep J 1988 Regional and urban development in the Netherlands: the retreat of government. *Geography* **73**: 97–104

Venzi L 1988 The evolution of the agrarian structure in Italy. In *The Dynamics of Agrarian Structures in Europe*. FAO, Rome, pp 87–114

Verrips J 1975 The decline of small-scale farming in a Dutch village. In Boissevain J, Friedl J (eds) *Beyond the Community: social process in Europe*. Department of Educational Science of the Netherlands and European Mediterranean Study Group of the University of Amsterdam, The Hague, pp 108–23

Vickerman R W 1986 Economic implications of the Channel Tunnel. *Economic Review* **3**: 2–6

Vickerman R W 1987 The Channel Tunnel: consequences for regional growth and development. *Regional Studies* **21**: 187–97

Vickerman R 1990 Regional implications of the Single European Market. *Built Environment* **16**: 5–10

Vickerman R (ed) 1991 *Infrastructure and Regional Development*. Pion, London

Vidal de la Blache P 1903 *Tableau de la géographie de la France*. Hachette, Paris

Vielvoye R 1985 Future gloomy for refining industry in Western Europe. *Oil and Gas Journal* **83**: 41–6

Vincent J A 1980 The political economy of Alpine development: tourism or agriculture in St Maurice. *Sociologia Ruralis* **20**: 250–71

Wallace W 1990 *The Transformation of Western Europe*. Pinter, London

Wallace W, Spät M 1990 Coming to terms with Germany. *The World Today* **46**(4): 55–6

Walter I 1975 *International Economics of Pollution*. Macmillan, London

Wannop U 1990 The Glasgow Eastern Area Renewal project. *Town Planning Review* **61**: 455–74

Warnes A M (ed) 1982 *Geographical Perspectives on the Elderly*. Wiley, Chichester

Watts H D 1990 Manufacturing trends, corporate restructuring and spatial change. In Pinder D A (ed) *Western Europe: challenge and change*. Belhaven, London, pp 56–71

Webb M 1985 Energy policy and the privatization of the UK energy industries. *Energy Policy* **13**: 27–36

Weidenfeld W, Janning J 1991 *Global Responsibilities: Europe in tomorrow's world*. Bertelsmann Foundation, Gutersloh

Wild T 1992 From division to unification: regional dimensions of economic change in Germany. *Geography* **77**: 224–60

White P E 1976 Tourism and economic development in the rural environment. In Lee R, Ogden P E (eds) *Economy and Society in the EEC*. Saxon House, Farnborough, pp 150–9

White P E 1984 *The West European City: a social geography*. Longman, London

Whitelegg J 1988 *Transport Policy in the EEC*. Methuen, London

Whysall P, Benyon J 1981 Urban renewal policies in central Amsterdam. *Planning Outlook* **23**: 77–82

Widgren J 1990 International migration and regional stability. *International Affairs* **66**: 749–66

Wilkinson P F 1978 The global distribution of national parks and equivalent reserves. In Nelson J G, Needham R D, Mann D L (eds) *International Experience with National Parks and Related Reserves*. University of Waterloo, Canada, pp 603–24

Williams A 1981 Portugal's illegal housing. *Planning Outlook* **23**: 110–14

Williams A M (ed) 1984 *Southern Europe Transformed*. Harper and Row, London

Williams A M 1987 *The Western European Economy: a geography of post-war development*. Hutchinson, London

Williams A M 1991 *The European Community*. Blackwell, Oxford

Williams A M, Shaw G (eds) 1991 *Tourism and Economic*

Development: West European experiences 2nd edn. Belhaven, London

Williams A V, Zelinsky W 1970 On some patterns in international tourist flows. *Economic Geography* **46**: 549–67

Williams C H 1980 Ethnic separatism in Western Europe. *Tijdschrift voor Economische en Sociale Geografie* **71**: 142–58

Williams R 1990 Supranational environmental policy and pollution control. In Pinder D A (ed) *Western Europe: challenge and change.* Belhaven, London, pp 195–208

Wils A B 1991 Survey of immigration trends and assumptions about future migration. In Lutz W (ed) *Future Demographic Trends in Europe and North America.* Academic Press, London, pp 281–99

Winchester H, Ilbery B W 1988 *Agriculture and Change: France and the EEC.* John Murray, London

Wise M 1987 European, national and regional concepts of fishing limits in the European Community. In Blake G (ed) *Maritime Boundaries and Ocean Resources.* Croom Helm, London, pp 117–32

Wood P A 1976 Inter-regional migration in Western Europe: a re-appraisal. In Salt J, Clout H D (eds) *Migration in Post-War Europe: geographical essays.* Oxford University Press, Oxford, pp 52–79

Wood P A 1987 United Kingdom. In Clout H D (ed) *Regional Development in Western Europe.* Fulton, London, pp 257–83

Woodruffe B 1990 Conservation and the rural landscape. In Pinder D (ed) *Western Europe: challenge and change.* Belhaven, London, pp 258–76

Woods S 1987 *Western Europe: technology and the future.* Croom Helm, London

Wynn M 1980 Conserving Madrid. *Town and Country Planning* **49**: 53–4

Wynn M (ed) 1984 *Housing in Europe.* Croom Helm, London

Wynn M, Smith R 1978 Spain: urban decentralization. *Built Environment* **4**: 49–55

Yuill D, Allen K, Hull C 1980 *Regional Policy in the European Community.* Croom Helm, London

Index